T0234070

Texts, Textual Acts and the History of Science

Archimedes

NEW STUDIES IN THE HISTORY AND PHILOSOPHY
OF SCIENCE AND TECHNOLOGY

VOLUME 42

EDITOR

JED Z. BUCHWALD, *Dreyfuss Professor of History, California Institute of Technology, Pasadena, CA, USA.*

ASSOCIATE EDITORS FOR MATHEMATICS AND PHYSICAL SCIENCES

JEREMY GRAY, *The Faculty of Mathematics and Computing, The Open University, Buckinghamshire, UK.*
TILMAN SAUER, *California Institute of Technology*

ASSOCIATE EDITORS FOR BIOLOGICAL SCIENCES

SHARON KINGSLAND, *Department of History of Science and Technology, Johns Hopkins University, Baltimore, MD, USA.*
MANFRED LAUBICHLER, *Arizona State University*

ADVISORY BOARD FOR MATHEMATICS, PHYSICAL SCIENCES AND TECHNOLOGY

HENK BOS, *University of Utrecht*
MORDECHAI FEINGOLD, *California Institute of Technology*
ALLAN D. FRANKLIN, *University of Colorado at Boulder*
KOSTAS GAVROGLU, *National Technical University of Athens*
PAUL HOYNINGEN-HUENE, *Leibniz University in Hannover*
TREVOR LEVERE, *University of Toronto*
JESPER LÜTZEN, *Copenhagen University*
WILLIAM NEWMAN, *Indiana University, Bloomington*
LAWRENCE PRINCIPE, *The Johns Hopkins University*
JÜRGEN RENN, *Max-Planck-Institut für Wissenschaftsgeschichte*
ALEX ROLAND, *Duke University*
ALAN SHAPIRO, *University of Minnesota*
NOEL SWERDLOW, *California Institute of Technology, USA*

ADVISORY BOARD FOR BIOLOGY

MICHAEL DIETRICH, *Dartmouth College, USA*
MICHEL MORANGE, *Centre Cavaillès, Ecole Normale Supérieure, Paris*
HANS-JÖRG RHEINBERGER, *Max Planck Institute for the History of Science, Berlin*
NANCY SIRAISI, *Hunter College of the City University of New York*

Archimedes has three fundamental goals; to further the integration of the histories of science and technology with one another: to investigate the technical, social and practical histories of specific developments in science and technology; and finally, where possible and desirable, to bring the histories of science and technology into closer contact with the philosophy of science. To these ends, each volume will have its own theme and title and will be planned by one or more members of the Advisory Board in consultation with the editor. Although the volumes have specific themes, the series itself will not be limited to one or even to a few particular areas. Its subjects include any of the sciences, ranging from biology through physics, all aspects of technology, broadly construed, as well as historically-engaged philosophy of science or technology. Taken as a whole, *Archimedes* will be of interest to historians, philosophers, and scientists, as well as to those in business and industry who seek to understand how science and industry have come to be so strongly linked.

For further volumes:
http://www.springer.com/series/5644

Karine Chemla • Jacques Virbel
Editors

Texts, Textual Acts and the History of Science

 Springer

Editors
Karine Chemla
SPHERE (ex-REHSEIS;
CNRS & University Paris Diderot)
Université Paris 7, UMR 7219
Paris Cedex 13
France

Jacques Virbel
IRIT
CNRS
Saint-Félix Lauragais
France

ISSN 1385-0180
Archimedes
ISBN 978-3-319-36673-9
DOI 10.1007/978-3-319-16444-1

ISSN 2215-0064 (electronic)

ISBN 978-3-319-16444-1 (eBook)

Springer Cham Heidelberg New York Dordrecht London
© Springer International Publishing Switzerland 2015
Softcover reprint of the hardcover 1st edition 2015
This work is subject to copyright. All rights are reserved by the Publisher, whether the whole or part of the material is concerned, specifically the rights of translation, reprinting, reuse of illustrations, recitation, broadcasting, reproduction on microfilms or in any other physical way, and transmission or information storage and retrieval, electronic adaptation, computer software, or by similar or dissimilar methodology now known or hereafter developed.
The use of general descriptive names, registered names, trademarks, service marks, etc. in this publication does not imply, even in the absence of a specific statement, that such names are exempt from the relevant protective laws and regulations and therefore free for general use.
The publisher, the authors and the editors are safe to assume that the advice and information in this book are believed to be true and accurate at the date of publication. Neither the publisher nor the authors or the editors give a warranty, express or implied, with respect to the material contained herein or for any errors or omissions that may have been made.

Printed on acid-free paper

Springer International Publishing AG Switzerland is part of Springer Science+Business Media (www.springer.com)

A Cloé,
pour son usage mesuré du langage

Contents

Contributors

Florence Bretelle-Establet SPHERE (ex-REHSEIS; CNRS & University Paris Diderot), Université Paris 7, UMR 7219, Paris Cedex 13, France

Yves Cambefort SPHERE (ex-REHSEIS; CNRS & University Paris Diderot), Université Paris 7, UMR 7219, Paris Cedex 13, France

Karine Chemla (林力娜) SPHERE (ex-REHSEIS; CNRS & University Paris Diderot), Université Paris 7, UMR 7219, Paris Cedex 13, France

Agathe Keller SPHERE (ex-REHSEIS; CNRS & University Paris Diderot), Université Paris 7, UMR 7219, Paris Cedex 13, France

Christine Proust SPHERE (ex-REHSEIS; CNRS & University Paris Diderot), Université Paris 7, UMR 7219, Paris Cedex 13, France

Anne Robadey SPHERE (ex-REHSEIS; CNRS & University Paris Diderot), Université Paris 7, UMR 7219, Paris Cedex 13, France

Michel Teboul (戴明德) SPHERE (ex-REHSEIS; CNRS & University Paris Diderot), Université Paris 7, UMR 7219, Paris Cedex 13, France

Jacques Virbel IRIT, CNRS, Saint-Felix Lauragais, France

Chapter 1
Prologue: Textual Acts and the History of Science

Karine Chemla (林力娜) and Jacques Virbel

Abstract Scholarly texts usually combine a variety of elements: sentences, diagrams, tables and other specific types of inscriptions, layouts and notations. After a long period when History and Philosophy of Science mainly focused on the verbal aspect of these sources, in recent decades these disciplines have actively begun to investigate other dimensions, such as diagrams and symbolic writings. This book is an attempt to return to the verbal components of these documents and explore them from a new perspective.

The collective work that led to this book started in the fall of 2002. In the context of the seminar "History of Science, History of Text," Karine Chemla (REHSEIS, CNRS & University Paris Diderot), a historian of science, and Jacques Virbel (IRIT, Toulouse), a linguist specialized in text linguistics, launched a series of workshops with the title "Textes de consignes et d'algorithmes. Approches historiques et linguistiques (Instructional Texts and Algorithm Texts. Historical and Linguistic Approaches)." This resulted in formation of a group that was enthusiastic about the collective work carried out in this framework. Before long it was focusing on the issues addressed in this volume. The group members included specialists of China, Europe, India and Mesopotamia. They were linguists as well as historians of mathematics, lexicography, zoology or medicine. The collective work continued for several years, during which colleagues learned about the theories and sources of each other's disciplines, and each elaborated his or her own approach in the framework of this interdisciplinary enterprise. As a result, the book brings Linguistics and History of Science into close interrelationship, with the aim of helping the two fields to advance together. Each chapter of the book has been subjected to close scrutiny by the participants of the seminar and we are thus jointly responsible for any omissions or errors. We were able to complete the book thanks to Anthony Pamart, and to the generous hospitality of Silke Wimmer-Zagier and Don Zagier in Bonn, and Lorraine Daston and the Max Planck Institut fuer Wissenschaftsgeschichte in Berlin during summer 2012, as well as that of the Fondation des Treilles during summer 2013. We would like to express our heartfelt thanks to all of them. Bruno Belhoste, Ramon Guardans, and Skuli Sigurdsson have carefully read this introduction whole or in part, and we are grateful for their constructive comments. We would especially like to thank Karen Margolis for her immeasurable part in completing the introduction. Many thanks also to Sarah Diému-Trémolières, Shubham Dixit, and Neelu Sahu for their contributions during the publication process.

K. Chemla (林力娜) (✉)
SPHERE (ex-REHSEIS; CNRS & University Paris Diderot), Université Paris 7, UMR 7219, 75205 Paris Cedex 13, France
e-mail: chemla@univ-paris-diderot.fr

J. Virbel
IRIT, CNRS, 1, rue du Commerce, 31540 Saint-Felix Lauragais, France
e-mail: virbel@irit.fr

© Springer International Publishing Switzerland 2015
K. Chemla, J. Virbel (eds.), *Texts, Textual Acts and the History of Science,*
Archimedes 42, DOI 10.1007/978-3-319-16444-1_1

1.1 Acting by Means of Scholarly Texts: A Research Program

Scholarly texts usually combine a variety of elements: sentences, diagrams, tables and other specific types of inscriptions, layouts and notations. After a long period when History and Philosophy of Science mainly focused on the verbal aspect of these sources, in recent decades these disciplines have actively begun to investigate other dimensions, such as diagrams and symbolic writings.[1] This book is an attempt to return to the verbal components of these documents and explore them from a new perspective.

1.1.1 Assertions and other Types of Utterances

When History and Philosophy of Science have dealt with the discursive part of scientific texts, both disciplines have laid great emphasis on the assertive dimensions of the statements they contain and the logical structures of their arguments. This is how one can express that, when reading the words that have come down to us from the practitioners of science, these disciplines have endeavored to determine in the most rigorous and fruitful way the bodies of knowledge available to actors, or created by them, on the basis of how actors described part of the world as they understood it, or reported on other topics of inquiry. However, as John L. Austin has emphasized, language can be used for other purposes. (Austin has since been followed by many others, including John Searle.) In addition to allowing its users to make assertions, language allows us to "do things with words."[2] For instance, language can be used to change the "world." Austin explains that this is the case with the famous sentence "'I do (sc. take this woman to be my lawful wedded wife)'—as uttered in the course of the marriage ceremony."[3] In the proper circumstances, uttering the sentence modifies my status and others' as well as my relationship to people and institutions. Clearly, the sentence does not assert a state of the world, but actually brings it about: the speaker performs an act by means of words. It is an act of communication, performed in a given context. Language can also be used to get someone to do something. For instance, I can suggest that a person shouldn't go any further by uttering the sentence "There is a bull in the field."[4] More generally, Austin and his followers have attempted to determine all possible ways of using language and to provide a classification for them. In the classification of

---

[1] We can illustrate this trend by mentioning just a few books that deal with ancient and modern science, as well as various academic disciplines: (Galison 1997; Netz 1999; Klein 2003; Mancosu et al. 2005; Giaquinto 2007).

[2] This refers to the title of Austin's posthumous book, which marked the opening of this field of inquiry: *How to Do Things With Words* (Austin 1962).

[3] Austin (1962, p. 5).

[4] Austin (1962, p. 32).

Austin's "speech acts" that John Searle has proposed and that has become standard, the utterance "I do," mentioned above, falls within the category of "declarative" use of language, whereas the latter, "There is a bull in the field," is referred to as "directive."[5] We shall return to these terms below, as well as referring to the other categories that have been identified.

What can be said about these other uses of language in scientific texts? In which ways can History and Philosophy of Science usefully consider them? Moreover, can Speech Act Theory be applied directly in order to address these questions, or should it be developed specifically to suit this purpose? These, briefly, are the questions this volume intends to consider.

Such questions are not devoid of meaning: it suffices to point out that some scientific writings contain texts for mathematical algorithms or medical prescriptions. In other words, as soon as we look for uses of language that are not merely assertive, we find examples in our sources: texts for algorithms and prescriptions are texts *that make* readers *do* things. By extension, let us call such texts "directive."[6] *How* these texts make readers do things is a question that has not been systematically examined so far. And yet, as we show in this volume, although inquiring into this question is by no means obvious, it casts a new light on actors' engagement with texts. At the very least, conducting a detailed survey of the various uses of language in scholarly writings will shed light on their textual complexity beyond the assertive function with which they have mostly been associated, or which has been sought for in them. This book aims not only to detect such uses, but also to analyze how actors performed these acts using language. As we will discover, the examination of these questions reveals features of scholarly writings that have been neglected so far.

1.1.2   How to Do Things with Texts

The two examples above—texts for algorithms and prescriptions—illustrate directly that Speech Act Theory can certainly be a source of inspiration, and that it has to be extended to enable us to reach our goal. Indeed, as the term "speech act" indicates, the theory was mainly conceived on the basis of an observation of circumstances in which language is used orally. We can also point out that historically, the study of speech acts has mainly been based on ordinary uses of language. It is true that Austin sometimes evokes "written utterances"[7] or considers the written version

---

[5] (Searle 1979, p. viii). See Chap. 2 by Virbel in the present book. The validity of Searle's classification has been questioned. See, e.g., (Vernant 2005). We shall not discuss these debates here; we only wish to call attention to a domain of research and modes of analysis that should prove fruitful for the History of Science.

[6] For the sake of simplicity, we only focus on one of the things these texts do: they enable users to carry out computations or manage medical treatment. As we shall see, sometimes this is not the only thing they do.

[7] (Austin 1962, pp. 8, 60–61). This expression seems to have an interesting history whose exploration falls outside the scope of the present book. We employ the expression "written utterances" below to refer to sentences as used in our sources.

of the speech act.[8] However, the main thrust of his theory is to account for *oral* events, whereas the phenomena we want to consider occur in written documents.[9] In fact, they occur in the framework of specialized uses of language. This is the first reason why Speech Act Theory needs to be reconsidered in the process of applying it for our purpose. Given this, in relation to our goal we shall use the expression "discourse act," rather than "speech act."

There is a second reason why Speech Act Theory must be extended. The theory was mainly devised to describe how *sentences* work in actual utterances.[10] However, for our purpose, as has already become clear, we have to consider not only how sentences, but also how *texts*—we mean *actual* texts like that of an algorithm or that of a prescription—are used to carry out directives or declaratives. In some of the language acts he considered, John Searle used texts as examples, notably when he discussed the case of the shopping list devised by G.E.M. Anscombe in 1957.[11] The same list of items, he emphasized—that is, the same text—can be used to perform (at least) two different acts. It is to be interpreted as a "directive" when used by the person who does the shopping: the list then dictates what has to be bought. It can also be an "assertive," if it is written down by a detective reporting on the shopper's actions: the list documents the customer's acts. However, seen from the viewpoint of these acts, we can observe that the phenomenon would be the same whether the list consists of a single word or a whole text. In this respect, the textual features of the list are "weak."[12] One may also argue that Searle considers "actual" texts when he addresses the issue of the "logical status of fictional discourse."[13] Note that in this case the point where Searle does address the textual dimension of the fiction is

---

[8] Austin (1962, pp. 57, 74–75).

[9] This sentence is a rough description of our set of sources, which is to be taken as a first approximation. We are well aware, as we show below, that some of these written sources adhere to oral activity and even sometimes record texts composed and/or practiced orally. In this respect, Sanskrit documents are always a source of warning and inspiration, since some of the written documents deliver texts that were once, and still are, practiced orally. See (Filliozat 2004), and our remarks below. This illustrates the usefulness of considering scholarly texts on a broad basis. The present book deals more generally with textual issues by relying on sources produced in many different places in the world and in different time periods. In this respect, as in many others, we are working within the framework of the options elucidated in (Chemla 2004).

[10] A similar comment occurs in (Vanderveken 2001) as an introduction to an attempt to describe speech acts not singly, but in the context of a conversation.

[11] (Anscombe 1957, pp. 56–57; Searle 1979, pp. 3–4).

[12] The example has a family resemblance with a whole class of texts —the "enumerations"— to which we devote Part II of this volume. Jacques Virbel analyses this example in Chap. 6, when he considers pragmatic features of enumerations (Sect. 6.4). In this context, he concentrates on the other dimension of the list, the enumeration. In the process he looks at the issue of how the list differs specifically from a single term.

[13] Searle (1979, pp. 52–75).

that of its coherence. This leaves open the question we want to consider: how does one do things with texts?[14]

Formulating this question allows us to perceive directly, even though indistinctly for the moment, how History and Philosophy of Science can benefit from raising these questions. An inquiry into this range of language phenomena is likely to provide tools to discover, in written sources, evidence about aspects of the practices, and not only the knowledge, of actors. Further, we shall see that this type of investigation rapidly leads us to terrain where scientific practice intersects with questions of law, and reflects social environments.

1.2 Discourse Acts and Textual Acts—The Example of Instructional Texts

The first part of this book presents a preliminary exploration of various types of acts carried out by means of "textual utterances" in the context of scholarly writings.

Our research did not start from scratch. In the last two decades, a group of linguists, logicians, neuroscientists and computer scientists, which Jacques Virbel was part of, has been developing a multidisciplinary approach to the textual acts carried out by means of instructional (or procedural) texts.[15] Texts giving instructions are meant to enable users to do things. As mentioned above, scientific documents also contain such types of texts, if we think of algorithms or prescriptions. It seemed natural to begin cooperating on this kind of "textual act." This is also the starting point for the present book.

---

[14] Barry Smith has raised similar questions in (Smith 2010 (August 23–26); Smith 2012). As mentioned above, Vernant has suggested reshaping the classification of discourse acts used by Searle. His main reason is to introduce "quotation" as a kind of discourse act. To do this, Vernant has introduced the more general concept of "metadiscursive acts," or acts that perform an action on the discourse itself. In performing such acts, actors also do things with texts. We shall return below to a phenomenon of that kind. In the field of science studies, Brian Rotman, in the context of attempting to understand how mathematics persuades, has noted the capacity of mathematical texts to "give commands" (Rotman 1998). Our focus in this book will be different. (Fortun 2008) also invokes Speech Act Theory to understand the part played by "promises" as speech acts in the recent "rapid rise of the science and business of genomics." The book analyzes how promises allow actors to raise funds and find biological materials. Fortun's analysis focuses more on the actual promises than the textual acts as such.

[15] Instructional texts were among the first types of texts approached as textual acts in linguistics. Results of this research work, done by Virbel and colleagues at IRIT (Institut de Recherche en Informatique de Toulouse—Toulouse Institute of Computer Science Research), were published in (Pascual and Péry-Woodley 1995; Pascual and Péry-Woodley 1997a; Pascual and Péry-Woodley 1997b; Grandaty et al. 2000; Virbel 2000). Their efforts and achievements convinced the historians of science who have contributed to the present book that we could draw on their expertise and the tools they had devised to tackle new problems in our discipline. This was the starting point of the present volume.

1.2.1   Speech Act Theory

In Chap. 2, Virbel begins by providing elements of Speech Act Theory that will yield
a theoretical basis for the whole book. Using the example of instructional texts, he
proceeds by illustrating possible extensions of the theory in order to describe "textual acts." This example will constitute one of the main foci of the volume.

Let us start by mentioning some of the principal basic ideas and concepts of
Speech Act Theory that Virbel introduces and that are taken up throughout this
book. The central notion is that of the "illocutionary act," which refers to the communication act performed by an utterance. Here we must introduce a distinction for
the sake of precision. When a speaker utters a sentence, he or she performs an act
that involves the body and produces a sound phenomenon. Speech Act Theory does
not dwell on this dimension of the act of communication (known as a "locutionary
act"); instead, it concentrates on the linguistic communication performed by the utterance. It is this dimension of the discourse act we are referring to when we use the
expression "illocutionary act." Following Searle, we take as our starting point the
assumption that the "illocutionary act" constitutes "the minimal unit of linguistic
communication."

Any illocutionary act has a "purpose:" this is the first key feature that characterizes it. The "purpose" captures the actor's intended action in performing a "discourse act." For instance, the point of the declarative "I do," mentioned above, is,
by virtue of the utterance, for the speaker to enter the state of matrimony. Or the
purpose of the utterance "I forbid you to go into the field" is to cause someone (not)
to do something. The theory refers to such a purpose as "the illocutionary point"
of the discourse act. The importance of this feature of the act is shown by the fact
that it provides a tool to identify illocutionary acts, and establish a first fundamental
classification of them. On the basis of a distinction between different types of illocutionary points, Searle surmised "five general categories of illocutionary acts."
We have already encountered three of them at the beginning of this chapter: the "directives," the "declaratives," and the "assertives." The five general categories also
include the "commissives," whose point is to commit the actor to do something, and
the "expressives," whose purpose is to communicate our "feelings and attitudes."[16]

These categories do not derive from an *a priori* approach to the topic, but from
an observation of actual uses of language. Once they were constituted, however,
Searle introduced the concept of "direction of fit" to explain the structure of the set
and the differences between the types of act. The concept encapsulates differences
among types of relationship between the world and the words that the various acts
establish.[17] It is best illustrated by the example of the shopping list. When it is used
for buying, the text of the shopping list performs an act that makes the world fit the

---

[16] Searle (1979, p. viii). In Chap. 2 of the present volume, Virbel provides illustrations for all these
acts (commissives in particular in Sect. 2.3.3.1).

[17] Searle (1979, p. 3–4), where the notion is introduced using the shopping list example. See also
Chap. 2 in this volume, Sect. 2.2, § 9.

text. When, however, it is produced by a detective, the text makes the words fit the world. All these notions play a role in the approach to "textual acts" outlined in this book.

The identification of these different types of illocutionary acts raises two sets of issues.

To begin with, how can we capture the differences between the acts themselves? We have already introduced a first element that enables us to do so. This is the "illocutionary point." It allows us to distinguish, for instance, between an "order" ("I forbid you to go into the field") and a "commitment" ("I promise you I won't go into the field.") More generally, the "illocutionary point" provides a criterion to define general types of discourse act. However, if the distinction between the various types of acts were only based on this criterion, it would be rather crude. We also want to be able to distinguish, for instance, between an "order" ("I forbid you to go into the field") and a "suggestion" ("There is a bull in the field.") In relation to this issue, and following Austin's first investigations, aside from the "illocutionary point" Searle suggested considering more broadly the "illocutionary force" with which an act is carried out. This force comprises a set of general features that enable one to characterize a discourse act and thereby distinguish between two acts. The "illocutionary point" is one of the components of this force. It contains others, and it is part of Searle's achievement to have developed the analysis of its various components. We need not go into detail here; it suffices to refer to the function of the "illocutionary force." Chap. 2 provides the description of the "force." Let us simply mention that its components notably include various types of conditions of success (essential conditions, preparatory conditions, sincerity conditions, propositional content conditions).[18] The components also include ways of describing the "modes of achievement" and "degree of strength" of the force.[19]

The first set of issues evoked in the previous paragraph looked at discourse acts from the viewpoint of *what* they perform. The second set of issues raised by the identification of illocutionary acts focuses rather on *how* these acts are carried out. The question is: how can we describe the differences between the utterances used to perform the illocutionary acts? In the terms just introduced, we can reformulate the question as: how are the illocutionary forces actually performed? In this vein, a study of the markers of illocutionary forces has been developed; Virbel alludes to this in Chap. 2. On the other hand, it is easy to see that the *same* utterance can carry out *wholly different* illocutionary acts. John. R. Searle 1969, pp. 70–71 discusses the phenomenon with the elementary example of, "It's really quite late." Depending on the context, the actor saying, "It's really quite late," can perform different acts. For instance, an act that corresponds to what the sentence means at face value, that is, the statement of a fact. By means of the same utterance, however, the actor can perform "an objection," "a suggestion, or even a request," or "a warning." In

---

[18] These conditions are explained in Chap. 2, Sect. 2.2, § 4, 7, 8, pp. 51–54. They are also illustrated for the case of commissives at the beginning of Sect. 2.3.3.2.

[19] This element of the illocutionary force was introduced later; see (Searle and Vanderveken 1985, pp. 12–20).

the last four cases, we speak of an "indirect illocutionary act." The term, "indirect illocutionary act," refers to the use, in order to perform an illocutionary act, of an utterance that, when employed directly, performs another illocutionary act (in the example just quoted, that of stating the fact that it is late).[20] The act performed can depend on who utters the sentence as well as to whom it is addressed. For instance, the sentence, "Dinner is served at 8 pm," uttered by the hotel manager is a directive when said to the chef; a commitment when addressed to customers; and an assertive in the context of an interview with a journalist.

It is obvious that these phenomena also occur with texts. We have already mentioned the case of the shopping list, in which exactly the same text can carry out different "textual acts" depending on the context. We can also refer to the text of the "program of a conference." As a text produced by organizers for participants, the program is a commitment. But from the perspective of the program that the organizers hand over to a chairperson or a speaker, the same text performs a directive, and when recorded in a report for a research assessment agency, the text performs an assertive. We shall see that "written utterances" and texts contained in scholarly sources also testify to these phenomena.

In turn, such phenomena raise a new question: when a single utterance can carry out different illocutionary acts, how do users of language understand what is meant? Here, Austin introduces a fairly useful distinction to clarify the question. The issue of how users construe the discourse act does *not* bear on the understanding of the propositional content of the sentence. In fact, it bears on another dimension of the understanding: that which determines the illocutionary act performed. To distinguish between the two types of understanding, Austin introduces a technical term. He designates by the term "uptake" the specific type of understanding that allows the "hearer" to grasp the intended act, in contrast to his or her understanding of the propositional content. Again, we shall see that this concept is useful to describe "written utterances" in scholarly documents.

However, clarifying the question does not solve the puzzle: how do users of language understand what is meant? More generally, how can one anticipate and describe which utterance is likely to be used for a given illocutionary act? As Virbel explains in Chap. 2, in recent decades it has been possible to clarify these questions in an impressive way. In the event, the logical analysis of illocutionary forces referred to above provided essential tools that enabled researchers to address these questions.[21]

In the case of the scholarly texts discussed in this book, which were written in many different languages and contexts, the questions related to utterances all raise important and tricky issues. To begin with, how can we establish the illocutionary force of "written utterances," that is, the nature of the illocutionary act they carry

---

[20] For an analysis of the various ways of carrying out, e.g., directive illocutionary acts, see Sect. 2.3.2.3 and 2.3.4.3 in Chap. 2 of the present book.

[21] Searle and Vanderveken (1985), in particular Chap. 2. See also, in the present book, Chap. 2, Sect. 2.2, points 10 to 12, pp. 55–61, for a theoretical explanation, and Sect. 2.3.2.3, p.72, for an illustration by example.

out? Moreover, and perhaps more importantly, how do the ways in which "illocutionary acts" are carried out by means of actual utterances reflect the contexts in which they were produced or, more precisely, the local scholarly cultures to which they adhere? Some chapters of the book address these issues. More generally, as we shall see, the notions of Speech Act Theory that are presented in Chap. 2 and were developed for a study of "utterances" provide tools for dealing with the sentences that occur in scholarly texts and drawing interesting conclusions from them.

1.2.2 Instructional Texts

As the second part of Chap. 2 explains, these notions were also essential in enabling Virbel and his colleagues to sketch a linguistic and pragmatic treatment of "textual acts:" instructional (or procedural) texts. These texts are important for the present book.

The first step in extending Speech Act Theory to the treatment of instructional texts is to consider these texts as a means to carry out illocutionary acts. In this case, the acts of "giving instructions" are, at least partly, of a directive type. As such, the illocutionary point of their authors —and not that of a speaker, the "S" of Speech Act Theory—is clear: they aim to "help" a reader or a user—and not a hearer, the "H" of Speech Act Theory—to do something. In this respect, the illocutionary point is materialized precisely by what the actor does, that is, by the actions he or she performs in compliance with the text. This is how, even for documents from the past, these acts can be studied as such: the historians' task is to find methods of reconstructing the acts readers, after familiarizing themselves with such documents, could perform.

In fact, through instructional texts authors often perform not only directives, but also other types of illocutionary acts. For instance, depending on the kind of instructions given, the acts of "giving instructions" can combine, in various ways, a directive and a commissive of some kind. The latter dimension captures the relationship of commitment established between the author and the user through the instructional text.

Virbel's presentation allows us to distinguish three modes of contribution of Speech Act Theory to the study of acts of "giving instructions." We shall now outline them, as they illustrate a general method that can be applied in the case of similar extensions.

Firstly, Speech Act Theory provides tools to describe the type of utterances to be encountered in instructional texts. In this book, in relation to the aim they pursue, Florence Bretelle-Establet (Chap. 3) and Karine Chemla (Chap. 9) make an inventory of this kind for medical and mathematical texts.

Secondly, Speech Act Theory provides a model for the study of new discourse acts. In particular, Virbel begins his study of the various types of illocutionary acts instructional texts can carry out by borrowing Searle's approach to Speech Acts and considering "different types of differences between different types of instructional texts." Note that these types of differences offer a multitude of research paths to historians of science interested in the study of procedural texts, as well as other types of text.

Thirdly, Speech Act Theory provides a description of the various types of discourse acts to which verbs such as "advise" or "prescribe" correspond. For the sake of simplicity, we shall call these acts "verb acts." Drawing on the study of "verb acts" described in Vanderveken 1990, Virbel shows how they form grids that allow him to situate any act of "giving instructions" as a discourse act by looking at its similarities and differences with respect to other discourse acts.

Virbel's method is systematic. Looking at "verb acts," he considers those to which instructional texts have some relation. Verbs that name "directives" include "order," "prescribe," "suggest," and so on. Those that name "commissives" include "promise," "guarantee," and so on. Then, for each type of illocutionary act, he offers a classification of these verb acts by illuminating the key components of the illocutionary force that allow us to capture the differences and relations between these acts. By this method he can develop a robust backdrop against which he can situate any specific act of "giving instructions."

Virbel begins by doing this with directives (See Sect. 2.3.2). He then turns to the area of commissives in order to situate within it the kind of "guarantee" that, depending on the context, the act of "giving instructions" involves (See Sect. 2.3.3). This analysis enables him to describe the combinations of directives and commissives that the acts of "giving instructions" constitute and to interpret the various possible meanings of the expression "with the help of" that characterizes instructional texts (See Sect. 2.3.3.2).

Virbel then considers assertives (See Sect. 2.3.4) as they are performed in instructional texts. On the one hand he highlights how, in this framework, assertions can combine with declaratives, in diagnostic texts, for instance. The reader will be given examples from medical books studied by Bretelle-Establet (Chap. 3), and publications on zoology described by Yves Cambefort (Chap. 4). On the other hand, Virbel focuses particularly on original forms of indirection that one encounters in instructional texts, for example modalities according to which assertions of the reasons for doing or not doing something perform directives. Such discourse acts occur in the medical books analyzed by Bretelle-Establet (Chap. 3) and in the texts for algorithms composed in ancient China, as examined here by Chemla (Chap. 9).

In conclusion, the purpose of approaching "giving instructions" as a language act has led Virbel to develop Speech Act Theory further, notably with the shaping of local classification of "verb acts." Conversely, Speech Act Theory inspires here an approach to text as the carrying out of acts. In particular the "illocutionary point" of the author(s) when writing a text, as well as other elements that have proved essential in describing a language act, appear to provide analytical tools to develop a linguistic and pragmatic approach to texts. In the present book, Agathe Keller (Chap. 5) and Chemla (Chap. 9) concentrate on procedural texts, as Virbel does. However, Bretelle-Establet (Chap. 3) and Cambefort (Chap. 4) extend the approach presented in Virbel's contribution beyond instructional texts. They, too, draw on the concepts and methods of Speech Act Theory to address other forms of textual acts. The succeeding chapters of Part I consequently describe specific acts carried out by means of texts, while at the same time the authors consider the issues at stake for History of Science.

1.3   How to Make Readers Do Things: An Inquiry into the Variety of Scholarly Writings

The chapter by Bretelle-Establet is devoted to Chinese medical writings. The key question she wants to address, and for which she finds the description of "textual acts" fruitful, can be formulated as follows: a huge number of medical books has been composed and published in the whole territory of China since the seventeenth century. How can we grasp differences between them without using observers' categories, which are in fact fairly superficial? Indeed, to understand contrasts between some of these books, historians have often employed opposites such as "scholarly" and "popular," and have attempted to classify them into genres. Are these categories meaningful?—and, if so, which sense can we attach to them? To tackle these questions, Bretelle-Establet puts forward a subtle and powerful strategy which promises to yield results for a much wider set of sources than those she bases her approach upon. She defines several criteria that allow us to perceive a variety of textual features in these books and that can be used to characterize the different ways in which authors have textualized medical knowledge.

1.3.1   The Definition of a Corpus and its Key Features

Bretelle-Establet's strategy begins with the definition of a corpus. She concentrates on books produced since the seventeenth century in the southernmost provinces of China, that is, at the margins of the empire and, like most of the writings that have come down to us from that time period, outside the central institution of medicine. Some of her selected books were compiled in prosperous urban areas, whereas others were used in poor, rural regions where physicians were scarce. In addition, Bretelle-Establet chooses to focus on books written by authors whose social status differs quite considerably. The contexts of production of the selected books vary in these two main respects. Furthermore, the authors' motivations for producing these books, and hence their intended readership, were diverse. Some authors wrote a textbook that could be used in the context of medical studies; others wrote for family use and to enable relatives to self-medicate; and there were others who wrote to facilitate emergency care in urgent cases where no physician was available. Bretelle-Establet notes that all these books aim to enable readers to do things, whether the action be learning or healing. This means our framework here is the study of texts carrying out "directives." However, in the terminology introduced by Searle and explained by Virbel in Chap. 2, there is a difference between the "illocutionary points" of the books—what they intend to make readers do.

On the basis of this corpus, Bretelle-Establet develops her questions as follows: in which respects are these books different and how can we correlate the differences highlighted with the geographical environment in which the books were written, or the social status of the authors, or the purpose of the book? Bretelle-Establet focuses on how each of these books deals with the same nosological entity, which she translates as "sudden disorder." These are the actual "texts" she uses as a basis for her analysis of how the "textual acts" are carried out.

In the first part of her analysis, where she concentrates on the semantic features of the texts, Bretelle-Establet examines the types of information provided, and, where applicable, she compares the information of the same type found in the various selected writings. It comes as no surprise that the texts differ in this respect. The reason is that the understanding of the ailment and the theories expounded do not concur. These differences, Bretelle-Establet emphasizes, illustrate that medicine in China was by no means a unified body of knowledge and practice, a fact that historians have endeavored to understand better in recent decades. However, the texts also differ in the *nature* of information of a given type that is provided, and this fact reveals an essential issue for the present volume. For instance, the various kinds of clinical signs the texts provide to help users diagnose an ailment demonstrate that the authors did not assume the same types of competency in their intended readers. The competency and knowledge required for readers to use a text have not been a topic of systematic inquiry in the History of Science—yet this issue proves to be very fruitful, as the present volume shows. In terms of "textual acts," these aspects characterize the conditions an author assumes with respect to his or her readers in order for the act to be successful. In the case of the directives analyzed, these types of competency and knowledge are part of what John Searle and Daniel Vanderveken have described as their "preparatory conditions."[22] We shall return to these issues below.

In this first stage of her analysis, Bretelle-Establet does more than simply comparing the types of information given by the various texts: she tries to determine whether differences in this respect can be correlated with factors characterizing the contexts in which the books were written. Here she obtains her first surprising result. The types of information contained in a text correlate less with the author's social status than with the declared aim of the book and its intended readership. To begin with, this conclusion suggests that the opposition between "scholarly" and "popular" texts is apparently not meaningful in this context. More generally, it invites reflection on the use of observers' categories in descriptions of historical writings. The following, however, is more important for the purpose of our book: if we reformulate the conclusion in the terms of the formula "who writes for whom and to make them do what?" what Bretelle-Establet shows is that, for this part of the analysis, the "whom" and the "do what"—the illocutionary point

---

[22] Searle and Vanderveken (1985, pp. 16–18), Searle and Vanderveken (2005, pp. 123–124).

of the texts—prove to be the key factors that allow her to account for distinctive features of texts.

1.3.2   Books Meant for Different Readings in Different Kinds of Places

The second part of Bretelle-Establet's analysis bears on syntactical properties of the writings. She identifies several features as meaningful for identifying differences between the texts selected, distinguishing between syntactical properties at the level of the whole text (e.g., the choices made in organizing the information[23]) or at the level of propositions. One conclusion she derives from the examination of all these features is notable for the present volume: differences with respect to syntactical properties can all be correlated to the distinct types of use for which the books were intended. Some writings were suited to pragmatic reading which could be done during a journey and enabled the reader to act, even quickly, whereas other books were more appropriate for slower reading, maybe in the reader's own study, and were designed for understanding first and acting only afterwards. All this can be captured in a whole collection of formal features of the texts. These remarks highlight a key fact: the illocutionary point of a text is reflected, on the one hand, in syntactic features of the text—both global and local—and, on the other, in the specific type of reading for which the text is suited. Once again, among all the possible explanatory factors, it is the illocutionary point of a book that appears most important in accounting for distinctive textual features.

Among the syntactical features at the level of propositions that Bretelle-Establet examines, one is particularly important for us here: the distinct uses she shows authors make of "textual connectors" and "discourse markers." Both these entities correspond to what in classical Chinese are called "particles," and in principle they are dispensable. The former category of particles is mainly used to make links between successive propositions explicit, whether these links are logical, temporal, or relate to other kinds of connections. The latter category is even more dispensable; it is a linguistic resource an author can use to express attitudes with respect to a given sentence. On the one hand, Bretelle-Establet argues that the number and variety of particles used probably reflect the author's attempt to write in a scholarly style. One of the longest texts, which is full of particles, was written by a polymath. It is probably not accidental that, of all the authors considered in Bretelle-Establet's chapter, he enjoyed the highest official status and compiled his book for the study of medicine. On the other hand, Bretelle-Establet suggests, the choice to limit the use of such particles could relate to the intention of keeping a book to a handy size. It is revealing that the two books that explicitly mention the issue of portability use barely any

---

[23] What is important about this feature is that it reflects Florence Bretelle-Establet analyzes texts rather than sentences. In the same way as for the parts of a sentence, different ordering of the sentences in a text makes different pieces of information salient and creates a hierarchy between them which correlates to the point of the texts.

discourse markers, if at all (see Table 3.5 in the chapter under consideration). This particular syntactical feature endorses the general conclusion mentioned above that Bretelle-Establet draws from her overall study of these features.

1.3.3   How do Authors Attempt to Achieve their Aims?

As we shall see, the reason why "discourse markers" are particularly important here is because they are related to the third part of the analysis developed by Bretelle-Establet. In this part she addresses the issue of *how* authors "guide their readership to follow their first intentions." The texts, as mentioned above, aim at making readers do something. It is in order to analyze *how* authors achieve their aim in different ways by means of textual features that Bretelle-Establet makes use of Speech Act Theory, but this time at the level of the sentences that make up the texts. In other words, she needs to examine the acts performed in the "written utterances" to carry out her analysis of the "textual acts."[24] The first noteworthy result is the significant number of "written utterances" performing "directives" in all these texts (see Table 3.7), a confirmation that the authors specifically intended to make readers do things. Bretelle-Establet also shows the variety of types of assertives as well as directives that these texts contain. To use the terms introduced by Virbel in Chap. 2, we find directives of the following types: direct, implicit, and indirect, by stating conditions of success or reasons for performing (or not performing) the action. Bretelle-Establet illustrates these categories in the translations of the texts given in the appendix of the chapter. More generally, "directives" are a type of "discourse act" which constitute a primary focus of this volume. We shall encounter these distinctions again.

Here we should note an important and general point that Bretelle-Establet emphasizes for the case of her medical sources: the intention of enabling readers to do things is not in contradiction to developing theoretical explanations. It has been often argued in the History of Science that writings aimed at enabling action were meant for lowbrow users limited to doing things without understanding them. This thesis has often been advanced with respect to algorithms in mathematical sources. We can see in the case of medical writings, as we shall also see in the case of algorithms, that this assumption does not fit with the evidence we have—a fact which becomes evident when we observe precisely how directives are carried out. This is one of the questions at issue in our study of "written utterances."

To return to medical books, what is most striking is Bretelle-Establet's conclusion that we can distinguish between different kinds of "textual directives" and identify the various types of goals assigned to books if we observe the distinct

---

[24] The exploration of the relationships between these two levels is an open question raised by applying Speech Act Theory to texts. Should we approach this question by looking at how "written utterances" are combined to make the "textual act," or is there a specific textual level? We leave this question open. It has been touched upon in (Nef 1980; Smith 2010 (August 23–26), Smith 2012; Vanderveken 2001).

distributions among various types of discourse acts performed by the "written utterances" contained in each text.

So far, the issue of how authors determine ways to achieve their goals for the reader was approached using the discourse acts carried out in their texts. The issue can also be viewed from another angle, that of the authors' textual self-fashioning. The last set of questions Bretelle-Establet raises that are worth noting for their general implications deals with this very point.[25] Bretelle-Establet begins by remarking that the various texts differ in the way they make readers feel the presence of the author. This is the point where "discourse markers" are important. They represent one of the textual techniques some authors use to make their presence strongly visible, as opposed to those who produce "highly impersonal texts." Interestingly enough, as Bretelle-Establet shows, only texts where the presence of the author is marked contain quotations by authors from the past. These texts establish and display their authors' authority in a specific way and by means of specific acts. Given this, Bretelle-Establet demonstrates how the same specific techniques display not only the presence of the author, but also his or her commitment with respect to the book's content. By contrast, Bretelle-Establet suggests that impersonal texts establish their authority in a different way, and this fact echoes the specific way in which they achieve the intended effect on the reader. (In the technical terms of Speech Act Theory, we would say this is how these texts fulfill their perlocutionary goal.)

To sum up, rather than offering a theory of genres for classifying all medical texts produced in China, Bretelle-Establet provides an analytical grid that enables historians to rely on textual features of the writings to capture differences and similarities between texts in a much more differentiated way. In particular, she offers a reflection on the various features by means of which a text reveals how an author performs the act of making his or her prospective readers do things. In our view, this is one of the key issues in a close examination of the texts inspired by Speech Act Theory.

---

[25] This gives us the opportunity to qualify the statements made at the beginning of this introduction. In recent decades, several books have focused on a rhetorical approach to scholarly texts. See, e.g., (Loveland 2001, pp. 17–23), which provides a useful survey of recent publications on the topic; (Ceccarelli 2001; Gross 2006). These rhetorical approaches also explore a dimension of scholarly texts which is not purely assertive. A review of research paths developed in this context would exceed the scope of this book. In relation to what Bretelle-Establet explores here, we can simply note that several of these books have used rhetorical analysis as a useful tool for understanding how an author attracted readers or persuaded them of the validity of his or her arguments. However, they still focus mainly on the assertive aspects of the text. Here, Bretelle-Establet considers the author's self-fashioning through textual shaping by means of other textual acts.

1.4   Christening Organisms: Declaratives as Textual Acts in Zoology

Cambefort's chapter is devoted to zoology and focuses on the examination of another type of "textual act," a declarative. Cambefort analyzes declaratives in his study of the way in which zoologists use publications following certain rules to institute a name for an organism (or more precisely, a group of organisms), as well as instituting and describing this group—a "taxon," e.g., a species, to which the organism belongs. Interestingly enough, zoologists themselves refer to such texts as performing a "nomenclatural act." In this case, then, the idea of a "textual act" is an actors' category.

By means of its very performance, a successful "nomenclatural act" changes the state of the world. This is why, using the criteria John Searle introduced to distinguish between speech acts, Cambefort suggests that the act the text performs belongs to the category of "declaratives." Before it is carried out, some animals are unknown. The species—in the case where the taxon is a species—does not exist, and nor does the name as Cambefort explains it. Once the act is performed successfully, a species has been created, the organisms observed are stated to belong to it, and the name will be attached to them both forever, or at least as long as zoologists retain their nomenclatural practices unchanged. In a sense, the act introduces something in the world—a taxon, to which a previously unknown being now belongs—and something in our language that is attached to it by virtue of the "declarative."

1.4.1   The Historical Shaping of a Textual Act—Its Textual and Legal Features

The creation of both the taxon and the name also has a legal dimension. This is made explicit in, and governed by, an *International Code* drawn up by the International Union of Biological Sciences and regularly updated. It stipulates the formal conditions governing all aspects of the "textual act." Many interesting issues are at stake here. Firstly, in dealing with the rules governing the texts as such, we immediately encounter legal and social dimensions of scientific practice. We also see facets of the texts that we would miss if we considered the publications carrying out this act as simply descriptions. Attempting to describe the institutions created to perform such "textual acts," and the consequences of these acts, is a way to perceive more precisely what is at issue in these texts. It is also a method of understanding how the publication of texts relates to other features of scientific practice. Lastly, the regular updates of the *International Code* express the part of the scientific work carried out by biologists that is devoted to working out formal rules for publications. We shall return to this point later. It testifies to how designing textual acts has been an important component of scientific activity.[26]

---

[26] This was discussed in (Chemla 2004).

These are some of the questions at issue for historians of science. It is also interesting to consider these texts from the viewpoint of the acts they perform and ask how they compare to other types of declaratives. In exploring the specificities of this type of "textual act," Cambefort underlines differences which reflect standards instituted by the discipline of zoology governing how such names and the taxa they designate should work when properly handled.

Firstly, conditions of success of the "textual act" include that only unknown organisms can be named, and the names must satisfy particular rules of formation adopted in the discipline. A proof that a condition is not fulfilled can be a reason to nullify a "nomenclatural act." This statement reveals that in the wide class of "declaratives," that which introduces a name and a taxon is one of the contestable "textual acts." Practitioners of zoology regularly dispute the validity of this kind of declarative. The scientific activities that are then done to confirm or invalidate the "textual act" highlight the role of these declaratives in zoological practice today. These activities are materialized in the form of the various types of publications, or "secondary nomenclatural acts," which contest, confirm, correct, or augment the "original description," or "primary nomenclatural act."

The reasons why a declarative can be contested reflect the various dimensions of the act. It can be invalidated because of the *name* introduced. In particular, if the name is shown to be already used for a similar purpose, another name has to be introduced. The importance of this aspect of the declarative is materialized by the tools created within the discipline, such as registers, to assist practitioners in designing names. It can also be contested on account of the *taxon*, for instance, if its novelty is in doubt. The recognition of this refutation as valid leads to canceling the declarative: the "textual act" is declared to have failed using a similar procedure to that which aims to perform it. In such cases, there is a rule determining which name should be kept for the taxon: the first printed name has absolute priority, while the names introduced later remain as "synonyms." This highlights the fact that a declarative of that type is invalidated in an unusual way: the part that deals with the creation of the taxon remains the only one affected by the failure, while the creation of the name in general is successful, with the name leaving a permanent trace in the attributes associated with the taxon in the specialist publications. It also reveals the emphasis zoologists have placed on defining rules for the management of names and taxa. They have elaborated the conditions for defining "naming" as a "textual act."

Secondly, present-day zoologists recognize the name and the taxon as effectively created only if they are "notified" in a specified way: the text carrying out the "declarative act" must be printed and published. Oral introductions and other modes of publication are not accepted as valid. The discipline of zoology has thus progressively defined formal conditions for a "declarative" to achieve success and, as Cambefort explains, the conditions are constantly under discussion. In Chap. 2, Virbel called attention to the distinctions between language acts, as well as between textual acts, with respect to their promulgation. We now see how the modality according to which this has been realized has been a topic of discussion and elaboration for zoologists. As part of a history of scientific practice, History of Science can set itself the task of inquiring into the historical shaping of these "textual acts" and their management principles. Cambefort goes on to describe the specifications of the textual artifact for achieving these "declaratives" today. It is also worth considering how each of these features was historically designed and came to be part of the "declarative," and also how the "textual act" as such was shaped accordingly.

Let us now mention some of these features to highlight issues of interest to us. To be successful, the "nomenclatural act" must not only create a name, but also provide a description of a specimen. It must also designate the specimen and specify where it can be found. As Cambefort formulates it, the "holotype"—the modern term for the specimen—is intended to remain the object of reference for the species name, and serves as a "proof" of the textual "act." Accordingly, the *Code* demands that the holotype be deposited in a public collection and freely available for consultation. We thus see how the discipline designed public collections in close connection with publications. Cambefort notes present-day debates about the rules governing the materiality of the holotype. This indicates the interplay of forces that brings about the transformations effected in zoological practice.

For any type of declarative, the status of the person performing the act is essential for it to achieve success (John Searle 1979, p. 26.) In recent decades, Cambefort indicates, zoology as a discipline has diminished to the benefit of other subfields of biology. Nomenclatural acts have partly become amateur endeavors, as they once were in former centuries. As the status of the author of the declarative changes, the role of his or her status in the success of the nomenclatural act becomes increasingly visible. It is interesting to examine, as a key feature of the "textual act," an author's attempt to establish his or her authority in the text to perform the act. Cambefort outlines means to fulfill this aim, including ways of naming, or acknowledgements. Interestingly, he suggests how the degree of emphasis in stating the authority is partly a function of the nature of the nomenclatural act performed.

1.4.2   Encountering a First Enumeration

The "nomenclatural act," Cambefort notes, as it is performed in a publication fulfilling all mandatory criteria, requires several other kinds of "textual acts" to be carried out. Two kinds of these play an important role: enumerations and "directives." For instance, the description of the specimen, as well as the diagnostic features characterizing the taxon, are textualized as enumerations. On the other hand, "instructions" are given to enable one to determine whether a given organism belongs to the taxon and to situate the taxon in relation to cognate taxa. Enumerations and "directives" are two kinds of textual acts to which the succeeding chapters return. From the perspective of how they are written down, the instructions, as illustrated on pp. 311–312 (see Fig. 1.1) of the first example given by Cambefort, display interesting textual features that are important for the theses of the present book. We shall now focus on them to introduce some notions that will subsequently prove useful.

The text of these instructions takes the form of an enumeration whose individual items are introduced by a number. This yields a combination of a directive and an

KEY TO SPECIES OF *CARINOSQUILLA* MANNING, 1968

1. Eyestalk with irregular dorsal carinae 4
— Eyestalk without dorsal carinae 2

2. Dorsal carinae on either side of midline of TS5 transverse. Inner margin of uropodal protopod spinose 3
— Dorsal carinae on either side of midline of TS5 longitudinal or oblique, not transverse. Inner margin of uropodal protopod crenulate .... *C. lirata*

3. Mandibular palp present. Telson prelateral lobe with sharp apex .... *C. multicarinata*
— Mandibular palp absent. Telson prelateral lobe with blunt apex .... *C. carita*

4. Ocular scales with bifurcate apices 5
— Ocular scales with apices entire, not bifurcate 7

5. Posterior margin of AS1-4 between submedian carinae lined with spines ...... *C. spinosa*
— Posterior margin of AS1-4 between submedian carinae unarmed 6

6. AS1-2 with unarmed submedian and intermediate carinae .... *C. redacta*
— AS1-2 with armed submedian and intermediate carinae .... *C. carinata*

7. Merus of raptorial claw with single longitudinal carina on outer margin 8
— Merus of raptorial claw with vermiform sculpture on outer margin 9

8. AS6 without posteriorly armed supplementary carinae between submedian and intermediate carinae. Distal segment of uropodal exopod entirely dark .... *C. australiensis*
— AS6 with one or two posteriorly armed supplementary carinae between submedian and intermediate carinae. Uropodal exopod distal segment dark on proximal third ....
.... *C. balicasag*

9. Dactylus of raptorial claw with five teeth. Dorsal carinae of AS6 and telson entire, not forming field of spines .... *C. thailandensis*
— Dactylus of raptorial claw with six or seven teeth. Dorsolateral carinae of AS6 and telson divided, forming field of spines .... *C. mclaughlinae* n. sp.

Fig. 1.1 Excerpt from "A new species of Carinosquilla..." Shane T. AHYONG. (Example 1, in Chap. 4, by Y. Cambefort)

enumerative act. Each item of the enumeration is composed of a pair of sub-items, the second sub-item beginning with a dash. At the end of each line, the reader finds either a number or an italicized name, which is the name of a taxon. The reader knows that each list of two sub-items represents an alternative. Formally, the text lists key diagnostic alternatives, which enable one to distinguish between taxa. It states them all as similarly essential for the desired aim. This is the first thing done with the text. We shall return later to this dimension of the act performed using the enumeration. In fact, as its title indicates, the text constitutes a "key" or "table." In this respect it carries out an assertive and a directive simultaneously, as with the shopping list or the program of a conference referred to above. However, it is easy to see that in the present case the method for handling the text of the enumeration differs from the use of the shopping list or the program. Accordingly, the "key" has a characteristic textual feature: the numbering of the items.

When using the "key," the reader knows he or she should begin reading at the first item. Whether the reader opts for a sub-item with the aim of determining the taxon of an organism under observation (text as directive), or reads the relationships between the taxa whose names appear in italics in the "key" (text as assertive), he or she knows that at the end of the first or the second line of item 1, the number indicates the number of the next item to be read. This principle holds true for all items, except when the italicized name of a taxon occurs at the end of a line. The use of the text as "key" thus leads to a reading of the enumeration in the form of content of a tree. The end of the lowest branches of the tree is found to contain either the name of the taxon for an organism or the list of the taxa classified, depending on the type of reading (text read as instructions or as description).[27]

What is important here is that, whether it is read as a directive or as an assertive, for appropriate use the text requires various types of textual competence. An essential feature of these types of competence is the knowledge of how to circulate in the text to work out its content adequately. The reader needs these types of competence to *do* his or her diagnosis *with the text*. Relying on the intended readers' textual competency is a requirement for "securing uptake." In this respect we are transposing Austin's expression to the use of texts (Austin 1962, p. 116.) Note that in this example, the circulation must be achieved in a precise way, in contrast to the circulation in the shopping list, which is more open. Despite superficial similarities, the two texts do not carry out directives in the same way. This correlates with the different types of textual competence the users require.

In some of the texts analyzed by Bretelle-Establet, particularly those suitable for reading slowly, textual circulation within the book was also an ability the authors relied on in organizing the information presented in the book. Illuminating the various types of textual circulation expected from readers for different kinds of texts written in different contexts is one of the important results presented in this book. The variety of such circulation and other types of textual competency that our sources require to "secure uptake" demonstrate the various cultural backgrounds forming the context of production of the texts under analysis, as well as their different purposes. In particular, as we shall see in Christine Proust's chapter, some mathematical tablets from Mesopotamia share several common features with the tables described by Cambefort. However, the tablets present distinct textual realizations of trees.

1.5   Texts for Directives in the Context of a Scholarly Culture: The Uptake Issue

Before returning to our focus on enumerations and the question of textual circulation in documents, we will continue our exploration of the important issues for History of Science in considering textual acts attested to in scholarly sources. The two examples discussed above illustrated different types of agenda for which focus-

---

[27] (McCarthy 1991) shows a similar interest in how writings can shape users' scholarly knowledge. McCarthy examines how the *Diagnostic and Statistical Manual of Mental Disorders* (1980) shapes psychiatrists' approach to illness, their gathering of information, their diagnosis, and how they communicate the acquired knowledge.

ing on textual acts benefits History of Science. In particular, Bretelle-Establet and Cambefort approach the question of "how authors do things with texts" from different perspectives. In Bretelle-Establet's case we have seen how a book's semantic, syntactic and pragmatic features reflected the textual directive the author aimed at performing. She also emphasizes that the nature of the directive carried out correlated with the expected type of reading and the intended environment for the book's use. Lastly, Bretelle-Establet revealed ways in which textual features demonstrate the various modalities by which authors established their authority in relation to the achievement of their goals. Establishing the author's authority also proved important in the case studied by Cambefort for the fulfillment of a "nomenclatural act." As we have seen, Cambefort correlated the modes of displaying this authority with the social status of the author in the context of a discipline where the balance between amateurs and professional actors was changing. However, there seem to be other issues at stake in Cambefort's chapter, relating to a discipline's historical shaping of valid textual acts, and the way textual acts actually shape practitioners' actions and knowledge.

These two examples show how awareness of the textual acts performed with texts enables us to perceive and interpret their specificities. They also provide two approaches to the question as to "how authors made readers do things with texts." The final chapter of Part I of the book (Chap. 5) examines documents that also attest to the performance of directives by means of texts. The chapter aims, however, at highlighting from a different angle the possible benefits to both History of Science and the study of "textual acts" from considering textual acts performed in scholarly documents. We focus on texts for algorithms as found in treatises written in Sanskrit in the fifth and around the tenth century. Establishing that these texts carry out directives, and how they do this, proves to be a real challenge. On the other hand, if we failed to attempt to describe the textual acts performed there, we would risk misinterpreting the sources and misrepresenting the authors' utterances. As Keller shows, this has often occurred in the past. What is at stake here is interpretation by means of identifying the textual act carried out in a source. To transpose Austin's specific term for the phenomenon to the use of texts once again, the problem is to "secure uptake."

1.5.1 Perceiving Textual Acts and the Issue of Interpretation

To introduce the problem and argue a solution, Keller relies on a corpus of five key texts, selected from Sanskrit mathematical texts composed between the fifth and presumably the twelfth century.[28] The composition method of the corpus is similar to that used by Bretelle-Establet in her study of the variety of Chinese medical texts from Southern China. The five key texts for Keller all deal with square root

---

[28] This is a coarse presentation. The reader will find in the chapter more precise information about the corpus.

extraction and even with the same algorithm to perform this operation.[29] The key point is that the textual contexts they belong to differ. Two of these texts are found in versified treatises, whereas the three others were composed as part of prose commentaries on these treatises. In addition, the first treatise, the *Aryabhatiya*, completed by Aryabhata at the end of the fifth century, is a theoretical astronomical book that devotes one of its four chapters to mathematics. In fact, many Sanskrit mathematical sources are chapters in astronomical treatises. By contrast, the *Patiganita*, composed by Sridhara in around the tenth century, is a book entirely devoted to mathematics and is presented as dealing with "practical mathematics."[30] Two of the commentaries analyzed by Keller in Chap. 5 bear on the *Aryabhatiya*. They are the seventh-century commentary written by Bhaskara I, and the twelfth-century commentary by Suryadeva Yajvan. The third, whose author and dating are both unknown, bears on the *Patiganita*.

We shall now focus on the case of the *Aryabhatiya* to explain the problem Keller tackles in Chap. 5. The treatise, like all those of the same type, is composed of *sutras* (rules) formulated in verses. Aryabhata devotes a single verse to square root extraction. The question is: what does the verse say on the topic? We don't know the context in which the book was written. We have no evidence as to how Aryabhata intended his verses to be read or how readers at that time understood them. However, there are two commentators, and their reply to the question is unanimous. In their opinion, Aryabhata "states a square root computation." He provides a text to enable readers to compute square roots. In our terms, the textual act carried out by Aryabhata's verse is a directive.

The next question is: how does he carry out this textual act? In fact, we can establish the algorithm that is the subject of the *sutra*. Keller's Fig. 5.2 summarizes its thirteen or fourteen steps. If we observe Aryabhata's verse, "stating the computation," we can't help being surprised. To elucidate this, we shall outline its content against the backdrop of Fig. 5.2, p. 187. In our view, the verse does not state the point of the computation. It contains no explicit indication as to where the procedure should begin or end. It provides no description of the analysis of the decimal expansion of a number according to a place-value system, on which the algorithm essentially relies. Indeed, key to the formulation of the *sutra*—and to the algorithm—is the opposition between "square places" and "non-square places," which is used without prior explanation.

Finally, the verse has two parts. We shall quote them here to convince the reader that their interpretation is by no means obvious.[31] The verse's first half performs two directives, one directly, one indirectly. It reads: "One should divide, repeatedly, the non-square [place] by twice the square-root." This corresponds to step 10 in Fig. 5.2. The second half contains an assertive regarding

---

[29] In fact, it is the same algorithm up to details that do not matter for the questions Keller addresses.

[30] The real meaning of this category is beyond the scope of the present discussion, but Keller's description of the text in the *Patiganita* devoted to square root extraction offers elements for discussing this issue.

[31] In her chapter, Keller explains in detail how the commentators make sense of these statements.

the status of partial results and, indirectly, another directive. It reads: "When the square has been subtracted from the square [place], the result is a root in a different place," which corresponds to step 5 in Fig. 5.2. Incidentally, we depend on the commentators to understand that the various occurrences of the term "square" in the verse sometimes mean the square of a value, and sometimes the name of positions in a decimal expansion—namely positions corresponding to powers of ten that are squares.

Clearly, if we rely on how we handle texts for algorithms nowadays, there is no way we could learn to carry out a root extraction on the basis of this cryptic text. Many of the steps made explicit in Fig. 5.2 seem to have been skipped. Keller captures the problem in its widest form when she identifies the "paradox of *sutras* which both prescribe and are cryptic." How did users understand what to do with the text and which actions did they derive from studying it? These are the first questions raised. However, their elliptical nature is not the only odd feature of these texts. Another key feature illuminates the fact that the text does not conform to our expectations with respect to ways of writing algorithms. As we have noted above, the first part of the verse refers to step 10, whereas the second refers to step 5. In other words, the text inverts the order of the only steps it mentions in the process of root extraction. To summarize, the text not only fails to formulate all the steps of the computation, but also lists those that are mentioned in an order that doesn't seem appropriate for action.

1.5.2 Observing Actors' Reading as a Key to the Identification of a Textual Act

The key fact is that the commentators fail to express surprise about the order in which the steps are formulated. They also apply reading techniques that illustrate how they derive the intended procedure from the text. The commentators use these techniques to make clear why they read a directive in the text, and they show how they capture this textual act. In other words, in this context they demonstrate how one could do things with this text. Noting which features of the text are meaningful for them and how they make sense of them is an essential task for us in learning to understand textual acts in this context. This brings us to another question: what accounts for our inability to read the directive?

Aryabhata's *sutra*, as well as Sridhara's, represent texts whose intention, at least partly, is to enable practitioners to perform computations. As we have seen, this is how commentators have interpreted them. However, none of these texts conform to the expectation historians have spontaneously developed with respect to texts of that kind. Indeed, it is commonly assumed that texts from the ancient world prescribing algorithms are ordered lists of operations practitioners followed step by step to execute a given computation. As Keller notes, this belief evokes the "descriptive fallacy," exposed by Austin when he coined the notion of speech act. With this, Austin designated the central mistake he aimed at exposing in his William

James Lectures: "the mistake of taking as straightforward statements of fact utter-ances which are *either* (in interesting non-grammatical ways) nonsensical *or else* intended as something quite different." (Austin 1962, p. 3). The fallacy Keller's chapter aims to expose is the idea that a text designed to enable a practitioner to carry out an algorithm must have had the form of a list of operations to be executed in the order in which they are prescribed in the text. Such an assumption caused his-torians to be bewildered by Sanskrit *sutras* and to propose ad hoc hypotheses to ac-count for *sutras'* unusual ways of formulating mathematics (see p. 194). The fallacy Austin exposed related to the nature of the speech acts carried out by utterances. The fallacy Keller is interested in relates to the nature of the texts used to carry out the textual act "giving an algorithm." What she brings to light in Chap. 5 is that algorithms can be given by textual acts of a type we will call "indirect directives." Just as it was a complex task to identify which utterances can carry out a directive, it appears that the texts used to perform directives sometimes display unexpected features. Describing these ways of performing directives is a tool for reading our sources. It is also an essential means to "secure uptake," if, as historians, we don't want to read these sources anachronistically, but aim instead at restoring their au-thors' illocutionary point.

The identification of these textual acts doesn't mean the end of our inquiry. There are still unanswered questions of interest for the project of the present book. Why did Aryabhata and Sridhara choose to state algorithms in this way? In particular, why did Aryabhata choose to invert the order of the two steps he explicitly men-tions, with respect to the order in which they occur in an actual computation? Once again, our only documentary resource for addressing these questions, aside from the *sutras* themselves, is provided by the commentaries. These documents of evidence suggest a twofold answer to the questions.

Let us start by considering the inversion. As Keller shows, Bhaskara reads a theoretical statement in Aryabhata's promotion of the division that occurs in step 10 of the algorithm at the beginning of his sutra, as well as in other related syntactical features. The inversion thus relates to another dimension of the act the text carries out. This fact illustrates how closely we have to read the sources if we want to identify the full dimensions of the textual acts. More generally, Bhaskara's commentary suggests that *at the same time* as Aryabhata refers to the algorithm in a way that allows the commentator to grasp it, he formulates a view on what constitutes the gist and inner structure of the algorithm.[32] According to this reading, the order Aryabhata formulates appears not to be the order of actions to be performed, but of a hierarchy of operations in root extraction—a hierarchy that distinct kinds of verbal forms express. In other words, what accounts for the way of formulating and organizing knowledge in the *sutra* is the fact that it car-ries out a combination of two textual acts, both indirectly: an assertive at the same time as a directive. The assertions that the *sutra* makes about the algorithm can be correlated with the genre of the treatise in which it occurs. The nature of the *sutra* reflects that we are in the context of a "theoretical treatise." Keller is able to show

---

[32] Incidentally, the steps his *sutra* propounds are entirely general and constitute the kernel of the computation without specifying any instrument that could be used to execute it.

in which respects a contrast can be established on these points with Sridhara's *sutra* and its commentary.

The conciseness of Aryabhata's—and Sridhara's—statement of the algorithm is a second specific feature that particularly interests us. It expresses more broadly the genre of the treatise to which the *Aryabhatiya*—and the *Patiganita*—belonged, and the constraints affecting the formulation of the *sutras* they contain.[33] In relation to their formulation, *sutras* were supposed to be maximally compact without a trace of redundancy. As for their meaning, commentators like Bhaskara expected to find in them the seed that exegesis had to develop to capture the meaning of the "written utterance." In other words, readers such as Bhaskara approached these texts with assumptions as to how these texts were making sense. Note that both compacting and unpacking required literary as well as mathematical knowledge. In this context, conciseness also reflects how *sutras* were brought into play: to be used, they were supposed to be appropriated with the help of a commentary and memorized. In this context, the *sutra* cited above was most probably meant to allow the user, who had memorized and studied it with the help of commentaries, to unfold all its layers of meaning when needed. This implies the combination of directive and assertive that we have shown was carried out by means of a text that displayed features and required action, both specific to the scholarly culture in which this form of communication developed. This is especially striking when we work with ancient sources handed down from the Indian subcontinent. However, it holds true as general rule. To confirm this, it suffices to mention the "nomenclatural acts" described by Yves Cambefort. The texts by means of which these acts are carried out were also written in the context of a scholarly culture that shaped ways of composing them. The interpretation of these texts requires that we inscribe them in the textual culture they belong to. This is supported by other illustrations presented below. All these cases show the importance of the background for the way discourse acts are carried out and for their uptake. This conclusion concurs fully with Searle's analysis of a discourse act.

1.6 Enumerations: A Key Textual Act in Scholarly Texts

So far in the present book, we have considered "textual acts" of a type comparable to the "speech acts," identified in Searle's taxonomy. To recapitulate, we have encountered assertives, directives, declaratives and commissives. However, there are other types of "textual acts" which are more specific to the level of the text, even though they are not limited to that level. We shall elucidate this claim by considering the textual object "enumeration," and the acts that can be carried out by means of it. Part II of the book focuses on that type of textual act and the issues it raises. Jacques Virbel introduces this part of the book with a chapter devoted to a lin-

---

[33] Keller (2006, pp. I:xvii, xliii–xlviii) gives an overview of Aryabhata's *sutras* and describes Bhaskara's exegetical techniques to make sense of them. The following remarks rely on that reference.

guistic analysis of "textual enumerations," illustrated by numerous examples. As in Part I, this introductory chapter provides the main tools required for the subsequent chapters.

1.6.1   Defining Enumerations

We already encountered an example of enumeration in our discussion on Cambefort's contribution to the book. In this case, the enumeration was clearly not limited to the framework of a sentence, but was composed of multiple phrases (see Fig. 1.1). Moreover, we then sketched how a directive was carried out through the enumeration. As Virbel explains, this fact is typical of the way enumerations can be combined with all types of discourse acts. More generally, enumerations are frequently found in scholarly texts, not least in instructional texts. The types of questions they raise include: Why do authors choose the form of an enumeration? Which kinds of textual act do enumerations perform? Which purpose do they fulfill? How are enumerations materialized in different sources? As we will see, consideration of their properties and purposes brings to light interesting features of our sources and opens fruitful domains of inquiry for History of Science.

In Chap. 6, Virbel begins by introducing criteria that can be used to define enumerations. In the subsequent chapters, these criteria play an essential role in the investigation of enumerations in scholarly texts. We shall mention some of them here, for the purposes of our discussion. The key criterion is, not surprisingly, that an enumeration contains a list of items, the items that are enumerated. From our perspective, what distinguishes lists from enumerations, or from other forms of enumeration, is the fact that lists simply contain items and nothing else. As Virbel emphasizes, this is where the range of enumerations could begin. The exact place at which we decide to cut into the continuum of similar textual phenomena to single out enumerations is merely a matter of convention.

Unlike lists, enumerations can contain other components. The set of items can be introduced by an initial phrase. Sometimes the initial phrase includes a classifier (also called a "hypernym," or "organizer") which makes the nature of the items listed explicit (e.g., the term "point" in the initial phrase, "here are the *points* characterizing an enumeration..."). Sometimes this phrase also indicates the number of items listed (e.g., the expression "four parts" in the sentence, "here are the four parts that make up Chap. 6 by Virbel...").[34] The set of items can also be concluded by a final phrase governed by the same options. An example is given below. Virbel analyzes the various features these elements can display in enumerations, focusing particularly on the markers that signal the presence of an enumeration, its beginning or the beginning of the items, and its end or the end of the items.

---

[34] In fact, the expression we took as an example for designating the items in the initial phrase should be quoted more precisely as "the four parts." Expressions of this kind may or may not assert the exhaustive character of the listed items. Virbel examines these assertives. Using examples found in classical texts, he shows that the statement of exhaustivity does not imply that the actual list of items corresponding to it conforms to it exactly. Interestingly, some lists of items differ from the information given in the initial phrase. Virbel discusses the issue of interpretation this raises.

The markers indicating items include lexical markers (e.g., "firstly," "secondly," etc.), symbols (e.g., numbers), signs (e.g., hyphens, indentations) and specific arrangements on the page.[35] Some of these markers are specific to modern texts of enumerations. We shall encounter other ways of shaping the texts of enumerations in ancient texts.

In fact, in scholarly texts as the example of Cambefort shows, enumerations that fulfill part or all of these criteria are multi-phrasal events. This indicates that, in addition to sentences, sections or paragraphs, texts have other types of components, and some of these go beyond the limits of sentences. Consequently, as Virbel suggests, the study of enumerations is one step towards an inquiry into the textual objects that can enter into the constitution of texts. It is to be hoped that the description and analysis provided here can contribute to the development of such a line of research.

Looking at the various methods of writing down enumerations reveals another theoretical option that is essential to how Virbel approaches enumerations. As we indicated briefly above, enumerations can take the form of fully developed "discursive formulations" in which the items are lexically marked (e.g., firstly, secondly...). However, they can also be textualized by means of "visual properties," including the use of typographical and dispositional markers that express their structure and characterize their spatial inscription on the surface of the written texts. Virbel considers both types of formulations as linked by means of linguistic transformations, and approaches the visual properties within his chosen theoretical framework, as the reduction of discursive formulations. The result is that the visual dimension of the texts of enumerations—and, in fact, of texts in general—is considered meaningful, and its study belongs within the same logico-linguistic description as other dimensions of the text.[36] In this context the choice of a spatial inscription of a text, whether an enumeration or not, is a textual act of a metalinguistic kind: it can be seen as a declarative, which brings out structural features in the text. In fact, reflections of the theoretical option outlined here can be found more generally in the present book.

1.6.2   Enumerations as Textual Acts

After an exploration of the syntactic features of enumerations based on the theoretical options summarized above, Virbel discusses semantic aspects of enumerations. The method he follows is similar to that deployed for the act of "giving instructions" we outlined in Sect. 1.2 of this introduction. The act of enumerating is approached through an analysis of its relation to verbs that name similar acts. These verbs can be classified into several groups. Linking "enumerating" with these groups highlights the various dimensions of meaning of the term: an enumeration forms categories, making them into a group and connecting the related items. These conclusions emphasize an essential fact that characterizes enumerations in

---

[35] In his study of "quotation," Vernant (2005) sketches the various ways "quotations" can be marked.

[36] This is an option that derives from the Textual Architectural Model to which Virbel subscribes and which he sketches in Chap. 6.

contrast to other related acts. Most importantly, as Virbel emphasizes, enumerating requires cognitive operations to be carried out and additionally makes other cognitive operations possible. These two issues are crucial for the History of Science. They also provide a foundation for Virbel's analysis of the logical features of "enumerations."

The final section of Chap. 6 is devoted to pragmatic features of "enumerations." The first point to note derives from reexamining the shopping list example. As evidenced by this example, the syntactic and semantic features of an enumeration are not sufficient to grasp the illocutionary force of the textual act carried out through the enumeration. This simple fact illustrates why a pragmatic approach to texts is absolutely indispensible. In Virbel's words, to secure uptake we must have "knowledge of the history of the document." We will have to examine how this can be understood in the case of scholarly documents.

More generally, one of Virbel's key remarks with respect to "enumerating" as an act is that it is a textual act that can be combined with any other act because "it allows any type of illocutionary act *to be performed in a particular way*." (Virbel's emphasis) We have already seen an example of this phenomenon with the enumeration quoted in Cambefort's example.[37] This derives from the fact that "enumerating" like "quoting," constitutes a discursive meta-act. Virbel goes one step further and highlights specificities of this meta-act. He emphasizes that one key dimension of the illocutionary point of such an act is to divide a matter into items and perform the co-enumerability of the items. From this perspective, we can see "enumerating" carries out a declarative. This feature of "co-enumerability" is essential to account for important facts that Virbel uncovers in his empirical study of enumerations. On the one hand, enumerations found in written sources are not always merely coordinating items: other types of relationship between items are shown to exist. On the other hand, the items enumerated are not always similar in form. We shall encounter examples of these phenomena below. The motivation for carrying out this meta-act can be to help the reader perform cognitive operations on the text that is materialized in this way. Once again, the enumeration quoted in Cambefort's chapter illustrates this point immediately. Another dimension that appears and will prove worthy of exploration is the assertion some enumerations carry out with respect to the items listed. Seen from the perspective of scholarly documents, enumerations are clues to intellectual acts that authors perform on their own discursive production. What do these clues tell us about their cognitive work? Why and how do actors enumerate? These are the key questions we will now address to evoke results that can be obtained in the History of Science by observing these acts as such.

---

[37] With respect to Austin's "expositives," which can relate to all types of discourse acts, Vernant (2005) suggests a similar conclusion.

1.7   The Production of Enumerations and their Interpretation

1.7.1   Identifying Enumerations: A Key for Interpreting Ancient Documents

The distinction between lists and enumerations introduced by Virbel is essential for the new interpretation that Michel Teboul offers in the present book for the meaning of the oldest dictionary of Chinese characters that has come down to us: the *Erya* or *Approaching Perfection*, which is dated to the third century BCE.

The *Erya* deals with Chinese characters entirely differently from the approach developed in the subsequent dictionary, *Shuowen jiezi*, completed by Xu Shen around 100 CE. The latter treats characters on the basis of a graphic analysis of the writing. In the *Erya*, the point is to elucidate the characters' meaning by listing them with characters of related meaning. Evidence for the connection between the meanings of these characters derives from earlier commentaries on the Classics. As a result, the nineteen extant chapters of the dictionary are composed of what has been interpreted until recently as lists of characters of related meanings. Each entry simply appears to give, successively, characters that are all positioned at the same level and semantically close to each other. This is the meaning of the term "list," that is used to refer to them. Often concluded by the character *ye* 也, which marks their end, these lists have been referred to as "semantic lists."

Briefly, the dictionary is a book merely made of lists, or of what has been understood as lists up until now. Consequently, understanding what is at stake with these "lists" is a central issue for its interpretation. The key questions Teboul addresses to the original text are simple: what is the format of these lists and what is the underlying structure of the dictionary? The main observation that enables him to offer a new answer to these questions is that the role of the last character in a "list"—the one placed right before the particle *ye* mentioned above—is not symmetrical to that of the other characters in the items sequence. This is an initial indication that the items listed are not all placed at the same level, or, in other words, are not all co-enumerable. But there is more to come.

This observation leads Teboul to discover that the sequence of characters is not a mere list, but an enumeration, in which the last character actually plays the part of the "classifier" or "hypernym." Accordingly, this character is not on the same level as the others, and, in fact, its function is wholly different. In other words, we have a dictionary composed not of lists, as scholars previously believed, but of enumerations. More precisely, Teboul establishes that each entry has a "headword"—in this case, we should rather say, a "tailword." And the characters listed before this "tailword" are, in fact, stated to fall equally under the scope of the "tailword." They are "types in the category of" tailword and, unlike the "tailword," they only are co-enumerable. The structure that proves to account for the meaning of each entry is thus entirely different from a list of items with equal status.

That result enables Teboul to offer a new interpretation of each enumeration and of the whole book as well. This illustrates the importance of focusing on enumerations as textual phenomena to deal with scholarly sources. Indeed, some authors notably opted for enumerations as a key mode of writing down knowledge. Here moreover, we have a clear-cut case in which a reflection on the structure of enumerations effects a shift in the interpretation of a book. Given the importance of this dictionary for the interpretation of the Chinese classics, the impact of Teboul's discovery shouldn't be underestimated. Yet it would remain mere speculation if there were no historical evidence supporting it. Teboul accounts for the validity of this new interpretation by providing evidence he found in the ancient commentaries on the *Erya*. The interpretation Teboul advances thus elucidates the structure of the dictionary and the meaning of its constituents. In doing so, it reveals more clearly how the book can be useful to researchers. It also clarifies the work carried out by the authors to produce enumerations. In our terms, the interpretation indicates how the dictionary's compilers carried out their lexicographic activity with texts. It reveals the work required to perform the act of enumerating.

Teboul can, in fact, take this one step further. In the case under consideration, enumerations are a method the dictionary employs to express relationships of meaning. In addition, Teboul goes on to suggest that the order in which items were arranged in each enumeration is meaningful. Incidentally, this observation offers an illustration of a remark Virbel made in Chap. 6 in his discussion of the "logical aspects of enumerations." Virbel addressed the issue of the "conditions of identity" of an enumeration and, in particular, raised the question of the role of the order of the items in the identification of the enumeration. With Teboul's interpretation of the *Erya*, we have a clear example in which the order of the items does, indeed, contribute to the identity of the enumerations and therefore to the meaning of the text. Last but not least, this new reading allows Teboul to offer solutions for vexed questions related to the *Erya*.

In this latter case, focusing on enumerations leads to a fresh interpretation of a classical work whose text lists enumerations. In the succeeding chapter, which is also devoted to ancient documents, examining enumerations illuminates the work done by actors to produce and read enumerations. It offers us a case in which the enumeration is at the level of the whole text.

1.7.2   Writing Enumerations: An Encounter with Complex Textual Acts

Chapter 8, written by Christine Proust, is devoted to an extraordinarily complex and remarkable enumeration recorded in cuneiform mathematical sources. The enumeration occurs in the context of specific types of texts called "series texts," which are characterized by the fact that, unlike many texts that fit onto a single tablet, they are usually recorded on dozens of tablets. Although no mathematical series text has survived in its entirety, colophons found in tablets belonging to series indicate the number of the tablet in the series. These numbers therefore provide information on the length of the series.

Several tablets belonging to mathematical series have survived. They all contain lists of statements of problems without recording any procedure for solving them. As far as mathematical writings are concerned, this characteristic distinguishes this type of text from "procedure texts" in which the statement of a problem is followed by a text of a procedure for solving this problem. Proust suggests that this may explain why historians interested in understanding the resolution methods to which cuneiform tablets bear witness have so far paid less attention to series texts than to texts dealing with solutions of problems.

The tablet on which Proust mainly focuses in Chap. 8, Tablet A 24194, is kept in Chicago. Discovered in illegal excavations, its origin and date are unknown, even though we may assume it was composed in the first half of the second millennium BCE. Series texts are characterized by their conciseness; however, in comparison, tablet A 24194 is extremely concise. This is easily seen by the fact that it records more or less 240 statements of problems on the obverse and reverse of a square tablet where each side measures roughly 10 cm (see the copy of the tablet in Appendix 1 of Chap. 8.) The question such tablets raise, and which is posed even more sharply by Tablet A 24194, is simple: what was the purpose of writing down these texts? Scholars like Neugebauer who studied them thought they were a repository of problems collected by teachers as a teaching aid. Proust convincingly argues that this interpretation scarcely matches the evidence contained in the tablet. Her own tentative answer to the question derives from a close analysis of Tablet A 24194. Let us consider some of its features.

In addition to recording a tablet number, the colophon of Tablet A 24194 says it contains 240 "sections." These sections are inscribed on the surface of the tablet in the form of boxes bounded by horizontal and vertical lines. The vertical lines define the columns that divide each face of the tablet, and the boxes are inscribed in these columns. They are separated from each other by horizontal lines drawn inside the columns. Proust points out that there are three sizes for the boxes containing sections: long, medium and short. Roughly speaking, the number of sections also corresponds to the number of problems stated on the tablet, with each statement usually recorded in a box. Each problem has the same structure, as far as its meaning is concerned. However, the formulation of the problem statement differs sharply between the boxes due to the compaction techniques used to write down all the items on the surface of the tablet. In other words, 240 similar items are listed on the tablets. They are given material expression by the format. With the descriptive terms introduced by Virbel in Chap. 6, this gives us an enumeration. The colophon, which concludes the tablet, can be considered as the final phrase that states the performance of the enumeration the text carries out.[38] In this case, the final phrase includes a classifier that indicates the nature of the items

---

[38] The tablet could also be considered as an item in itself in the enumeration that the series constitutes. However, we lack the evidence about this series to develop this line of inquiry.

listed according to the scribe of the text. It also indicates the number of items listed.

The interesting issue Proust addresses is to find evidence that could allow us to understand the work done to perform and textualize the enumeration. In this case, the scribe who wrote on the tablet didn't use indentation or punctuation. Nevertheless, many elements were brought into play, or even specially designed, to format the text of the enumeration physically, and these are precisely what Proust focuses on.

To begin with, a type of statement was shaped that allowed elision as well as classification. Proust distinguishes four levels to characterize the structure of problem statements. Each problem is stated in two sentences. Level 4 corresponds to the first sentence and is common to all problems whose statement is recorded on Tablet A 24194. This first sentence probably changed from one tablet to the next in the series. In the case of Tablet A 24194, it is stated at the beginning of the tablet and then elided throughout. This is the first example of an elision of something common to a set of items. The whole tablet makes systematic use of this resource.

Distinguishing between levels 3 to 1 allows Proust to describe the formulation of the second sentence that concludes the problem statement. We shall now describe the content of these second sentences and the organization of their enumeration. Level 3 corresponds to the formulation of a main expression P, level 2 to the formulation of a secondary expression S, and level 1 to the statement of a relationship linking P and S that constitutes the second sentence of the problem statement. This relationship is formulated by means of operations bearing on P and then on the result of the latter and S. Proust gives the details in Chap. 8; here we shall merely note the result that the second sentence of a problem statement has the following structure: P—S—relationship R. This is true both of its meaning and its syntax. It is on this basis that the actors can render the enumeration recorded in Tablet A 24194 in the form of a tree, similarly to the case in Cambefort's chapter (see Fig. 1.2, which shows a modern version of the tree; we shall describe below how it was realized on the tablet). The enumeration first leaves P and S unchanged, and changes the relationship sequentially with each successive item. It can thus be represented as P_1–S_1–R_1, P_1–S_1–R_2, P_1–S_1–R_3,.... Let us call such a set of items a sub-enumeration at the level of "leaves." Then, after the end of this sub-enumeration, P is kept unchanged, S is changed in the next item, and then again a sequence of items is listed in which only the relationship is modified from one item to the next one. We thus have a second sub-enumeration at the level of leaves (roughly speaking, it can be represented as P_1–S_2–R_1, P_1–S_2–R_2, P_1–S_2–R_3,....). A key feature, to which we shall return, is that the list of changes of the relationship is more or less the same from one sub-enumeration at the level of leaves to the next. Observed at this level, the text enumerates several sub-enumerations at the level of leaves (if we separate the sub-enumerations by a semicolon we can represent the items as: P_1–S_1–R_1, P_1–S_1–R_2, P_1–S_1–R_3,....; P_1–S_2–R_1, P_1–S_2–R_2, P_1–S_2–R_3,....) Let us call these higher-level enumerations "sub-enumerations at the level of lower nodes." Each of their items is a sub-enumeration at the level of leaves. At the end

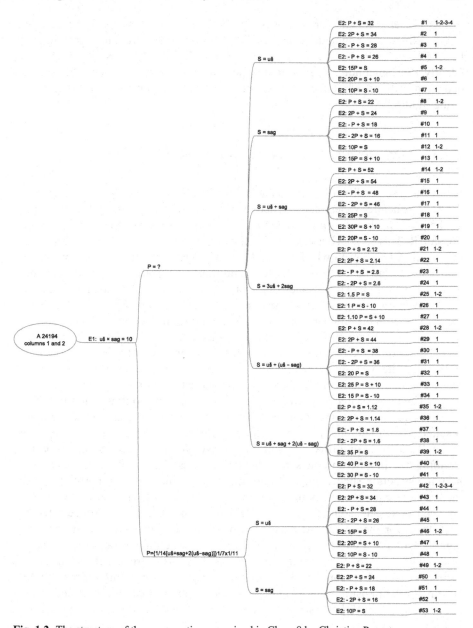

Fig. 1.2 The structure of the enumeration examined in Chap. 8 by Christine Proust

of this higher-level enumeration, P is changed, and with this new P, the text deploys again a "sub-enumeration at the level of lower nodes" (roughly speaking, we have: P_1–S_1–R_1, P_1–S_1–R_2, ….; P_1–S_2–R_1, P_1–S_2–R_2, …; P_2–S_1–R_1, P_2–S_1–R_2….) Clearly, this tablet holds a set of items whose content forms a sequence of embedded enumerations, while the whole tablet displays an "enumeration at the level of higher nodes."

1.7.3   Methods and Reasons for Enumeration

As we have seen, all the problem statements found in Tablet A 24194 have exactly the same structure. We emphasized above that the syntax of the second sentences of all the problems has the same structure as its content: it is formulated as: P–S–relationship R. This parallel between the meaning and the syntax of the sentence enables the system of elision to which Tablet A 24194 attests to work. Let us now move from considering the content of the items and their organization to considering how items are realized and the enumeration as a whole textualized. We are moving from the act of enumerating to the act of shaping the text of the enumeration. At this point the parallel with the enumeration cited by Cambefort breaks down.

In the textualization of the sub-enumerations at the level of leaves, P and S are elided if we omit the first item, which states the expression of S for all the others. In the other items, only the relationship is stated. Consequently, if we look at the second sub-enumeration at the level of leaves, the text looks like this: S_2–R_1, R_2, R_3,…. The last two items appear in the short sections. By contrast, the first items of these sub-enumerations are in medium-sized sections, in which P is elided, but S stated. As for sub-enumerations at the level of lower nodes, their first item, in which P and S are modified, are both stated. We are in the long sections (see, for example, the first item in the sub-enumeration P_1–S_1–R_1, R_2, R_3,….). We can see how the structure shaped for the statement of a problem closely correlates with the way the enumeration is materially formatted and information is distributed between boxes. In addition, Proust establishes how the expression selected at each level for the operations makes the whole system of formulation possible. Again, we have a correlation between local syntactic choices and the structure of the whole. In conclusion, all items recorded in boxes are strictly of the same form as regards their content. However, the sentences formulating them differ according to their position in the enumeration. This point illustrates a key feature of enumerations that Virbel insisted on: the items enumerated are not always similar in form.

This means that a type of statement was specifically created for these enumerations, while at the same time a type of text was shaped to textualize the enumeration. These two developments are interrelated. This is interesting in itself for the History of Science. But what does the foregoing analysis tell us about why and how the enumeration was produced? A first hypothesis could be that this type of text was used to classify problem statements that existed prior to the enumeration. However, that seems rather unlikely. The existence of similar lists of variations of the relationship that concludes the statement (R_1, R_2, R_3,....) in subsequent sub-enumerations seems to indicate that the enumeration does not record pre-existing problems. On the contrary, the text of the enumeration appears to have provided a framework for systematic production of problems statements. The form of the text was a tool for producing the enumeration, that is, its text as well as the items enumerated. This observation raises the issue of the knowledge needed to produce the enumeration. We have seen that it required knowledge about syntax of operations and forms of texts. Proust also emphasizes the mathematical knowledge needed to produce the items, as well as the textual competence to inscribe them in the tablet.

As concerns the purpose of producing the enumeration, Proust observes something that leads her to question the validity of Neugebauer's hypothesis. The "procedure texts" that have come down to us record only few problems of the type enumerated in Tablet A 24194, and these are only some of the simplest ones. It seems unlikely that the enumeration's goal was to provide teaching material. It would probably make sense to look for the enumeration's aim in the effort that led to the author writing it, that is, in the production of a set of statements that could be textualized in this highly skilled way. In this context extreme conciseness appears to be a key feature. On the one hand, it singles out Tablet A 24194 from similar tablets. On the other hand, Proust finds clues indicating that the scribe who wrote down the tablet actually placed great value on conciseness. In addition, most of the tablet's features that Proust highlights can be related to the ambition of utmost conciseness: the syntax chosen for the statement, the system of elision, and the principle of the text. If we assume that the purpose of creating the inscribed tablet is related to the attempt to achieve conciseness, then the production of the enumeration appears to have been an end in itself. In any case, it is an important feature of the knowledge that this tablet attests.

These conclusions reveal various facets of the knowledge required to carry out the meta-act of enumerating. They reveal the effort needed for the production of the text as such. This is what actors, i.e., scribes, do using texts. However, we can also look at the tablet from another viewpoint: the reader's. The reading of the enumeration also required various types of competency and knowledge, and the challenge it posed to readers was perhaps one of the driving forces behind the display of virtuosity the enumeration shows. Proust also considers the knowledge readers needed to possess to grasp an item of the enumeration, once these items were realized in their respective boxes. In addition to mathematical knowledge, which could be used to control the reading, the user of such a text couldn't do without some knowledge of the principles by which the enumeration was physically formatted. Nor could readers ignore the kind of circulation within the text required to grasp the statement of a recorded problem, e.g., in a short section. This remark brings us back to the issue of the kind of circulation within a text actors need familiarity with to make proper use of technical texts. Various types of competency of that sort are also discussed in the next chapter.

1.8   One Enumeration can Conceal Another: Strange Texts for Directives

Chapter 9 combines the tools of Speech Act Theory and the description of enu-merations to examine texts used to write down algorithms. This brings us back to textual acts of a directive type—even more precisely, to instructional texts.[39] The exploration is carried out using texts for algorithms from Chinese writings composed between the second century BCE and the seventh century CE. As we shall see here, the issue of how directives are performed, on the one hand, and of enumerating on the other, are intertwined in an original way.

The sources considered include a manuscript excavated from a tomb sealed in the second century BCE, and documents handed down through the written tradi-tion. Two classics are examined: *The Nine Chapters on Mathematical Procedures* (hereafter: *The Nine Chapters*), probably completed in the first century CE, and the *Mathematical Classic by Zhang Qiujian*, which dates approximately from the second half of the fifth century. Two commentaries on these classics are also use-ful: the commentary on *The Nine Chapters* Liu Hui completed in 263 and the commentary Li Chunfeng completed in 656 on the *Mathematical Classic by Zhang Qiujian*.[40] In effect, Chemla focuses on algorithms recorded in these sources that mostly deal with the multiplication and division between quantities combining integers and fractions. The key fact is that these documents contain different kinds of texts prescribing the same actions—i.e., the same operations to be done. One of the main aims of Chap. 9 is to understand what accounts for the differences between them.

We have already alluded to the shortcomings of the standard view about ancient texts for algorithms. This holds that texts of this kind allegedly list terms directly referring to operations in the order in which a practitioner should execute them. In other words, such texts would display a simple one-to-one correspondence between terms for operations and actions. In the case of Sanskrit sources, as we have seen, this representation did not match the facts. In particular, we showed then that the textual directives that texts for algorithms carried out in Sanskrit treatises were by no means simple combinations of discourse directives.

Chemla also shows that the widespread conception of texts for algorithms is a poor representation of the reality of texts in ancient Chinese sources. She describes how some of the textual acts of the directive type to which these sources attest are not merely combinations of discourse acts. This conclusion is a strong argument for the need to consider textual acts as such. However, this conclusion imposes itself on the basis of the Chinese material in a specific way. In Chap. 9, Chemla reveals two essential reasons why the reality of texts for algorithms is much less simple than

---

[39] In fact, Chemla focuses specifically on how these texts prescribe actions. However, texts for algorithms also occur in the context of proofs, where they do not carry out textual directives. Once the former issue is dealt with, it will be interesting to focus on the texts of algorithms that occur in the latter context.

[40] Again, these are broad descriptions. For more detail, we refer the reader to Chemla's chapter.

posited by the standard view. One of these reasons relates to how "written utterances" carry out directives. We shall now proceed to outline it.

1.8.1   Different Ways of Prescribing the Same Operations

The texts of algorithms Chemla considers usually start with the stereotyped phrase, "the procedure says…" This is followed by an apparently ordinary list of operations. In most cases, however, appearances are misleading. Despite this, one feature of the list conforms to an important property of enumerations identified by Virbel: the granularity of the items is more or less uniform. In other words, the flow of actions to which these texts refer was divided into operations of approximately the same "size." This corresponds to what Virbel called the "co-enumerability" of the items.

To begin with, Chemla leaves the text as a whole aside, and focuses on its items, that is, the discourse acts of the directive type which the text of an algorithm uses to prescribe operations (in Virbel's terms, the thing in the world to which the item corresponds). Indeed, the modes of prescription of the same operation show an unexpected variety.

One would spontaneously assume that operations are prescribed by means of a term referring to them, such as "multiply." This is, in fact, only one of many different ways of prescribing. Chemla shows that even this simple prescription can be less straightforward than it seems. In addition, we can identify several indirect ways of carrying out the directives. For instance, multiplication is sometimes prescribed by asserting the result to be obtained. We can also find—and this is the most interesting phenomenon for us—operations prescribed by the reasons for carrying out the operation.[41]

It is interesting to consider the competence the reader needs to understand such prescriptions. On the one hand, he or she needs to understand the mathematical situation in which such a discourse act is performed in a way that allows the directive to make sense. This implies that the comprehension of the situation is not merely an understanding of the values obtained, but also of their meaning. Consequently, encountering such directives in texts highlights cognitive operations performed by the successful users of these texts. On the other hand, readers need to understand the reasons stated. This is the meaning of the utterance, if it was carried out directly, and corresponds to Austin's "understanding;" but the nature of this "understanding" calls for analysis. The meaning corresponding to the "propositional content" of the prescription in such cases is part of a proof of why the procedure is correct. Lastly, to secure "uptake" the user must know how to determine the actual operations corresponding to these reasons. This outline analysis indicates how interesting it would be to examine Austin's opposition between "understanding" and "uptake" in the context of texts for algorithms more closely. In the last type of case examined, as in

---

[41] At the level of the reasons of the correctness of an algorithm, this way of prescribing is comparable to the previous case.

others, there may not be a one-to-one correspondence between the term in the text and the actions in the world.

The indirect mode of prescribing by stating the reasons for performing the operations is interesting for the historian of science in two main respects. Firstly, the fact that directives of that kind are used implies that these algorithms texts refer to the reasons why algorithms are correct at the same time as they prescribe actions to be done. Moreover, only practitioners who understand the algorithm at this level are able to use the text as it is formulated. Secondly, as Chemla brings to light, the *ways* in which reasons are stated in algorithms texts seems to have undergone a transformation between the second century BCE and the first century CE. In the earliest texts, the reasons are expressed at a material level, whereas later they are grasped formally. This transformation enables us to distinguish between ways in which reasons for the correctness are approached and formulated. Historians have not yet focused on this transformation, which only becomes perceptible if one observes how texts for algorithms carry out indirect directives. This shows the kinds of results we can anticipate by turning our attention to the ways of carrying out discourse acts in our texts.

1.8.2 Different Textual Acts for Algorithms

After exploring the local modes of prescribing evidenced by these Chinese texts for algorithms, Chemla turns to consider the texts as a whole. She identifies two kinds of texts shaped by practitioners for writing down algorithms. This element of analysis accounts for the main differences between texts for algorithms found in Chinese sources.

The first type brings nothing unexpected. The texts list prescriptions in the order in which operations are carried out. The second type of text, recorded between the second century BCE and at least the seventh century CE, is much more surprising. It corresponds to texts able to deal with the various cases to which the algorithm can be applied. The list of operations to be carried out varies according to the case. The text, which integrates the treatment of distinct cases, is formed in such a way that it yields the correct list of actions for each of the cases.

As Chemla shows, the mode of integration seems to have required prior work on the method of handling the different cases. Their treatments were made similar to each other as far as possible. In one instance the list of operations required for case 3 contains that for case 2, which in turn contains that for case 1. However, the integration also required work on the text itself. Chemla shows that such texts have a fairly specific structure. They begin with a list of operations that actually corresponds to a case chosen as fundamental. Then conditions occur in the text, followed by the list of operations to be prefixed to that of the fundamental case to achieve appropriate treatment of the case covered by the condition. In the example above, the text has the following structure: first, operations for case 1; then condition A, corresponding to case 2, followed by operations to be prefixed to the operations for case 1 to deal with case 2; then condition B, corresponding to case 3 and again followed by operations to be prefixed to the operations for case 1 to deal with case 3.

This type of text requires several remarks. First, clearly, such a text requires the user to carry out a specific circulation within its sentences in order to derive the operations to be performed from the text. This is where we again encounter the issue of circulation in texts, as a textual gesture practitioners had to learn to perform with texts to be able to work efficiently with them. In this context, for the text to provide the correct operations to execute in each case, the practitioner was clearly expected to handle the text, beginning with the conditions, i.e., to start in the middle of the text. The reason for starting from the conditions was to determine whether the actor dealt with case 2, 3 or 1. Using clues found in the texts, Chemla proves that the practitioner was *not* expected first to execute the operations listed at the beginning of the text. This is confirmed by examination of the commentaries. This feature of the text for algorithms is the second main way in which these texts do not conform to the standard view.

The textual artifact just described constitutes the tool used to make a text encompass different lists of operations according to the cases. If historians of science failed to investigate how texts prescribe, the interpretation of those texts would remain tentative and would lack arguments to support them. Moreover, they would miss an entire aspect of the work done by actors to write down these types of texts for algorithms.

Secondly, the way the text is written down determines that the fundamental case should be at the beginning. It is the case to which all the other cases will be reduced. The text organizes the arrangement of the cases from the most fundamental to the most complex. In fact, the text carries out an enumeration of a type completely different from the previous kind of texts, and different from the spontaneous assumptions of the standard view about texts as well. Its initial phrase states the fundamental operations, after which the text lists items that are each composed of a condition and the method for extending the list of fundamental operations to cover other cases. This enumeration carries out a textual act of the directive type for all cases. We can see from this how the textual act can't be considered as a mere combination of discourse directives.

This brings us to an unexpected conclusion, and this is where we find the mutual connection of the issues of how directives and enumerations are carried out. The first kind of phenomena revealed, the indirect prescription by means of stating the reasons, seems to occur mainly, if not exclusively, in the second kind of texts for algorithms, that is, the texts covering several different cases. The textual phenomena that require specific competence at local as well as global level to enable exploitation of the texts for algorithms seem to be concentrated in a specific kind of text. These texts were probably more theoretical. They were also more general. Perhaps the two types of texts represent traces of different professional groups that used texts for algorithms in China. Let us leave the question open for the moment. Whatever the case, Chemla's description of texts for algorithms in the present volume shows how groups of practitioners have shaped textual resources to formulate texts for algorithms and to use them. As this shows, a more systematic inventory of these acts could be of great interest to History of Science.

1.9   How could History of Science Profit from the Study of Enumerations?

Chapter 10, by Anne Robadey, continues our exploration of enumerations, but this time in the context of modern mathematics. Here, the focus shifts to another reason why enumerations can be of interest for historians of science: Robadey aims to highlight how enumerations attest to cognitive processes at play in scientific activity, and may sometimes be the only records to provide specific pieces of historical information.

Robadey concentrates on a corpus of texts that documents the shaping of Henri Poincaré's famous, specific approach to differential equations between 1878 and 1886, at the beginning of his career as mathematician. Poincaré's approach can be characterized by a set of related features. Let us describe some of them to highlight the points of general interest in Robadey's chapter. Poincaré, like some of his predecessors such as Briot and Bouquet, is interested in the definition of functions by differential equations, rather than in the properties of their solutions once they are expressed explicitly. Unlike his predecessors, however, Poincaré approaches them as curves and not as functions. Accordingly, geometry plays an important part in his work. He is also interested in the curves' global properties, and not only in their local behavior. Lastly, he is interested in the set of curves that are solutions, rather than in specific solutions.

Robadey's main focus in Chap. 10 is on Poincaré's major publication on the topic of curves defined by differential equations (the *Mémoire* "Sur les courbes définies par une équation différentielle," published in two successive parts in 1881 and 1882). More precisely, she focuses on the key Chapter II, which is structured as an enumeration. This distinguishes different types of "cases" for what happens to the curves that are solutions of a differential equation at a given point. Let us call this the "master enumeration." Again, in this case, the enumeration is a textual phenomenon which develops at the level of a whole chapter and is thus undoubtedly multi-phrastic. As in the case of Tablet A 24194 described above, the enumeration is actually a set of embedded enumerations. Robadey describes its material format, which makes use of typographical devices (e.g., italics), or dispositional devices (e.g., arranging titles of cases in a specific way). However, the material format doesn't spatially express the entire structure of the enumeration. This, Robadey notes, causes a problem for the interpretation of the last case, whose exact status in the set of embedded enumerations is not wholly clear.

1.9.1 The Cognitive Work Carried out in Enumerating

By highlighting a set of features of the enumeration, Robadey is able to show that in its smallest details its structure reflects Poincaré's mathematical approach to the topic.

A first level of the enumeration counterposes ordinary (and common) points that are easy to deal with, to singular points. The key phenomena Robadey points to occur in relation to the sub-enumeration of the latter, that is, in the treatment of particular cases, which is the main concern of Poincaré's Chapter II. The guiding principle of the master enumeration at that level derives from an analytical inquiry into the situation; the key concepts of this are given by a main theorem. However, in alternative enumerations of the same cases, based on the main theorem, Poincaré's moves reveal that he reads two distinct sets of cases there, each explored using a specific geometric approach to the situation. In fact, Robadey highlights Poincaré's recurring use of enumerations as a tool in his research. She also analyzes how Poincaré combines these two types of tools, analytic and geometric, to develop his analysis of the situation in the framework of the master enumeration. In this way, she illustrates the mathematical work and knowledge required for shaping an enumeration and working with it.

Poincaré does not enumerate to advance *a posteriori* an argument that is already available. The enumeration as text materializes Poincaré's actual process of mathematical research. It reflects the cognitive division of the world that Poincaré performs to work out the situation mathematically. At the same time it serves as a basis on which Poincaré carries out new operations, including new enumerations, which play an important role in his research. In other words, the enumeration offers a basis on which further cognitive operations can be carried out.

Lastly, and perhaps more importantly, the enumeration retains a key feature of Poincaré's research procedure. The previous remarks concern the *structure* of Poincaré's enumeration. Robadey then proceeds to focus on the *organization* of the items and Poincaré's specific *view* of them. Poincaré first analyzes three cases of the sub-enumeration, in the context of which he identifies types of "ordinary singular points." Only then does he turn to cases for which he emphasizes a key point: the differential equations in which they occur are *exceptional* compared to the previous cases. Poincaré thus enumerates in a specific way. He does not give equal weight to all particular cases. Rather, he differentiates between them in terms of importance and, in fact, treats particular case types in a hierarchy of decreasing importance. In addition, the word "importance" has here a specific mathematical meaning and refers to an assessment of the "degree of generality" of the phenomena dealt with.

This indicates that, at the same time as Poincaré is enumerating, he is supervising the items listed from a higher viewpoint, in this case a viewpoint that no longer focuses on points but rather assesses the generality of types of differential equations for which such points occur in comparison to other types. The criteria he uses to do this are not explicit. We shall return to this point below. What is important here, however, is that conducting the enumeration probably provides an essential basis for carrying out the assessment. Mathematical knowledge is again involved not only in the production, but also in the reading of the cases.

The decreasing importance of the cases is also reflected in how Poincaré handles them: the more "particular" the cases, the less developed his treatment is. Here we encounter a phenomenon about enumerations that Virbel emphasizes in Chap. 6: the enumerated items are not always parallel. This is the case here with respect to the things in the world referred to by the items of the enumerations, whereas in the case of Tablet A 24914 it applied to the text of the items.

Robadey shows how the items' titles, as well as Poincaré's incidental remarks on the enumeration in the *mémoire* analyzed, or the quotation of the *mémoire*'s results elsewhere, demonstrate Poincaré's precise awareness of the distinct degrees of generality of the particular cases. This perception, Robadey argues, is the outcome of a specific reflection on generality. Robadey succeeds in illuminating the main lines of this reflection as revealed and embodied by the enumeration. She advances hypotheses as to where Poincaré may have derived inspiration for this.[42] The essential issue here, as Robadey emphasizes, is that this is precisely what distinguishes Poincaré's approach from that of his predecessors on whose writings he relied and whose research he continued.[43] However, Poincaré offers no explicit development in relation to these degrees of generality. The reflection on generality is not "thematized," to use a concept introduced by Cavaillès. Were it not for the structure of the enumerations, there would be nothing in Poincaré's texts on the topic to testify to specific work on, and understanding of, generality and its degrees of differentiation. It is only by means of a careful analysis of the textual act of enumerating that the historian can approach Poincaré's reflection on generality, a task that Robadey fulfills excellently in Chap. 10.

1.9.2 Studying the Enumeration as a Key Tool for the History of Science

Studying the enumeration has so far revealed the work done by Poincaré to perform the enumeration and specific aspects of his enumerating practice. This enables Robadey to show very important results. In particular, she establishes the key role of the enumeration as such in Poincaré's successful research strategy.

Firstly, Robadey shows that the whole *mémoire* actually relies on the enumeration carried out in Chapter II. More precisely, to establish his new results, Poincaré needs to focus on a case which is general enough to be meaningful, but leaves aside particular cases that are both intractable and exceptional. The framework he adopts for this is precisely one he can define on the basis of the enumeration examined above. Consequently, the enumeration's properties as Robadey described them

---

[42] Robadey (2006, pp. 70–82).
[43] Robadey (2006, pp. 53–70).

prove *essential* for the success of the next step of Poincaré's program.[44] Examining the structure and organization of the enumerations, in addition to the use Poincaré makes of them, actually gives Robadey powerful tools for capturing a research method of Poincaré's. In fact, she establishes that Poincaré regularly makes use of the same method in several of his research works.[45] The method can be formulated as follows: Poincaré focuses on what is essential, and, in order to determine what is essential, he relies on enumerations that list items in decreasing degrees of generality. Nowhere does the method seem as visible as in this particular instance. In other words, describing the method in this context yields important clues about Poincaré's way of performing mathematical activity more generally. It also allows historians to find clues about his way of proceeding even when these clues are hardly visible. The textual features of the enumeration in the *mémoire* analyzed above reflect specificities of Poincaré's mathematical practice and provide ways for the historian to approach its operations. The reasons for this are clear. Poincaré uses the act of enumerating as a research tool in a specific way that leaves clues in the text. The text of the enumeration is produced as the result of an exploration of a specific type, which it materializes. Both the structure of the enumeration and the gradation of the items in terms of generality, yield plentiful information about Poincaré's mathematical work.

Secondly, the result of Poincaré relying on the enumeration to define the framework of the general case in which he operates is that his approach embodies rigorous general reasoning. This type of reasoning, Robadey stresses, must be distinguished from what Thomas Hawkins has described as the usual "generic reasoning," which mathematicians such as Cauchy and Weierstrass reacted against.[46] The enumeration carried out in his Chapter II enables Poincaré to define with great precision the framework he adopts for developing his reasoning. As we have seen, such enumerations and the method of approach they derive from are, in fact, a distinctive and recurring feature of Poincaré's mathematical practice. They are always correlated in a similar way with the use he makes of the enumeration: in distinct mathematical explorations, Poincaré focuses first—and sometimes only—on the essential, which should again be understood as the most general. The enumeration provides the basis for the definition of what is essential.

Lastly, Robadey emphasizes that this way of dealing with particular cases is not only specific to Poincaré, in contrast to his predecessors, but also characterizes Poincaré's approach to differential equations from his earliest writings in 1878 and 1879.[47] Shaping this approach carried out using the enumeration may well have been a key condition that allowed Poincaré to develop a wholly new and successful approach to differential equations. This leads Robadey to suggest a new periodization for this chapter of the history of differential equations. It awards a decisive role

---

[44] Robadey (2006, pp. 84–91).

[45] Robadey (2006, pp. 91–97).

[46] See the discussion by (Hawkins 1977b; Hawkins 1977a), as analyzed in (Robadey 2006, pp. 77–82).

[47] Robadey (2006, pp. 61–70).

to enumerations. These results illustrate clearly how the study of enumerations can provide valuable historical information.

In conclusion, identifying and studying Poincaré's enumeration in Chapter II of his *mémoire* enables Robadey to address key questions: the part played by the shaping of the master enumeration in Poincaré's further work in the *mémoire* under consideration; more generally, the correlation between the novelty of Poincaré's approach with respect to differential equations and the specific approach that enumerations represent; and lastly, a new periodization of the research done on differential equations. Robadey offers a detailed treatment of these questions, illustrating how History of Science could gain immensely from focusing on what our sources document indirectly.

This brings us to the conclusion of Part II. We shall conclude this introduction by outlining the research program inspired by the explorations conducted in this book.

In the case of enumerations, we have considered a textual object and the textual "meta-act" carried out with it, i.e., an act that operates on the text itself and leaves marks of the operations in the text. The results shown by research focused on enumerations could have been anticipated: we have seen evidence of the work carried out by actors using their own texts. This led us to focus not on what actors asserted, but rather on what they did with their own inscriptions. Actors have not merely used writings to set forth results or theories; in the course of their intellectual activities they have struggled with texts and inscriptions. As we have seen in the case of enumerations, concentrating more generally on textual objects and textual acts specific to the level of texts seems to be a way of finding traces of this other facet of actors' engagement with texts. Many textual objects appear equally promising and we intend to explore them in the future. They include titles and definitions, parentheses and parenthetical clauses, footnotes, and quotations, as well as sections actors distinguish in their texts and use to structure their texts.

However, these are not the only benefits that can be expected from such an inquiry. We have seen how reading what our sources document without asserting it proved essential for capturing historical facts not documented in any other way. *How*, and not only *what*, do texts document? This is a key question that our endeavor highlights.

One last important issue emerged from our investigations, and seems promising for the future. As we have seen, specific aspects of carrying out textual acts allowed historians to identify features of the contexts in which these acts were performed. The term "context" has been used here in relation to several layers of phenomena. In some cases it referred to the "immediate" context of production of a text, i.e., its producers, their illocutionary aim and their intended readership. It also included institutions, which we encountered through the rules governing the performance of certain acts. Last, but not least, the scholarly cultures in which textual acts were carried out were shown to leave their mark on the texts and to be rendered perceptible using textual studies. The information on the context that a careful examination of sources provides is highly meaningful for ancient historians, who usually work with only a small number of documents. However, several chapters in the book indicate that the development of approach methods useful for ancient history can also provide information on features of modern science that are not documented by sources.

References

Anscombe, Gertrude E. M. 1957. *Intention*. Oxford: Basil Blackwell.

Austin, John L. 1962. *How to do things with words. The William James Lectures delivered at Harvard University in 1955*. Oxford: Clarendon Press.

Ceccarelli, Leah. 2001. *Shaping science with rhetoric: The cases of Dobzhansky, Schrödinger, and Wilson*. Chicago: University of Chicago Press.

Chemla, Karine. 2004. History of science, history of text: An introduction. In *History of science, history of text*, ed. K. Chemla, vii–xxviii. Dordrecht: Springer.

Filliozat, Pierre-Sylvain. 2004. Ancient sanskrit mathematics: An oral tradition and a written literature. In *History of science, history of text*, ed. K. Chemla, 137–157. Dordrecht: Springer.

Fortun, Michael. 2008. *Promising genomics. Iceland and deCODE Genetics in a world of speculation*. Berkeley: University of California Press.

Galison, Peter. 1997. *Image and Logic. A material culture of microphysics*. Chicago: University of Chicago Press.

Giaquinto, Marcus. 2007. *Visual thinking in mathematics. An epistemological study*. Oxford: Oxford University Press.

Grandaty, Michel, Claudine Garcia-Debanc, and Jacques Virbel. 2000. Evaluer les effets de la mise en page sur la compréhension et la mémorisation de textes procéduraux (règles de jeux) par des adultes et des enfants de 9 à 12 ans. *PArole (special issue Langage et Cognition)*. 13:3–38.

Gross, Alan G. 2006. *Starring the text: The place of rhetoric in science studies*. Carbondale: Southern Illinois University Press.

Hawkins, Thomas. 1977a. Another look at Cayley and the theory of matrices. *Archives internationales d'histoire des sciences* 27:82–112.

Hawkins, Thomas. 1977b. Weierstrass and the theory of matrices. *Archive for history of exact sciences* 17:119–163.

Keller, Agathe. 2006. *Expounding the mathematical seed. A translation of Bhaskara I on the mathematical chapter of the Aryabhatiya*. Science Networks Vol. 30/31, (2 vols.). Basel: Birkhaeuser.

Klein, Ursula. 2003. *Experiments, models, paper tools: Cultures of organic chemistry in the nineteenth century*. Writing science. Stanford: Stanford University Press.

Loveland, Jeff. 2001. *Rhetoric and natural history: Buffon in polemical and literary context*. Studies on Voltaire and the eighteenth century, 3. Oxford: Voltaire Foundation.

Mancosu, Paolo, Klaus F. Jorgensen, and Stig A. Pedersen, eds. 2005. *Visualization, explanation and reasoning styles in mathematics*. 327 vols. Dordrecht: Springer.

McCarthy, Lucille P. 1991. A psychiatrist using DSM-III: the influence of a charter document in psychiatry. In *Textual dynamics of the professions: historical and contemporary studies of writing in professional communities*, ed. C. Bazerman and J. Paradis, 358–378. Madison: University of Wisconsin Press.

Nef, Frédéric. 1980. Note pour une pragmatique textuelle. Macro-actes indirects et dérivation rétroactive. *Communications* 32:183–189.

Netz, Reviel. 1999. *The shaping of deduction in Greek mathematics: A study in cognitive history*. West Nyack: Cambridge University Press.

Pascual, Elsa, and Marie-Paule Péry-Woodley. 1995. La définition dans le texte. In *Textes de type consigne—Perception, action, cognition*, ed. J.-L. Nespoulous and J. Virbel, 65–88. Toulouse: PRESCOT.

Pascual, Elsa, and Marie-Paule Péry-Woodley. 1997a. Définition et action dans les textes procéduraux. In *Le texte procédural: langage, action et cognition*, ed. E. Pascual, J.-L. Nespoulous, and J. Virbel, 223–248. Toulouse: PUET/PRESCOT.

Pascual, Elsa, and Marie-Paule Péry-Woodley. 1997b. Modélisation des définitions dans les textes à consignes. In *Cognition, discours procédural, action*, ed. J. Virbel, J.-L. Nespoulous, and J.-M. Cellier, 37–55. Toulouse: PRESCOT

Robadey, Anne. 2006. *Différentes modalités de travail sur le général dans les recherches de Poincaré sur les systèmes dynamiques*. University Paris 7 Paris Diderot, Département d'histoire et de philosophie des sciences. See http://tel.archives-ouvertes.fr/tel-00011380/. Accessed 6 April 2015.

Rotman, Brian. 1998. The technology of mathematical persuasion. In *Inscribing science: Scientific texts and the materiality of communication*, ed. T. Lenoir, 55–69. Stanford: Stanford University Press.

Searle, John. R. 1969. *Speech acts: An essay in the philosophy of language*. Cambridge: Cambridge University Press.

Searle, John. 1979. *Expression and meaning*. Cambridge: Cambridge University Press.

Searle, John, and Daniel Vanderveken. 1985. *Foundations of illocutionary logic*. Cambridge: Cambridge University Press.

Searle, John, and Daniel Vanderveken. 2005. Speech acts and illocutionary logic. In *Logic, thought and action*, ed. D. Vanderveken, 109–132. Dordrecht: Springer.

Smith, Barry. (2010, August 23–26). Document acts. In *Collective intentionality*. Basel: Switzerland. http://ontology.buffalo.edu/smith/articles/DocumentActs.pdf. Accessed 31 March 2015.

Smith, Barry. 2012. How to do things with documents. *Rivista di Estetica* 50:179–198. http://ontology.buffalo.edu/smith/articles/HowToDoThingsWithDocuments.pdf. Accessed 31 March 2015.

Vanderveken, Daniel. 1990. *Meaning and speech acts. Vol. 2, Formal semantics of success and satisfaction*. Cambridge: Cambridge University Press.

Vanderveken, Daniel. 2001. Illocutionary logic and discourse typology. *Revue Internationale de Philosophie* 216:243–255.

Vernant, D. 2005. Pour une analyse de l'acte de citer: les métadiscursifs. In *Citer l'autre*, ed. M.-D. Popelard and A. J. Wall, 179–194. Paris: Presses Sorbonne Nouvelle.

Virbel, J. 2000. Un type de composition d'actes illocutoires directifs et engageants dans les textes de type 'consigne'. *PArole (special issue Langage et Cognition)* 11–12:200–221.

Part I
Speech Acts and Textual Acts

Chapter 2
Speech Act Theory and Instructional Texts

Jacques Virbel

Abstract In this chapter, we shall consider, systematically, the types of speech act that are likely to arise in instructional or procedural texts and the forms they take. This will be preceded by a general presentation of Speech Act Theory. Thus, the chapter can be read autonomously and may possibly allow new applications to be formulated.

2.1 Presentation

Pragmatics, as a science of inter-human communications by means of language, seems able to bring a fundamental renewal to the study of texts, and in particular texts as a support to approach the history of science. This sentiment is all the stronger if the texts primarily in question are procedural, meaning that their aim is to achieve specific directive speech acts, as is the case for this chapter. This does not mean that we should deny the existence or the importance of other dimensions or aspects of texts seen as language objects: syntax, semantics, rhetoric, and bibliology for example, nor does it mean that we should ignore the role of generic rules in the writing or reading of texts, but it means on the contrary, as we will have the opportunity to see, that we should specify these components according to the perspective brought about by placing a priority on the pragmatic dimension.

This chapter has two objectives. Firstly, we would like to introduce a presentation of pragmatics that can be used (Part I). There are many extensive presentations of the principal pragmatic theories (Cohen et al. 1990; Levinson 1983; Tsohatzidis 1994), but we shall focus on the presentation of the most important features of one of the theories that is currently active in this field of research, Speech Act Theory

The author is grateful to Karine Chemla, Hélène Eyrolle, Julie Lemarié, Mustapha Mojahid, Colette Ravinet and Jean-Luc Soubie who have given many perceptive comments and much helpful advice on the writing of this paper.

J. Virbel (✉)
IRIT, CNRS, 1, rue du Commerce, 31540 Saint-Felix Lauragais, France
e-mail: virbel@irit.fr

© Springer International Publishing Switzerland 2015   49
K. Chemla, J. Virbel (eds.), *Texts, Textual Acts and the History of Science,*
Archimedes 42, DOI 10.1007/978-3-319-16444-1_2

(SAT: Bach and Harnish 1979; Hornsby 2008; Searle and Vanderveken 1985; Vanderveken and Kubo 2002)[1]. There is no sign of dogmatism in this approach, only the desire to render this domain more accessible. A presentation 'that can be used' will hopefully allow any readers who are not specialists of the field to fully appreciate the approach that follows and even formulate new applications for him or herself. This brings us to the second main objective of this chapter. We would like the reader to realize that this approach has a productive character by presenting text-centered analyses. These applications will be creative given that SAT was not in any way developed in the perspective of use in text analysis (Champagne et al. 2002; Lemarié et al. 2008; Tonfoni 1996; Vanderveken 2001).

In Part II, we shall present some analyses of textual phenomena which were carried out using as well as developing some aspects of SAT presented in the first part. Firstly, we shall analyze the concept of instruction and the types of text that instantiate this concept (recipes, medication dosage, DIY [Do It Yourself] instructions, game rules, itineraries, pharmaceutical instructions, etc.). The various kinds of prescription that an instruction can indicate to the user, like the various types of commitment which may establish between the user and he who produces and/or issues the instruction, are, for example, some of the questions that can be addressed in such a context.

2.2   Speech Act Theory

1. Natural language utterances do not only serve to describe states of the world (such as 'It's raining'): some allow the performance of specific types of acts, these are called illocutionary acts by Austin. Indeed, when a speaker S says 'I kindly ask you to leave' or 'I promise I will come around and help you tomorrow', he is not describing himself asking or promising something. He is literally making a demand or a promise by saying that he asks or promises. We can perceive the difference by comparing the above with a situation when S would say: 'When I want to be alone, I ask Max to leave me' or 'When I am a bit drunk, I always promise to Max I'll go around and help him'. These uses of language, which are performative, are thus different from the constative uses. This first distinction between the two was then put into perspective by Austin: describing a certain state or affirming what it is, is as much a speech act as a request or a promise, even if they are of a different type.

---

[1] List of symbols used: SAT=Speech Act Theory: S=speaker, author; H=hearer, reader; u=utterance, text; A=action named or described in utterance u; p=proposition; T: sentence uttered; ifid=illocutionary force indicating device; bIF=basic illocutionary force; CS=conditions of success.

In the case of procedural texts: H=consignee; A(s)=action A is done by S; A(h)=action A is done by H; CS+=added conditions of success.

We extend our thanks to Pierre Chaigneau and Richard Kennedy who have provided insightful remarks on the English version of this text.

2. The concept of a speech act was clarified by means of a standard distinction: the same utterance allows the speaker to realize a locutionary act (the actual physical utterance), an illocutionary act (the act of communication which can be realized through the utterance such as a request, a promise, an affirmation, etc.) and a perlocutionary act which refers to the effect of the illocutionary act on the receiver (joy, relief, fear, etc.). A major difference between the last two kinds of speech acts is the following: the performance of illocutionary acts involves some conventional resources of a given language: thus, a promise can be realized by this utterance: 'I promise I will come tomorrow' but also by others such as 'I will come tomorrow', 'See you tomorrow', 'You can count on me being there tomorrow', etc. The interlocutors are familiar with these expressive resources, and they can be listed. This allows us to say that the realization of illocutionary acts is in a way conventional. By contrast, the perlocutionary effect does not depend directly on any convention, whether it bears on the meaning of the utterance or the resources of expression that were used for the performance of the illocutionary act: the utterance 'I promise I will come tomorrow' can provoke gratitude, worry, or fear for hearer H.[2]

Following this distinction between three types of speech act introduced by Austin, Searle proposed a different classification, which excludes perlocutionary acts (which would belong to the field of psychology or ethnology for instance), and distinguishes between:

   a) Utterance acts (the fact of uttering the words, thus quite similar to locutionary acts)
   b) Propositional acts: referring and predicating
   c) Illocutionary acts (staging, questioning, commanding, promising, etc.)

3. The relationship between the illocutionary force and the propositional content is of a function-argument relationship: an illocutionary force F can be applied on a proposition p according to the formula F(p). Thus, the same proposition can be the argument for different illocutionary forces: for instance, assertion ('You're coming'), question ('Are you coming?'), directive ('Come'). These two components activate different comprehension procedures from a neuropsycholinguistic point of view. We can understand the illocutionary force of an utterance without (fully) mastering the propositional content, and conversely, not understand the force with which a proposition is uttered. Austin differentiated these two forms of comprehension by referring to 'understanding' as opposed to 'uptake'.

4. One of the consequences of the notion of speech act is that communication is considered to be the product of actions. This 'truism' masks one of the most significant advances. On one hand, the notion of action, linked for instance to the notion of event, allowed for considerable development in the fields of logic, philosophy of language, cognitive psychology, and even Artificial Intelligence. These developments can be directly applied in research on pragmatics. On the other hand, communication is an activity with some risks which we can foresee

---

[2] This led Searle to reconsider the notion of effect in Grice's conception of meaning, where the two kinds of act are combined, and to identify a truly illocutionary effect separate from the perlocutionary effect.

only partly: the success of an illocutionary act, just like the success of any action, requires the fulfillment of conditions of success (CS). There are many types of such conditions, and we need to distinguish between them and the conditions of satisfaction. The latter deal with the relationship (corresponding or not) between the propositional content of an utterance, and the 'state of the world' it represents.

5. Different markers of illocutionary force indicate which illocutionary force is realized. These include elements of syntactical structure, verbal and sentence modes, intonation, sometimes punctuation marks or even typographical attributes, and above all, specific verbs which name these forces (to advise, to command, to promise, to declare, etc.):

Sam smokes habitually.
Does Sam smoke habitually?
Sam, smoke habitually!
Would that Sam smoke habitually. (Searle 1969, p. 22)
Sam, I beg you to smoke habitually.

6. Following Austin's typology of illocutionary forces and/or verbs, many others were tested. For instance, Searle's typology, which is presented as a typology of forces and not verbs, is based on the fundamental fact that the basic unit of communication is the act of communication and not (in contrast to descriptive linguistics or logic) the utterance (which is the means of the realization of the act).

7. We can demonstrate the content of the conditions of success of an illocutionary act by referring to the act of promising as Searle did ('How to promise', Searle 1969, pp. 57–61 and, in a slightly modified form, in Searle 1971, pp. 46–51). Note that S designates the Speaker and H the hearer, p what is promised and T the sentence which is uttered.

"1. Normal input and output conditions obtain" ("conditions of understanding" and "for intelligible speaking")

"2. S expresses the proposition that p in the utterance of T"

"3. In expressing that p, S predicates a future act A of S" ("propositional content condition")

"4. H would prefer S's doing A to his not doing A, and S believes H would prefer his doing A to his not doing A" (preparatory condition)

"5. It is not obvious to both S and H that S will do A."

"6. S intends to do A" ("sincerity condition")

"7. S intends that the utterance of T will place him under an obligation to do A" ("essential condition")

"8. S intends

(i-1) to produce in H the knowledge (K) that the utterance of T is to count as placing S under the obligation to do A."

(i-2) "S intends to produce K by means of the recognition of i-1" by H

(i-3) "and he intends i-1 to be recognized in virtue of (by means of) H's knowledge of the meaning of T."

"9. The semantical rules of the dialect spoken by S and H are such that T is correctly and sincerely uttered if and only if conditions 1–8 obtain."

These conditions can be brought together into three groups according to their scope:

a. Conditions related to communication in general
   1, 5, 9
   Conditions 1 and 9 guarantee the existence of the conditions of effectuation of
   the communication (conditions of realization and completeness of the description), and condition 5 is a general condition of 'communicational economy' (S
   and/or H are not sure that the state of the world will change as they want it to
   change, if they do not say anything i.e. if they do not perform speech acts).

b. General conditions related to illocutionary acts (not specific to promises)
   2, 8
   Condition 2 is implicitly linked to the principle of expressivity (a language provides the resources necessary for expression), and condition 8 is a special interpretation of P. Grice's 'meaning' in his article bearing the same name (Grice
   1957).

c. Conditions specific to promises
   3, 4, 6, 7
   Condition 7, which is called 'essential', literally expresses the illocutionary point
   of the act of promising (we specify this notion below). Potential failure of the act
   should be attributed to S's intentions (for instance S wants to lie to H or cheat
   him). One should note that condition 4 is only a valid preparatory condition for
   promising and not for commissives, that is, the most generic acts that commit H
   and for which the desire of H that S should do A is not necessary. Consequently,
   one also notes that, beyond their similarities, acts such as 'to promise' and 'to
   commit to') are differentiated by the existence of a condition of success of a
   specific preliminary type. This idea was generalized and used by (Vanderveken
   1990) to create a systematic enumeration of the verbal performative lexicon of
   a language. Finally, one should note that condition 3 is necessary to characterize
   the commissives as well as the directives.

8. Searle's typology of illocutionary forces is initially based on a study of the 'different types of difference between different types of illocutionary acts', which is
   summarized as follows: (Searle 1979, pp. 1–8)

   "It seems to me, Searle writes, there are (at least) twelve significant dimensions
   of variation in which illocutionary acts differ one from another":
       "1. *Differences in the point (or purpose) of the (type of) act.* The point or
       purpose of an order can be specified by saying that it is an attempt to get the
       hearer to do something. The point or purpose of a description is that it is a
       representation (true or false, accurate or inaccurate) of how something is. The
       point or purpose of a promise is that it is an undertaking of an obligation by
       the speaker to do something. These differences correspond to the essential
       conditions in" the "analysis of illocutionary acts" (See supra paragraph 7)
       "The point or purpose of a type of" illocutionary act is called "its *illocution-
       ary point*," which is "part of but not the same as illocutionary force."
       "2. *Differences in the direction of fit between words and the world.* Some
       illocutions have" "to get the words, (more strictly, their propositional content)

to match the world" (e. g.: assertions), "others to get the world to match the words" (promises and requests). (see below for explanations)

"3. *Differences in expressed psychological states*" ("belief, desire, intention, regret", "want," etc.) "in the performance" of an illocutionary act, i.e.: "the *sincerity condition* of the act."

"4. *Differences in the force or strength" or commitment "with which the illocutionary point is presented.*" "Along the same dimension of illocutionary point or purpose there may be varying degrees of strength or commitment" ("I suggest" in contrast to "I insist that we go to the movies").

"5. *Differences in the status or position of the speaker and the hearer as these bear on the illocutionary force of the utterance.*" — "one of the preparatory conditions" ("if the general asks the private to clean up the room, that is in all likelihood a command or an order", and not "a proposal" or a "suggestion").

"6. *Differences in the way the utterance relates to the interests of the speaker and the hearer.*" For instance, from the point of view of S, boasts and laments concern S's interests and congratulations and condolences the H's one. "This feature is another type of preparatory condition" of the speech acts.

"7. *Differences in relations to the rest of the discourse.*" ("I reply," "I deduce," "I conclude," and "I object"). Some conjunctions, "as "however", "moreover", "therefore" also perform these discourse relating functions."

"8. *Differences in the propositional content that are determined by illocutionary force indicative devices*" —hereafter abridged into ifid — (i.e.: "propositional content conditions"); report (present or past) in contrast to prediction (future).

"9. *Differences between*" those acts "*that can be, but need not be performed as speech acts,*" even an *internal speech act* (for example, "I classify," "I estimate," "I diagnose," "I conclude") and "*those acts that must always be speech acts.*" (e.g.: promise).

"10. *Differences between those acts that require extra-linguistic institutions for their performance*" ("and generally a special position by the speaker and the hearer within that institution") "*and those that do not.*" (bless; excommunicate, christen, pronounce guilty, declare war).

"11. *Differences between those acts where the corresponding illocutionary verb has a performative use and those where it does not.*" (One can perform acts of stating or promise by saying "I state" or "I promise"; but "one cannot perform acts of boasting or threatening by saying "I hereby boast" or "I hereby threaten"").

"12. *Differences in the style of performance of the illocutionary act*" (announcing in contrast to confiding with the same illocutionary point and propositional content).

Searle created his typology using a restrictive choice of some of the above dimensions.

9. The direction of fit allows us to describe the relationship between the propositional content of the utterance and the situation (or state of the world). Thus, we can characterize the fundamental types of illocutionary acts. In classic logic, the situation-utterance relationship only pertains to the assertions and foresees two possibilities: true or false according to whether the propositional content of the utterance represents the state of the world. More precisely, the utterance needs to adapt its content to the state of the world in order for it to be true; this is obvious in the case of a mistake; if S says it is raining while it is not, he must change his utterance and not the weather (which is logically beyond his control). Yet, this situation, though belonging to the vast group of assertive utterances, is not generally applicable. There are four possible fit relationships between an utterance (u) and a state of the world (w) and, thus, four fundamental types of illocutionary acts:

$u \rightarrow w$: assertions

S must adjust the content of his utterance to the state of the world

$u \leftarrow w$: directives, commissives

In these two categories, conversely, the state of the world changes to fit the content of the utterance

$u \leftarrow \rightarrow w$: declaratives (I pronounce you husband and wife)

Declarations (I hereby declare this meeting open) have a double direction of fit: by changing the state of the world according to u, S realizes one act which would be simultaneously like an assertion and a directive

$u - \varnothing - w$: expressives (Let me congratulate you on your victory)

In expressives, there is no direction of fit because it is assumed that the proposition which represents the propositional content of u is true.

This basic structure can be, and has been, enriched in many ways. It is useful now to distinguish between directives and commissives based on whether H's or S's actions change the state of the world according to the content of u. In the same way, we can track among the directives the cases where S wants to make H perform these particular acts that are speech acts—this is the case for interrogatives (questions). We will also see that it is possible to differentiate numerous types of declaratives or assertives, if the world situation in question is that of a language or the text in which the acts are realized (meta-speech acts).

10. Speakers and readers of a given language can recognize which illocutionary acts are realized by utterances because these utterances bear 'illocutionary force indicating device' (ifid) markers which can be:

Syntactical

Get out! What time is it?

Or lexical

I (state + assure you + swear + ...) it is raining

I (ask + order + advise + beg + invite + implore + ...) you to leave

I (promise + guarantee + ...) I will come tomorrow

I declare this meeting open

I (congratulate) you for winning

I am (happy + thrilled + ...) + ...) you won

Yet, the truth about the expressive resources of a language is far more complicated. Indeed, the same ifid allows the realization of very different acts. For
instance, the imperative mode can also be used to realize acts such as:

Advice: S: Excuse me, I want to go to the railway station
        H: Take the number 3 bus
Conditional instructions for use:
To listen to the recorded conversations
Press the REW button
Press the MP button
Press the STOP button to stop playback
Permission: S: Can I open the window?
            H: Oh, open it, then.
Insistent assertions: Note that he is often right!
Good wishes: Get well soon (visiting someone in the hospital)
Audience-less cases: Start, damn you (to a car)
Threats and dares: Go on. Throw it! (to someone who's getting ready to
                     through a snowball)

Conversely, an illocutionary force can be realized using different ifids which are
usually indicative of other illocutionary forces:
• Could you leave?
• You can leave now.
• I would like you to leave.
• I would like to be alone.
• You'd be better off outside.
• It's a great day (make the most of it!)

In order to explain this, we use the concept of indirect illocutionary acts (Bach
1994; Bertolet 1994; Holdcroft 1994; Searle 1979) and differentiate, following Grice (Grice 1989), between the meaning of the utterance and the speaker's
meaning (or: the meaning the speaker aims at); the coincidence of the two is
certainly an important case, but special. Considering one or more utterances (u,
f, etc) as the representation of the meaning intended by S, we can enumerate the
following possibilities:

$u \rightarrow u$ literal act: S says u and means u;

$u \rightarrow *u$ irony: S says u and means the exact opposite;

$u \rightarrow -u$ fiction: by saying u S means to say less than what he is saying, at
least in relation to the conditions of truth of his utterances;

$u \rightarrow u+f$ indirect act: S says u and means u plus another act (f) (Could you
leave?);

$u \rightarrow f$ metaphor, implicature (on this term see below): S says u but
means something different (Sally is a dragon);

$u \rightarrow \emptyset$ useless politeness or phatic function (Hello?) (Jakobson 1960): In
saying u, S has nothing (specific) to say.

Table 2.1 Conditions for Directives and Commissives

	Directives (request)	Commissives (promise)
Preparatory conditions	H is able to perform A *You can leave.* *Can you leave?*	S is able to perform A H wants S to perform A *I can come in.* *Can I come in?*
Sincerity condition	S wants H to do A *I want you to leave.*	S intents to do A *I want to come in.*
Propositional Content condition	S predicates a future act A of H *You will leave tomorrow.*	S predicates a future act A of S *I will come tomorrow.*

One can see in this list that sentences like 'Could you leave?' are interpreted as indirect illocutionary acts: by saying u S does not just want to ask H whether he can leave but also wants to ask H to leave. In other words, S realizes a directive act ('Leave!') by means of another act: a question about the possibility to do so. A complete analysis shows that all indirect illocutionary acts of this type consist of assertions and/or questions on the preparatory conditions, condition of sincerity or propositional content condition of the corresponding direct illocutionary acts. The table 2.1 lists these conditions for directives and commissives and offers examples of corresponding indirect illocutionary acts using these sentences (from Searle 1979, p. 44).

11. Taking the content of some of their conditions of success as a basis of other direct acts, such as assertions and questions, is not the only form of indirect realization of illocutionary acts. In effect, these conditions of success are only one type of pre-condition for an action; actions have other constituent elements such as, for example:

- Reasons or motives or causes or explanations
- Initial state of the world (to change or conserve)
- State of the world to be achieved (goal)
- Pre-conditions (conditions of success) and pre-constraints
- Elementary actions whose composition leads to A and the logic of planning the actions
- Post-conditions and post-constraints

We can prove that that all these elements can be used equally in questions and assertions to perform illocutionary acts:

>Assertions or questions on the existence of a (good) reason to (not) do A
Do you want me to come?

S can realize an indirect directive by asking if there are, or by asserting that there are, good reasons to do A. As in this particular case: one reason could be that H wants or wishes A, so S may ask whether H wants, or wishes, A.

>Assertions or questions about the reasons to do (or not do) A
You should rest.
You'd better leave.

Must you leave right now?
Wouldn't it be better if you left now?
Why not stop here?
Why won't you stay in one place?
There is no reason for you to stay.

>Assertions or questions about H's desire to do A, or his agreement to do it
Would you be willing to write a reference letter for me?
Would you like to hand me that hammer over there?
Will you mind being less noisy?
Would it bother you to make it next Wednesday?
Would it be a problem for you to pay me next Wednesday?

>Assertions or questions on the content of a (good) reason to (not) do A
Is it reasonable to eat so much spaghetti?
It wouldn't hurt you to go out!
Wouldn't it be nice to leave town?
You're stepping on my feet.
I can't see the screen if you keep your hat on.
S: Let's go to the cinema tonight.—H: *I need to* study for my exams.

>Assertions or questions on the content of pre-conditions
Do you know where the salt is? *Can you see* the salt? *Do you see* the salt?
(To see a target or know where it is is normally a pre-condition to picking it up)

>Assertions or questions on the content of the initial or final states of the world
You're stepping on my feet.
I can't see the screen if you keep your hat on.
It's hot in here.

>Assertions or questions on the content of parts of the action
Could you grab the salt?
Can you (reach+stretch) over and (grab+pick up) the salt and (pass+put) it
here?

(All of the above amount to the decomposition into more elementary actions of
the global action 'pass the salt)

>Assertions or questions on the content of post-conditions
S: Can I come in?—H: Close the door when you do.
S: Can I get myself a beer from the fridge?—H: Just put some more in.

(H indirectly answers S's question by asking him to respect one of the post-conditions of the script[3] 'to come in' or 'to take a beer from the fridge').

12. Recognizing indirect illocutionary acts, and generally all non-literal aspects of speech, may seem problematic if we only consider the formal aspects of utterances. How are we to decide, on this basis, if 'You can pass the salt?' is a uniquely direct question (for instance a chiropractor who wants to know if his patient is making any progress), or if someone at the table is asking to be given the salt? Obviously, the context itself allows one to decide, but we know this is a very complex notion. Grice proposed a tool which allows the necessary formalization for this need in the context of his research on the cooperative character of conversation: conversational maxims and a specific form of rea-soning: implicature.

The conversational maxims are divided into four groups (which we state here before illustrating their application) (Grice 1989, pp. 26–27):

A Quantity
    1-" Make your contribution as informative as is required for the current purposes of the exchange."
    2-" Do not make your contribution more informative than is required."
B Quality
    1-" Do not say what you believe to be false."
    2- "Do not say that for which you lack adequate evidence."
C Relation
    1-"Be relevant": respect the flow of the conversation, handle focalizations and changes in topics, etc.
D Manner (of saying): be clear ("be perspicuous")
    "1- Avoid obscurity of expression
    2-Avoid ambiguity
    3- Be brief (avoid unnecessary prolixity)
    4-Be orderly"

Let us note that maxims A and B pertain to propositional content *stricto sensu*, whereas C pertains more to the structure of the speech (conversation), and D pertains to the use of the expressive resources of the language.

On conversational maxims, let us emphasize the following points:
 - they belong to a wider ensemble of maxims ("aesthetic, social, or moral in character"....) for instance: "Be polite"; Do not lose face unnecessarily; etc.
 - they are a special case of maxims governing cooperative action
 - they are not well-formed formulae but general principles that are used and applied in various ways to ensure cooperation
 - they allow an original form of inference: implicature (Grice 1989, pp. 32–37).

---

[3] In Artificial Intelligence and cognitive psychology, a script (or frame) is a conceptual representation of procedural knowledge on physical or mental stereotyped actions.

When all the conversational maxims cannot be respected together there is implicature if one maxim is ostentatiously violated; in this case, this infringement occurs in favor of another maxim which is respected.

Examples:

S: "I'm out of petrol."

H: "There is a garage round the corner."

(H infringes the maxim of Relation and the first maxim of Quantity; yet, he is applying the second maxim of Quality; so he implicates: it is possible that the garage is open and that it sells petrol)

S: Max "doesn't seem to have a girlfriend these days."

H: He has been paying a lot of visits to Paris lately.

H infringes the maxim of Relation to observe the first maxim of Quantity and the maxim of Manner.

S: Where are the car keys?

H: Somewhere in the living room.

H infringes the first maxim of Quantity to observe the maxim of Relation and the maxim of Manner.

S: Can you vouch for student X?

H: X is very punctual, never late for class, always smartly dressed, fluent in English, and very polite.

(H infringes the maxim of Relation and the third maxim of Manner to observe the first maxim of Quality)

The application of this principle may affect the expression of the performance of illocutionary acts (and not only the propositional content, as in the examples above). This is how we can account for the following example:

S: Can you pass the salt?

I am not having lunch with my chiropractor who would like to check on my progress. Thus, the conversational flow is interrupted: the maxim of Relation is infringed for this reason.

Furthermore, S is realizing an illocutionary act (a question) of which a preliminary condition of success is obviously infringed given that the answer is evident, while the preliminary condition of a question is that S does not know the answer. It is another case of infringement of the maxim of Relation.

If S infringes this maxim so overtly and deliberately, we may assume that he does so to respect another (or others). Asking a question on a condition of success of an illocutionary act is a way of indirectly performing the said act: asking someone if he can do A, may be an indirect way of asking him to do A. Given the relationships between S and H, by infringing a maxim, S is pointing out that he wants to respect the maxim of Quantity (the speech act must be 'as illocutionary' as needed but no more).

This necessary and sufficient 'intensity' in the choice of the realization of the directive act must be compared to other possibilities:

'I order you to give me the salt!' (certainly too intense)

'Give me the salt!' (maybe too intense)

'It's not very tasty. It may taste better with some salt' (maybe too weak)

'I'm off my no-salt diet (and I'm making up for lost time)' (presumably too weak)

13. There is one last important point to discuss. We have already seen how directives can be distinguished from assertives and commissives. Nevertheless, one question remains. How to characterize what differentiates forces within the same type? For instance, what is the difference between illocutionary forces realized by verbs such as 'to demand' and 'to order', 'to advise', 'to ask', etc.? The technique we should apply has already been suggested above for the difference between 'to commit to' and 'to promise' (See supra paragraph 7).

    D. Vanderveken (Vanderveken 1990) proposed that a method of systematic enumeration of the main illocutionary forces expressed by a verb (in English and French) should be defined for each of the five fundamental types of illocutionary forces. His approach can essentially be described as follows:

    a. each type of fundamental illocutionary act is defined by a basic illocutionary force (bIF), which corresponds to the illocutionary point of that type of act (the point represents the purpose of the illocutionary act and determines the direction of fit and the presentation of the propositional content);

    b. each bIF is defined by the content of the other three types of conditions of success: preparatory, sincerity, and propositional content;
    Based on this, a particular illocutionary force can be defined by the addition of new conditions of success of any of the three types.

    c. Furthermore, two other components of the illocutionary force may result in variations: the mode of achievement of the illocutionary point and the degree of strength of the expression of the condition of sincerity. The values of these two components are neutral or equal to zero for the bIF, and the individual illocutionary force may include (on top of the specification of the conditions of success mentioned above) variations in these two components.

Thus, any given illocutionary force is defined by the content of the six components of illocutionary force:

- the essential condition, which defines the fundamental type of the illocutionary force (assertive, directive, commissive, declarative, expressive), in relation to the value of the direction of the world-utterance fit of the illocutionary point;

- the contents (which are supplementary to those of the bIF) of the other three types of condition of success: preparatory conditions, conditions of sincerity, and conditions of propositional content;

- the particular modes of achievement of the illocutionary point and the degree of strength of the expression of the condition of sincerity, which are by definition neutral for the bIF.

The procedure is said to be recursive in the sense that a given illocutionary force can be defined by using another illocutionary force, of which it "inherits" all or part of the characteristics by the addition or the substitution of new features. It can be noted that the six components of illocutionary force retained by D. Vanderveken (the four conditions of success: essential, preparatory, sincerity, propositional content; the mode of achievement of the illocutionary point; and the degree of strength of the expression of the condition of sincerity) are a subgroup of the 12 dimensions initially identified by Searle (see paragraph 8 above) by which illocutionary acts can be differentiated.

2.3   A Speech Act Theory Based Approach for Instructional Texts

The aim of this chapter is to evaluate, by means of a detailed analysis, the possible contributions of Speech Act Theory to research on instructional (or procedural) texts.

This objective can be justified in two ways:

Firstly, texts of this kind happen to be in a situation of relative paradox (Heurley 1994): their extended use in both technological and professional fields, as well as domestically and in personal areas and the importance of their impact (technical, legal, financial, medical, etc) are in contrast with the relative modesty of the knowledge of their properties, efficiency, etc. Nonetheless, instructional texts raise many important theoretical issues: as text types, their position with respect to stories or descriptions is not completely acknowledged; as texts which can lead to an execution of an act, they lend themselves, from a psycholinguistic point of view, to an original modality of examination of the comprehension through observation of the performances of subjects in different situations. Another aspect, linked to modeling, appears again. The instructions describe the content and the effects of the actions that are executable, learnable, transmissible, modifiable, improvable, etc. Thus, instructions should be able to support (maybe even more than other text types) the demanding operations of control: well-formedness (adequate for a model), consistency (no contradiction in the prescriptions), completeness (all cases are considered), coherence (lexically and syntactically unambiguous for instance), etc.

It appears that referring to Speech Act Theory (SAT) is quite natural in this perspective: instructional texts are explicitly given as different types of realization of directive acts. A more in-depth observation reveals that they can also realize acts of the commissive or assertive type and even of declarative type in some contexts. A priori, it should be useful to differentiate and characterize a direct order, a piece of advice and a recommendation; a commitment with no conditions from the prescriber and a commitment conditional on a given attitude by the user of the instruction text; a mere assertion on the state of the world and a prediction; etc. The crucial point here is that understanding an utterance is not limited to understanding the propositional content but also comprises understanding the intentions that the utterances express (understanding in contrast to uptake, to use Austin's words).

Conversely, this research confirmed the idea that if SAT (sometimes denounced as generalization so far removed from the effectiveness of utterances as to be of little use in concrete situations) offers an enlightening conceptualization, it can itself be enlightened in a particularly detailed implementation.

Three preliminary clarifications must be made on the scope of this work:

1. It is clear that the component of an instruction, as far as speech acts are concerned, is only one of the dimensions on which one can base taxonomic criteria. In order to define the contribution of this dimension, we preceded this study with an inventory of dimensions that are likely, in our view, to be the basis of a taxonomy.
2. In general, procedural texts are seen by their authors and readers as aids. In particular, the definitions of the instruction in a professional context stress the importance of their value as 'aid' to an agent. The notion of 'aid', which seems clear, turns out to have multiple forms and to be very relative. It has multiple forms in that it can relate to all conceptual aspects, such as the number of details (granularity) included in the breakdown of a global action into constituent actions, and all expressive aspects: linguistic and visual means of realization. It is relative to knowledge, expertise, situations, etc. of the agent (a partial analysis of the concept of aid can be found in Sect. 2.3.3.2).
3. Finally, one last clarification must be made on the objects of the following analyses. What we are aiming at is to shed light on and characterize, if possible, the specifications of the types of speech acts that can be present in instructional texts. This will be done considering, as does Searle whom we follow here too, that the *basic unit* of communication (which is an activity) is not the utterance but an action, the elementary speech act, which is realized by means of an utterance. Consequently, we shall not address, in this study, a completely different question: the multiple forms of syntagmatic composition that these speech acts can adopt in a given instructional text. Thus, we shall not discuss the exhaustive taxonomy of these texts as forming a single category, but only the speech acts within them.

2.3.1 Different Types of Differences Between Different Types of Instructions

The title of this section is a transparent reference to and an adaptation of the title Searle (1979) gave to a section of his famous article 'A classification of Illocutionary Acts' (this later became the first chapter of *Expression and Meaning*).

This reference indicates not only a source of inspiration for the content but also a parallel approach. Just as Searle started with an identification of the dimensions or aspects for which illocutionary acts can be distinguished from each other, in order to continue with the identification of the dimensions that can be used as a basis for the taxonomy of these illocutionary acts, we are trying in this section to define a frame which aims to play the same role for 'giving instructions' type acts.

An initial empiric inventory includes 17 of aspects that are likely to be, subject to later analysis, potentially contrastive.

For the sake of clarity, these aspects are grouped into three classes as they relate to:

a. Illocutionary forces that can be realized in the context of an instruction
b. Semiotic and linguistic aspects
c. Aspects related to the activity

a. **Illocutionary Forces**
  1. **Directive illocutionary forces**
     We perceive intuitively that giving an order, or a command, or a piece of advice, or making a suggestion, etc. each performs an illocutionary act of the same fundamental (directive) type, but with different modalities, for example, the relationship between the speakers, the obligations assumed by one and/or the other, etc. vary from case to case.
     Directive illocutionary forces fall under groups that cover at least 40 types defined by verbs (in English and French).
     The question here is where 'to give or provide instructions' is placed in this group. This is discussed in detail in Sect. 2.3.2.
  2. **Commitments (guarantees, etc)**
     Given on the one hand the function of aid of instructions and, on the other, the social, technical, ethical, contractual, and other contexts where instructions may occur, it is possible that instructions can be, explicitly or implicitly, a form of contract, so that the person who provides the instructions commits up to a given point (for instance about the success of the consignee and his right to expect success). This aspect of commissives can also include variations of strength and/or effect (Sect. 2.3.3).
  3. **Assertions (forecasts, diagnostics, etc)**
     Like any realistic (serious and literal) utterance, instructions include assertions of the simplest form, for instance references to the state of the world in question. Furthermore, due to their nature of aid to the realization of activities, instructions seem to be meant to contain future assertions on new states of the world, for instance, which are necessary or possible and realized by said activity, by the nature or properties of the goal, etc. This type of assertion about future states includes differentiated illocutionary forces (conjectures, predictions, prophecies, etc.), the illocutionary definitions of which vary (Sect. 2.3.4).
     It should be added that assertions (future or not), like questions, can carry out indirect directive acts, evoking a condition of success, a state of the world to modify or obtain, or a reason (not) to do the action.
  4. **Explanations**
     One other type of assertion is likely to appear in an instruction: explanations, which can also have varied effects and/or logical status. We present below (Sect. 2.3.4.3)) a kind of explanation involved in the realization of indirect illocutionary acts.

5. **Promulgation**

In order for instructions to obtain their full status as reference texts, they must be rendered public in a broad sense: the potential recipients must at least know that the instructions exist and the modalities for their consultation and use (the minimum seems to be: where are they?). And vice versa, the issuer of the instruction can consider, once the instruction was properly promulgated, that no-one affected by it can claim ignorance.

Ensuring that a text is taken into account is a question that does not only apply to instructions; it refers to illocutionary acts that are both assertive and declarative but have many modalities (to promulgate, to notify, to declare, to warn, etc.) which can differentiate instructions from each other (Sect. 2.3.5).

b. **Semiotic and Linguistic aspects**

6. **Mode of Utterance**

The mode of utterance is without doubt a significant aspect of the instruction. Variations on this dimension become clear when studying certain cases of strictly identical instructions whose enunciative range are nonetheless different. For instance, two different texts on drug dosage with the exact same content have different effects, according to whether they can be found on the packaging of a drug or in the instructions inside the pack (in which case the text is a recommendation or advice), or if it is found on the prescription given by the general practitioner to his patient (in which case it is clearly a direction). Obviously, this aspect matches that of the type of illocutionary act performed.

Likewise, the (exact same) 'No smoking' notice in a public location does not have the same status before or after the promulgation of a law about public smoking (it is a kind of local rule in the first case and the reminder of an existing law in the second).

Thus, it appears that, like any other utterance, instructions can be, for instance, original, a quotation, a reminder, etc.

7. **Semiotic and linguistic modalities**

There are many such modalities. We can talk of at least:

- the linguistic modal forms of expression (imperative, indicative, infinitive, nominalizations, etc.);
- developed discursive expressions in contrast to reduced discursive expressions accompanied by typographic and positional traces (hereafter 'typo-dispositional');
- discursive expression in contrast to diagrammatic, logogrammatic, figural, etc.

8. **The textual 'genre(s)' of instruction**

Knowing if instructions correspond to a fundamental text type (in the manner of descriptions, stories, dialogues, etc.) is an important problem, in the sense that a positive answer would shed light on useful distinctive characteristics that could be used in a definition. This question has already received various responses in the literature.

On the other hand, the notion of text 'genre' or 'type' has recently been profoundly renewed, and a study on instructions could benefit from this.

It appears that generic logic, which aims to answer the question 'what does it logically mean that a text t is of a specific genre g?' strongly relativizes the semantic or structural aspects, which are in general the only ones taken into account in classic linguistic approaches. The aspects related to reference modes (Schaeffer 1989) seem to be more pertinent.

With this term "reference modes", Schaeffer introduced a key idea. The meaning of a term designating a genre captures different features of the communicational process, of which the text is a means. When we say that a text is a "sonnet," we refer to some syntactical properties of the message, whereas when, we say it is a comedy, we refer to its perlocutionary effect. This remark allows Schaeffer to clarify what is meant by stating that a text t belongs to a specific genre g. Schaffer suggests that we refer to one feature of the communicational process or another, according to the meaning of the term with which one designates the genre. Another feature that emerges from this approach concerns the granularity of the concept of genre or that of type of utterance, which relates more generally to the problem of categorization.

9. Instruction = textual script and/or textual score?

The notions of 'script' (as in the work of playwrights) and 'score' (in the musical sense) belong to the formal theory of notational systems which was developed within Goodman's theoretical semiotics (Goodman 1968 (1976)). Using an entirely reformed notion of reference Goodman differentiates, among other things, on a general level, between 'scripts', which are syntactically articulated but semantically dense[4] (including texts among others), and 'scores', which are syntactically and semantically articulated (music scores, numerous schemas and diagrams among others). In the case of instructional texts, it appears, we believe, that we are situated in an in-between area bounded by two clear poles. At one of the poles are found would find instructions (clearly the greatest number), the semantics of which is not articulated (and thus not formalized); at the other, those with entirely articulated semantics which can also create a score (generally speaking: a notation representing the execution).

Such an approach brings to mind other related questions, such as:

- to what extent can scores be paraphrased or adapted into more natural speech without losing their properties? (this seems to be largely the case, specific as it may be, for theatrical texts)
- as far as instructions in an intermediary position are concerned, how can we characterize the composition of what would simultaneously be a text and a score?
- can the notions (Genette 1997) of transitive texts (i.e. instructions) and intransitive texts (i.e. stories), compared to the notions of transitive reading (i.e. the text is to be performed) and intransitive reading (i.e. the text is to read only), shed light on these questions?

---

[4] The concepts of "dense" and "articulated" refer to a property of differences in structure among units as they are continuous (in the passage from one unit to another, there may be a third) or discrete (between two structurally contiguous units, there can be no a third intermediate unit).

10. **Location**

Where can instructions be found? It would seem that not any type of instruction can be found in any place. Some are detached from the objects of the action or place of action (for example, game rules or recipes which can be found in game manuals or recipe books); others are attached (for example, instruction in the case of fire or electrocution which are placed in the place of their possible application); others can be mixed (the aforementioned game rules and recipes could be found on the packaging or boxes containing the game or of the ingredients for the recipe). This aspect is obviously related to the 'pragmatics' of instructions.

One other modality of localization, which is related to the last, without being the same, is that of the character of the instruction of being 'on display' or not, given that 'exhibited' texts have particular properties (writing style, visual characteristics etc.; Petrucci 1986 and Harris 2000). Thus, the aspects of localization of instructions strongly relate to other questions: for whom are the instructions issued? Is the relation between the consignee and the localization adequate?

c. **Aspects related to the activity**

This chapter is more difficult to circumscribe than the previous ones, as it deals with aspects external to the text itself and specific to the effectuation. Thus, these aspects are only evoked here from the point of view of the content of the instructions (and not as an analysis of the effectuation), that is to say, as long as they are supposed to affect *the very text* of the instruction in various ways, for instance because they account for a differentiation which is assumed to be significant.

11. **The types of actors**

It seems unavoidable that the question 'Who gives instructions to whom?' needs to be asked in this study, as the answers obtained will necessarily provide important distinctions between types of instructions. On the other hand, this question seems to be almost impossible to study, given that the possible variations of the answers seem infinite. A remark that goes in the same direction would consist in noting that, in the context of this question, the expression 'to give' (which is only useful as a support verb to designate the illocutionary act 'to give instructions') also needs to be specified: who conceives, writes, corrects, checks, transmits, diffuses, updates, etc., the instructions. We know that in certain cases inadequacy between the context of the writing and the context of reception can be annoying (for example, the user manual for an appliance written by the engineers who designed the machine and not by someone practiced in operating the machine). This group of properties, which is very far from any conventional context, can only be studied in the case of fully defined examples.

12. **Rules defining activities**

The notion of 'rule' is particularly polyvalent and needs to be clarified. Especially since it appears that all activities that can be the object of instructions are not equivalent from this point of view (for example, recipe in contrast to game rules or instructions in the case of fire).

We need to consider at least two differentiations: constitutive rules in contrast to normative (regulating) rules; detachable rules in contrast to non-detachable (or incorporated) rules.

13. **The 'consignee ← → world' direction of fit**

Even though, clearly, any action simultaneously changes the state of the world and the state of the person who acts, we should nevertheless be able to distinguish between two cases for interpreting the principle change of state sought: change of the world (for example, construction of an object using existent materials with the help of a manual) in contrast to change of the state of the agent (for example, someone walking in a park and respecting the *Parks and Gardens Regulations*).

This aspect is to be linked to the qualification of the most fundamental objective of the action; for instance: transform the world, safeguard the current world state or even restore a previous state—it is also necessary to take into account the fact that the world in question can be the 'exterior' world or the 'interior' world of the consignee (this last case is typically that of medicines).

14. **Modes of reference to the action**

The general modes of reference (direct reference in contrast to indirect; reference through denotation, description, or deixis) assume particular aspects in the case of action, underlined by the cooperative component of references. For instance, in the case of instructions, there are original forms of indirection which refer to the properties of the objects of action like 'Top', 'Bottom', or 'Handle with care', on the packaging of objects that need to be lifted, moved, transported, stacked, opened, etc.

15. **The types of onset**

As texts, instructions as texts have a certain lifespan (duration of validity) but what activates them by an agent is a particular occurrence of an event. From this point of view, it would be useful to distinguish between multiple types of onset; for instance, independence of the will of the 'consignee' (e.g. in case of fire), direct dependence on will (e.g. using a recipe book), or even mixed (e.g. how to get a travel visa).

16. **Types of failure**

From Searle's work and taking that of Austin, based on the concept of conditions of success as a starting point, it has been demonstrated that it is possible to establish a system for the types of failure of illocutionary acts. This guarantee comes from two sources: from the role that the conditions of success play in the definition of success of an illocutionary act: these conditions are necessary one by one and all together sufficient so their satisfaction implies the success of the illocutionary act; and from the fact that there is a given number of ways to not satisfy them (i.e. to contradict them). This approach brings a fundamental status to the notion of failure: instead of being conceived as a sort of termination or limit of success, and constituting a post script in the study of success, failure is established within the analysis of success as a privileged means of its very definition.

It is tempting to consider the possibility of an equivalent approach for non-discursive actions. It is clear that advancement in this domain would have multiple effects on the entire problematic of instructions: from the 'art' of writing to the detailed specification of multiple forms of 'guarantee'. Such elements can be found, for instance, in (Norman 1981) and (Reason 1990).

Whether or not to take into account, in the drawing up of instructions, the reasons for and types of potential failure, or even mentioning them directly, is without doubt an important criteria for differentiation.

17. **The role of results**

In a somewhat parallel manner to failure, the knowledge, real or estimated, by the consignee, of the nature of the result attainable thanks to the instruction, plays a fundamental part in many aspects of the instruction (and a fortiori in its execution). The nature of this knowledge can come from many factors: the agent's know-how and expertise is one of them, but the nature of the task is another.

Again, the possibility of evoking, or not, this aspect directly in instructions must be a factor of diversification, all the more so as the evocation of the expected result can be part of the explication of the action.

This initial provisional study is undoubtedly still incomplete, not least in the obvious fact that the aspects, isolated here in order to present this analysis, are not independent but instead present a great number of interdependencies. This study, nevertheless, allows us to represent where and how a contribution of SAT can be placed in the study of instructional texts.

2.3.2 Directive Illocutionary Forces

2.3.2.1 The Position of Instructions among Directive Utterances

As a beginning, we should repeat here the content of directive illocutionary forces in terms of conditions of success, according to Searle's analysis (Searle 1969):

Essential condition
• S attempts to get H to do A
Preparatory condition
• H is able to perform A
Condition of sincerity
• S wants H to do A
Condition of propositional content
• H will carry out the action A in the future

An examination of the content of the conditions of success of the basic directive illocutionary force directly suggests an important point about the position of instructional text. If the preparatory condition of a directive is that H can do A, and if the condition of sincerity is that S wants H to do A (i.e. S wants the preparatory condition to be satisfied), then, the function of aid of the instructional text (utterance: u) finds a natural interpretation: S wants H to be able to do A thanks to u and, correlatively, H is believed or supposed to be able to perform A thanks to u.

This interpretation leads to the introduction, for the illocutionary act 'to give instructions', of a specification of these two conditions of success, in which the instructional text acquires a precise illocutionary status: S does not only express (condition of sincerity) the fact that he wants H to do A, but also that H should do A thanks to u; and it is not only presupposed (preparatory condition) that H can do A, but also that he can do it thanks to u. A first result of this analysis would thus be that, in the case of instructions, the preparatory condition and the condition of success have a stronger relationship than in more general cases, in which they are independent. In the interpretation proposed for instructions, S wants to ensure the success of the preparatory condition, which is that H does A, thanks to the instruction itself.

These formulations may seem defective: one could indeed note that it is in the essence of any illocutionary utterance u to be the means of realizing successful illocutionary acts (Thus, I try to make someone leave the room thanks to the utterance u: 'Get out of here'). The fundamental difference is that this aspect of u (to be the means of) is managed by conventional types of rules: it is by the intrinsic conventionality of the illocution and the communicational reasoning that the consignee will be made to leave 'thanks to' my utterance 'Get out of here'. The form of aid of the instruction lies in a rule of a normative, and not a constitutive, type. The only thing that can be considered to be conventionally constitutive, in the proposed analysis, is that instructions constitute aid of whatever kind, of which the form, the reach and the content have nothing to do with communicational conventionality but with normative type rules.By the fact, we can note an original form of composition of conventional and normative rules.

2.3.2.2   Classification of Directive Forces

Using this specification of the conditions of success of the directive act in the case of instructions, we propose to verify whether it is possible to push the limits of this specification further, that is, to identify distinct illocutionary directive forces in more detail. The approach will consist in an attempt of identification using an inventory of the entire English illocutionary directive forces named by verbs and supplementary illocutionary criteria as completed by (Vanderveken 1990) (See paragraph 13 in Sect. 2.2).

There seems to be a possible classification of these illocutionary forces into four main groups:

a. the group of requests (to beg, to solicit, to beseech, etc)

b. the group relating to asking questions (questions are a particular group of directives as what they seek to accomplish is to make H realize speech acts in response)

c. the group derived from 'to tell to' (orders, commands, demands, etc) with some cases of negation (to forbid, to ban)

d. the group of advice (suggestions, recommendations, etc)

The first two groups do not seem to be significant for instructions, because making requests or asking questions does not seem to be a characteristic of instructions (unless it is possible to demonstrate that these illocutionary forces have a specific definition in the context of instructions by contributing to the function of aid in a special way).

A 'structuralist' method is applied here, as well as for the other illocutionary forces that we will examine later on: it amounts to considering, for a given component (for instance, the mode of achievement of the illocutionary point), all the values that it can assume for all the listed verbs in order to come up with pertinent subgroups in our context and, thus, obtain pertinent verb groups.

In this manner, from the point of view of illocutionary directive forces likely to be used in "giving instructions", it appears that two values of the 'mode of achievement of the illocutionary point' component play an important part:

- the reference or not by S to a position of authority (commonly of S but sometimes of H), whether this superiority is institutional (hierarchical position of S over H), intellectual, technical (expertise) or even a simple position of power (cf. the difference between 'to order' and 'to command')
- the existence or not, from S's point of view, of the possibility of H's refusal

Distinctions that are introduced by these two modes of achievement of illocutionary points which are relevant in the context of the act of giving an instruction overlap quite extensively with groups of verbs discussed above and their relationship to the act "give an instruction":

- requests and questions (derived from 'to ask') have an option of refusal and do not refer to a position of authority
- orders and commands refer directly to the refusal of such an option and to a position of authority
- the group of advice inherits the neutrality of the bIF: 'to suggest', 'to advise' or 'to recommend' refer more to H's interests, or even S's, as seen by S, and do not indicate anything of an option of refusal or a position of authority, which depend in this case on S's status and not directly on the illocutionary force he uses.

One could think that these two values of the mode of achievement of the illocutionary point are not independent and that, on the contrary, the option of refusal is only possible if a position of authority is not invoked; and that conversely, the absence of an option of refusal is only possible by the invocation of a position of authority. Nevertheless, the examples of 'to prescribe' (where an option of refusal exists even if a position of authority is evoked) and 'to tell (someone) to (do something)' (where an option of refusal does not exist even if no option of authority is evoked) indicate that the two other cases in the relationship of the two values are attested.

For the same case, the verbs can be classed according to how strongly the condition of sincerity is expressed, starting with the weakest; for instance: to suggest, to advise, to recommend; to order, to bid to do, to command. We must note that negations, such as 'to prohibit' or 'to advise against', are not negations of illocutionary acts but illocutionary acts based on negations: to prohibit does not mean not to order to do but order not to do; to advise against means advise not to do.

In conclusion to this analysis: irrespective of the finer variations introduced by the degree of strength of the expression of the condition of sincerity, instructions can realize four different types of directive illocutionary forces:

a. prescriptions
b. commands
c. advice
d. demands

Prescriptions are as a whole perfectly represented by medical prescriptions, that is, the written instructions issued by a doctor, and can be distinguished by the double invocation of a position of authority of S and an option of refusal by H.

Commands pertain to the case where, by contrast the option of refusal does not exist. It would undoubtedly be useful, in the context of an analysis of the particular classes of instructions, to distinguish between the types of position of authority (expertize and institutional positions for instance). We must note that this illocutionary force is linked to the S's and H's institutional positions but is also separate: an officer can give orders in the context of his position, but he can also give advice or recommendations for one aspect or other of the general order, in which case his position of authority is in a way suspended or neutralized.

Demands are differentiated from commands by the fact that there is no invocation of a position of authority.

Finally, advice is differentiated from the other three types by the absence of a position of authority and the existence of a refusal option.

2.3.2.3 Indirection of Directive Forces

The above analysis could prove to be useful to characterize certain types of possible failure, concerning the interpretation of the type of illocutionary directive force, when the latter is not explicitly performed through the use of a directive verb: for instance, an utterance that signifies a command for S may be interpreted as advice by H (and vice versa).

Indeed, summarizing the above for indirect illocutionary acts, any illocutionary act (thus, directives as well) can be realized:

- directly (explicitly) thanks to:
  - a lexical marker (lexical ifid)

I (order + command + advise + suggest + beg + etc.) you to bring me the book
  I lent you.
  – or a syntactical marker, when it exists:
    Bring back the book I lent you!
The use of the imperative form marks the realization of a generic or undifferenti-
ated directive (as opposed to the use of a lexical marker as for the first case);

- indirectly realized by many means which need a form of illocutionary reasoning:
  – by asserting the propositional content of the illocutionary act:
    You will bring me the book I lent you.
    This mode presupposes and realizes at the same time the success of the essen-
      tial condition.
  – by asserting or questioning the preparatory conditions or the conditions of
    sincerity:
    You can bring me the book I lent you.
    Can you bring me the book I lent you?
    I want you to bring me the book I lent you.
  – by presupposing H's acceptance
    (Will you + want to + will you accept to + etc.) bring me the book I lent you?
  – by asserting or questioning the state of affairs in the world that S wants H to
    modify or achieve, or the reasons for H to do or not do A:
    I have nothing left to read.
    I would gladly reread the book I lent you.
    I gave you that book some time ago.
    I don't like lending my books for long.
    Max would like to read the book I lent you. Etc.

This last group is represented in particular by the case of instructions assuming the
form of a directive realized indirectly by an assertion not of the action to undertake
but of the state to achieve. A very common example is those of 'guidelines for
manuscript preparation' where one expects to read:
'The main body of the text will be written using Times New Roman size 12'
instead of how to achieve this result (which would be more likely be found in a
word-processing software manual).
There are significant stylistic variations in the use of these language resources.
For instance, in *Guidelines for manuscript preparation,* one can find the following[5]:

Assertions: Optimum resolution for photographs is 300 dpi
Assertions using the future tense:

            All formatting will be performed by the typesetter

---

[5] The same diversity can be observed for direct illocutionary acts :imperative : Cite as follows …
Type captions on a separate sheet Please mark clearly …directive : Permission to reproduce mate-
rial is entirely the author's responsibility Authors are permitted to …Contributors are advised to
…UCP accepts electronic files …

With should: Contributions should be sent ...
                  Contribution should be submitted ...
With may: Standard works may be cited ...
                  Copies may be sent ...
With can: We cannot accept ...
With must: Accepted manuscripts must be ...
                  Authors must ensure that ...
                  Tables must be numbered ...

What unites all these cases of indirection, which are quite different, is that in each case S, by saying u, means u and refers to another utterance f, thus, the illocutionary force of the directive *stricto sensu* remains implicit and can lead to misinterpretations. We can give two simple examples: what is exactly the illocutionary status of 'Instructions in the case of fire' or of 'Emergency exit' that can be found in many buildings?

2.3.3   Illocutionary Forces of Commissives and Position of Guarantee

2.3.3.1   Classification

By following the same procedure as above we note that:

a. Some of the verbs studied by Vanderveken (1990) and carrying out commitments may be left out in the study of instructions, mainly because, as for some assertives, they presuppose the existence of a dialogue (bid, outbid, make a counter-offer, consent, accept, etc.), which is not the usual case for written instructions.
b. the preparatory conditions include, for some verbs, a relationship between H and the action A expressed by the utterance u (for example, H wishes or doubts)
c. the mode of achievement of the illocutionary point is more particularly marked in relation to:
   – a given force (publicly, solemnly)
   – two forms of performativity: absolute (i.e. no conditions) or depending on one or more conditions
These conditions are varied and relative:
   – to H (for example, to sign a contract)
   – to a third party (for example, to stand as guarantor for)
   – to a state of the world not necessarily dependent on the activity of S or H (for example, to guarantee)
These remarks allow us to classify quite easily the various illocutionary forces of commissives, denoted in English by a verb, in some classes allowing us to place the guarantees:
   • firstly, promises are distinguished from guarantees because the former are not dependent on any condition;

- among guarantees which are dependent on a condition, we can distinguish between security (or a deposit), for which the guarantee does not depend on H, and the opposite possibility: the guarantee depends on H;
- lastly, we can distinguish between offers, for which the success of the commissive depends on H's acceptance, and contracts, for which success directly depends on H's actions.

2.3.3.2   Directive-Commissive Composition and the Logic of the Commitment: The Notion of 'Thanks to' (or of 'Aid')

Insofar as an instruction contains a dimension of commitment by S, it is not without interest to identify the point or points, besides the logic-illocutionary type of the contracted obligation, which this commitment affects:

In the case of SAT, an act of commitment has the following conditions of success:

- essential condition: S assumes the obligation to do A(s)
- preparatory condition: S is able to perform A(s)
- condition of sincerity: S intends to do A(s)
- condition of propositional content: S accomplishes A(s) in the future

It appears that, considering that in instructions S wants H to be able to accomplish A(h) thanks to u, the action A(s) which S wants to accomplish (commitment level) is that H be able to accomplish A(h) thanks to u.

This is an extremely strong condition which is completely dependent on the content of 'thanks to' of which we need to define the effects.

A first version of the interpretation of 'thanks to' could be expressed by a conditional formula such as:

F1: S commits himself to: if H conforms with utterance u, H can accomplish A(h)

The problems arising from this representation are the following:

- the content of u is not explicitly stated and cannot be so because as such it is always particular
- what happens if the reader does not conform with the instructions? (we are not in the case of a truth-functional implication, but that of a causal relationship)

We can analyze this situation, through this formula, in some representative examples of instructions:

- *Assembly instructions* (for flat pack furniture for instance): is the 'commitment' referring to the furniture's 'correctness' (is there a reference model?) or to its 'non incorrectness'? What is the nature, or the 'status', of the object constructed, in case one does not conform with the instructions?
- *Drug dosage*: the commitment cannot refer to recovery, but perhaps rather to the maximum efficiency (or the minimum inefficiency?), or maybe even that respecting it will guarantee that no harmful effects will arise?
- '*In case of fire*': the commitment does not refer to the reader's capacity to put out the fire but to the effectiveness of behavior (if we conform with the

instructions we are guaranteed that we will be the most cooperative in the situation, even if we do not understand this directly through our actions). Additionally, it refers to the fact of not incurring blame or legal action (a concept similar to what in France is failure to provide assistance to a person in danger).

- *Park and Gardens regulations*: in principle, this type of rules seems to guarantee the most enjoyable use of the park for all (do not walk on the grass, respect other 'users', etc.). There is, nonetheless, a negative component: should we not respect the rules, we are 'guaranteed' the risk of (potential at least) legal action or a fine.

- *Game rules*: in this case, the concept of commitment has a different meaning (to the point that we could ask if it has the same meaning): if we follow the game rules, we are guaranteed to play the game 'correctly' or legally (but not that we will win); if not, we simply do not play this game.

- *Instructions in the case of an incident while using* a machine, appliance, etc.: this case seems to be similar to that of 'In the case of fire'.

- *Maintenance instructions* (regular chimney-sweeping, car oil change, etc.): other than the contractual or legal aspect, instructions of this type seem to function mainly negatively: there is no commitment on the smooth running of the engine if we change the oil often, but there is the possibility of serious problems if we do not. Likewise, for an annual chimney-sweeping what is guaranteed is not that the chimney will function properly, even if this can be an important prerequisite, but that should an incident occur which could involve the user's responsibility, the insurance company will pay for any possible damage only if the sweeping has been done, or, rather, that sweeping can be proved to have been done.

- According to a remark by J. Vuillemin (Vuillemin 1984), we need to note the special status of the instructions for the construction of figures given in the context of proofs of geometrical theorems. The imperatives they contain have a specific status: what they indicate is either possible or not (thus, contributing to the demonstration). It is not meaningful to talk for them of success or failure (we are not talking about the inability to draw).

The above remarks suggest at one and the same time a relative uncertainty on the exact content of the 'positive' engagement and, by contrast, a certain precision on the content of a negative 'commitment': even if the content of the instructions does not indicate clearly what the commitment of following them is, not following them may or must bear consequences CSQ that could or should be enacted.

So we opt for a different logical form for sentences expressing commitment. In this second formula F2, the commitment has a kind of specific content (defined via the consequence CSQ) instead of a content referring to the goal of the instruction itself. The content follows a hierarchical structure related to the 'positive' and 'negative' consequences: what is clear is that non-conformance with the instructions may or must provoke a consequence CSQ, which is either not possible or not necessary in the case of conformance with the instructions.

F2: S commits herself to:

- if H does not follow the instruction, it is possible or necessary that there will be a consequence (CSQ)
- if H follows the instruction, it is possible or necessary for CSQ not to happen (or it is neither possible nor necessary for CSQ to happen)

F2 gives the element 'thanks to' particular content in relation to the position of failure vis à vis success.

We must conclude this study on the position of the commitment in instructions by noting that there is no inherent relationship between directives and commissives (i.e. commissives can be achieved independently of directives). And the conditional aspect we have mentioned above is only a particular case for commissives.

We can imagine that, at one pole of a continuum, contexts exist where S's main, if not only preoccupation in issuing an instruction is not to help H, but to free S of his responsibilities: thus, the goal is the correctness of the instruction as such, as difficult to apply or unsuitable as it may be. At the other pole, the concern would be to give H as much aid as possible. It is not evident that in this case maximum effectiveness implies that instructions must be (exhaustively) correct[6].

2.3.4   *Assertive Illocutionary Forces*

2.3.4.1   Classification

In a first approximation, we can distinguish between two main groups of verbs likely to designate an assertive illocutionary force:

- assertions p per se
- assertions which include 'moral' positions or more generally positions in relation to p of S and/or H

This distinction appears quite clearly in the examination of the values of the preparatory condition and the condition of propositional content.

We only consider the verbs of the first group here.

For this group, the examination of the different values of the preparatory conditions (which are, as we said, presupposed), indicate that practically all are related to the existence or non-existence and the particular values of a relationship:

- either between S and p: for instance
  - S has solid evidence of the truth of p (e.g. to conjecture)
  - S has solid reasons to believe p (to predict)
  - S can provide reasons for his belief (to support)
  - etc.

---

[6] This situation is well known for reference in relation to Grice's maxims: if I'm living in a tiny village near Toulouse and a foreign colleague asks me where I live, the most cooperative answer would be 'in Toulouse,' which is, strictly speaking, wrong, or 'near Toulouse,' which is imprecise.

- or between H and p: for instance
  - H knows or should know of p (to remind)
  - H does not know about p (to inform)
- or else between S and p and H and p: for instance
  - H has doubts about p and S wants to convince H about p (to guarantee, to assure)

The relationships between S and p on one hand and between S and H on the other are important to differentiate two assertion situations previously examined above: that where S's objective is to communicate to H the content of his beliefs and no more, and that where, on the contrary, S aims to modify the content of H's knowledge (and communicate his knowledge to H).

From the temporal point of view (condition of content), some forces are related to the present and the past but not to the future (to relate, to testify), while others can only refer to the future (to predict).

The mode of achievement indicates two axes: the force (from weakly to strongly) and the necessarily public or not character[7].

2.3.4.2 An Original Composition between Assertive and Declarative Acts: Diagnostic-Type Discourse Acts

In his analysis of declarative acts, Searle identified a group of declarative acts that comprise a composition with an assertive act (due to this they are called declarative-assertive acts). Indeed, in 'I declare this session open', 'I hereby declare you husband and wife', 'I name this ship the "Queen Mary"', there is only a 'pure' declarative act, whereas in 'Guilty' (by a judge) or 'Penalty' (by a referee) there are simultaneously:

- an act of assertion consisting of an authorized person saying what the fact are: what the accused did was a crime or what the player did was a foul in the penalty area;
- and a declarative act itself creating a state of the world: there is guilt and a penalty by the fact that the judge or the referee says so.

This reminder allows us to note that in the context of instructions another form of composition between these two types of illocutionary acts can be found. Indeed, there is a group of verbs which usually indicate assertive illocutionary forces such as: to describe, to characterize, to qualify, to identify, to diagnose, to class, which imply syntactical structures such as 'someone V something as adjective/substantive' (for example, The doctor V this sickness as (being) meningitis). This example is an assertive and not a declarative act: it is not the fact that the doctor declares the patient

---

[7] Another criterion, not pertinent for us here, shows that assertive acts can also be distinguished between those which do not necessarily involve dialogue (to affirm, to declare,…) and those which are normally only used in the case of a dialogue (to object, to contradict,…)

sick that makes him sick, nor that the doctor declares that he has meningitis that makes the sickness meningitis. Yet, there are contexts in which an authorized person makes certain assertions which create *ipso facto* institutional decisions. Thus, the diagnosis of certain sicknesses (meningitis, rabies, etc.) by certain doctors has as the effect of creating new situations and enacting certain public health decisions[8]. Likewise, it is not the meteorologists who create hurricanes and storms by their weather reports; but if the office which establishes these diagnoses has the recognized qualifications, this automatically creates an official state of emergency.

2.3.4.3   On the Mode of Indirection of Directives by the Assertion of Reasons for (not) Doing the Action

We saw (Sect. 2.2. Paragraph 11) that a way of realizing a directive illocutionary act indirectly is by creating an assertion (or possibly a question) on the reason(s) for doing or not doing an action. Thus, 'It's cold in here' can be the reason why we would like someone to close the window and is also an indirect realization of a request. We could ask how such an indirection can have a proven efficiency, since it seems to be a lot more difficult to master than asserting or questioning the conditions of success ('Can you close the window?'). The modalities for expressing such reasons are numerous and diverse (Virbel and Nespoulous 2006): from the simple mention of their existence and the obligations they imply ('You ought to be more polite to your mother'), to the expression of the content of the reason ('You're standing on my foot'), passing from the willingness given (by S) to H to do A: 'Would you be willing to write a reference letter for me?'

From all the reasons, we can refer to the following:

1. **The reasons participate in an implicit argumentative structure**
   They are an answer to a question with 'why' supported by a directive (direct or indirect):
   S: pass the salt/ could you pass the salt? H: why? S: because this food is tasteless/ it would be better with more salt/ I want to add some/ I can't reach it.
   We can see here the approach followed by Austin J. (1962 (1976)) and Anscombe E. (1957) who observed and exploited the power of revelation of the dialogue structure < question with 'why'/ answer with 'because'> on the reasons to do something. The reasons for (not) doing something which fall within such a structure are explications of the action.
2. **Indirection by reasons to (not) do A exploits to the maximum the background knowledge**, such as frames, scripts, standard plans, just like the indirection referring to the planning of the action and contrarily to the indirection by the conditions of success.

---

[8] Furthermore, a diagnosis is also related to a 'prescription' type act; that is of the directive type (Sect. 2.3.2.2)

A rational action (performed by a rational agent) includes not only, under specific circumstances, reasons to do it but also a given method to follow. The use of modeling of plannable action in Artificial Intelligence is very useful in this case (Bratman 1987; Camilleri 2002; Zaraté et al. 2005). This form of modeling of the action presupposes that for an action to be executable, preconditions must be fulfilled (and if they are not, they can themselves be planned), the main action can be analyzed in smaller executable actions (subgoals), the post-conditions are fulfilled, and, finally, that the effects of the action on the world can be identified. Thus, to change the wheel on a car, one must (preconditions) be able to use one's hands, have adequate light conditions, have a jack, have a spare wheel, etc.; we can break down the global action 'change the wheel' into some smaller actions whose sequence results in changing the wheel; we must fulfill post-conditions such as put away the jack and the crank handle, etc; finally, the state of the world receives a new description: the previously spare wheel is fitted as one of the car's operational wheels and, thus, becomes usable, and the wheel previously in use cannot be used as a spare wheel. This reference allows us to understand how 'Can you reach the salt?' can be an indirect request not only by a question on a condition of success but also by reference to the action by means a smaller action of the plan 'pass the salt', since 'reach' is a smaller action of the corpus of the global action 'pass'. The form of indirection used here is the reference to an action by one of the parties: 'Can you (stretch + reach with) your arm and (grab + take + bring + place) (next to + near me) the salt?' or to the preconditions:'(Can you see + do you see + do you know) where the salt is?' (to see a target or know where it is, is usually a precondition to grabbing it). In adequate contexts, indirection can also refer to post-conditions: 'S: Can I come in? H: Close the door behind you' (H indirectly allows S to come in by uttering a directive that indicates the satisfaction of a post-condition of the plan 'enter a room').

Thus, it is possible to realize indirect illocutionary acts by referring to a model for the action (that to do in the world). Such models belong to the background knowledge shared by the speakers, and are, thus, relative to the propositional content. One should note here that this propositional content is subject to the same indirect illocutionary acts: Can you reach the salt? Can you pass the salt? Can you see (where) the salt (is)? I'd like you to reach for the salt.

Thus, it appears that invoking a reason to do A with an indirect directive is a peculiar case of a general situation: we can realize an indirect demand by referring to any element of an execution plan belonging to the background knowledge of the participants.

3. **Dissymmetry between (types of) actions and the reasons for accomplishing them can lead to privileging the expression of the reasons:**
   It is easy to observe empirically that **an action can have many reasons to be performed**: S: 'Slow down' H: 'Why?' S: '(It's starting to rain + there's a toll booth ahead + we're taking the next turn on the right + the car's suspensions isn't in great shape + we're not in a hurry + it's a bit foggy + there's a speed limit + …)'. While, in certain circumstances, one reason for doing an action refers to one or a

small number of actions: 'We're taking the next turn on the right' → slow down (+signal your intention).

A satisfying analysis of this situation can be found in the theories of action developed in works by E. Anscombe (1957), J. Austin (Austin 1962 (1976)) and D. Davidson (Davidson 1980). It is impossible to present these theories here, even in summary, so we will only present those points which are directly pertinent to our study.

Firstly, an event can be described in many concurrent or combined ways, just like an object. Consequently, just as we can refer to an individual by indicating his name, and/or that he is our neighbor, and/or that he is married to Lea, and/or that he is a journalist who writes for such and such a newspaper etc., we can refer to an event by saying that Max set his house on fire, or that he set fire to his bedding, or that he put a match to his bed etc. Yet, there is a special category of events—the actions—that have a specific relationship with the reasons the agent had to do A. One reason is the combination of a desire, and beliefs on the state of the world, and the possibility for A to satisfy this desire: thus, Max wanted, for instance, to get the insurance money and believed that setting the house on fire was an adequate means for achieving this and believed that setting the bedding on fire with a match was an adequate means of setting the house on fire. The important point here is that mentioning a reason to do A is another way of describing A (and not the description of a different event independent from A). 'Max set the house on fire to get the insurance money' is a description of the *same* action as 'Max set the house on fire by burning his bed'[9].

Secondly, not only can a reason be described in many ways, but many reasons can justify or explain an action: Max could also wish for the fire to spread to Luc's house next door because he has hated Luc for a long time. Whatever these reasons may be, there is one reason (or a set of reasons), the 'primary reason' according to D. Davidson, which explains *this* action under *these* circumstances; it is *the* reason: there could be a thousand reasons for a driver to slow down, but, in *this specific case,* Max slowed down because he was coming up to a toll booth. The element of intentionality is evidently a determining factor here: the same event will be an action (intentional because of the primary reason) according to one description, but will not be under another: I may want to publish a book on pragmatics, but if by misfortune it has a great number of typos, it is not possible to (and I cannot) say that I wanted to publish a book full of typos.

A cognitive hypothesis may then be advanced from this analysis: the nature and the power of the relationship between the constituents of the group < action, primary reason, intentionality > are such that it is equally, if not more (less cognitive effort for the analysis of the information), efficient to indicate the reason ('We're coming up to a toll booth') in order to indicate the action to do rather than indicate the directive ('Slow down') and (risk) provoking the activation by H of the search for a (primary) reason to justify this action, among others. This 'risk' of

---

[9] In Frege's terms, it can be said that the two sentences have the same referent but not the same meaning.

Table 2.2 Analysis of declarative illocutionary forces

	C. Sincerity	Mode of achievement of illocutionary point		
	Binding	Officially	Validly	Solemnly
To endorse			+	+
To homologate	+		+	
To ratify	+			+
To promulgate	+	+		+

activation is directly linked to the principle of cognitive charity (believing that an utterance is truthful for reasons of economy) and the attitude of attribution of mental contents.

This hypothesis is undoubtedly dependent on annexed hypotheses. For instance:

- the cost of investigation is lessened when circumstances are managed by rules such as 'scripts' or 'frames' which are part of the background knowledge of the speakers: when coming up to tolls we must slow down, start looking for our ticket, and prepare a way to pay, and (supposedly) that is all. Thus, 'We're coming up to a toll booth' is not only the reason to slow down but also the reason to prepare for the two other actions that are linked to the script 'to go through tolls'.
- we can also suppose that this efficiency can be greater if the circumstances of the situation include a danger or a risk of greater or more immediate crisis: 'You're on the wrong side of the road' in contrast to 'Turn the wheel' or 'Go further to the left (or right!)'.

This is a strong hypothesis, which we believe to be supported by Action Theory and SAT, whose credibility can be evaluated through various experiments (Virbel and Nespoulous 2006).

2.3.5 Declarative Illocutionary Forces

Using the same procedure, we chose four declarative verbs which we deem to be pertinent for our case among the verbs examined by Vanderveken: to endorse, to approve, to ratify, and to promulgate. They are differentiated by a characteristic linked to the condition of sincerity: p becomes binding or not, and three characteristics of the mode of achievement of the illocutionary point: the declarative action may be accomplished or not officially and/or validly and/or solemnly (for this three terms according to definitions for each context) (Table 2.2).

2.4 Conclusion

We would like to highlight two points to close this study.

Firstly, we hope that we have illustrated the idea that SAT allows us to ask new or re-state existing questions in the scientific study of texts, and specifically, as in this study, instructional texts. What allows us to move beyond the linguistic or logical description of the objects that texts are is the writers's intention for the actions that they want the readers to perform and/or help the readers do. Yet, as we have seen, this intentionality is not described in psychological terms, but in terms of a conceptualization of the action (here, as a speech act), in a specified theoretical frame where we can find concepts such as the conditions of success and satisfaction (a given composition of both contributes to define the 'felicity' in Austinian term): if S asks H to do something (A), both S and H know that, from both of their points of view, it is a sincere demand, S wants H to do A and S believes that H can do A; if S recommends that H should do A, it is normal, for both S and H, to believe that S has some kind of authority (moral or of expertise, but not of power) over H, and that the latter has an option of refusal without creating a conflict. These conceptualizations allow some aspects of a wider question to be addressed: the various forms of cooperation in inter-human communicational activity, as P. Grice showed.

Nonetheless, we are aware that this demonstration is only partial. There are points that we did not have the chance to develop. For instance, we showed that we can distinguish between four types of directive illocutionary acts in instructional texts, and also four (this is coincidental) types of commitment acts. Yet, we have not studied cases where these intersect, which would undoubtedly allow for studies on new questions. Nevertheless, maybe only within the framework of defined types of texts (for instance recipes and car manuals): for instance, is there a reason for S to realize a directive act of the command type combined with a commissive of the promise type, given that the success of a promise does not depend at all on H's behavior? Another area we have not been able to examine here is SAT's contribution to the theory of other speech acts (not only illocutionary), specifically referential and predicative acts. The latter are very important in the choice of the granularity of the description of an action. On the other hand, there are important aspects of instructions on which SAT, in its current formulation, is unlikely to shed light. This is particularly true for the expression of conditional preconditions of an action ('If the light flashes red, insert credit card again', 'The battery low indicator will light up when the battery in the handset is low', 'Return the off/stby/talk switch to stby after you call'). The choice of syntactical structure, as well as the order in which the elements of the schema <conditional precondition—action> are presented plays an important part in the efficiency (comprehension, memorization, etc.) of such instructions. We note, for instance, that presenting the precondition first, and then the action does not realize the same act as the inverse, but, to the best of our knowledge, for now, we are better placed to address these questions in terms of Rhetorical Structure Theory (Kosseim and Lapalme 2000) than SAT.

The other point we would like to highlight in conclusion is in our view important in relation to the history of scientific ideas; however, we will only present it here briefly, as it would normally require lengthy development. While, as we believe, SAT is an interesting theory that can contribute to research programs on texts, we also maintain that this approach suggests interesting, and, to the best of our knowledge, original developments in SAT. For instance, it is because we felt it was necessary to qualify the illocutionary acts that were most pertinent in the case of instructions, that we were led to the classification we have presented and which adds a certain amount of detail to SAT. Likewise, the composition between directives and commissives that we revealed in the case of instructions is specific compared to other forms of this composition like, for instance, the plural imperative form ('Let us go to the cinema') which combines a directive and a conditional commissive (S commits to going to the cinema with H if H accepts to go); or in bets ('I'll bet you $ 10 that...'). The beginnings of the SAT—*William James Lectures* in 1955 (Austin 1962 (1976)), *Meaning* (Grice 1957)—took place at practically the same time as those of cognitive sciences (MIT 1956) and Artificial Intelligence (Dartmouth 1956), and the relationships between the three fields are numerous: conceptual borrowing, applications and experiments on SAT coming from the philosophy of language, (Tsohatzidis 1994) as well as linguistics, cognitive psychology and neuropsycholinguistics (Champagne et al. 1999, 2003) and even from domains such as formal logic, discourse act planning in AI and cognitive engineering (Bratman 1987; Camilleri 2002; Kautz 1990; Zaraté et al. 2005). Thus, we must note that textology has its place in a vast movement, which we can reasonably take as a mark of maturity.

References

Anscombe, Gertrude E. M. 1957. *Intention*. Oxford: Blackwell.

Austin, John L. 1961. *Philosophical papers*. Oxford: Clarendon Press.

Austin, John L. 1962. *How to do things with words*. The William James lectures delivered at Harvard University in 1955. Oxford: Oxford University Press (second edition, 1976).

Bach, Kent. 1994. Semantic slack: What is said, and more?. In *Foundations of speech act theory. Philosophical and linguistic perspectives,* ed. L. Tsohatzidis, 267–291. London: Routledge.

Bach, Kent, and Robert Harnish. 1979. *Linguistic communication and speech acts*. Cambridge: The MIT Press.

Bertolet, R. 1994. Are there indirect speech acts?. In *Foundations of speech act theory. Philosophical and linguistic perspectives,* ed. L. Tsohatzidis, 335–349. London: Routledge.

Bratman, M. E. 1987. *Intention, plans and practical reason*. Cambridge: Harvard University Press.

Camilleri, G. 2002. Dialogue systems and planning. In *Text Speech and Dialogue, 5th International Conference TSD 2002*, Brno, Czech Republic, Book Part III: 429–436.

Champagne, M., J. Virbel, and J.-L. Nespoulous. 1999. Comprehension of nonliteral speech acts: Is there a need to active literal meaning first? In *European Conference on Cognitive Science* Siena Conference Proceedings, ed S. Bagnara, 349–352.

Champagne, M., A. Herzig, D. Longin, J.-L. Nespoulous, and J. Virbel. 2002. Formalisation pluridisciplinaire de l'inférence de certains types d'actes de langage non littéraux. *Information—Interaction—Intelligence*. Numero Spécial *Modèles Formels de l'Interaction* (Hors série): 197–225.

Champagne, M., J. Virbel, J.-L. Nespoulous, and Y. Joanette. 2003. Impact of right hemispheric damage on a hierarchy of complexity evidenced in young normal subjects. *Brain and Cognition* 53: 152–157.

Cohen, Ph., J. Morgan, and M. Pollack, (eds.) 1990. *Intentions in communication*. Cambridge: The MIT Press.

Davidson, D. 1980. *Essays on actions and events*. Oxford: Clarendon Press.

Genette, G. 1997. *The work of art: Immanence and transcendence*. Trans. G. M. Goshgarian. Ithaca: Cornell University Press

Goodman, N. 1968. (Rev. ed. 1976). *Languages of art. An approach to a theory of symbols*. Indianapolis: The Bobbs Merrill Company. Revised edition: Hackett Publishing Company.

Grice, P. 1957. Meaning. *The Philosophical Review* 66:377–88. Reprinted in (Grice 1989).

Grice, P. 1989. *Studies in the way of words*. Cambridge: Harvard University Press.

Harris, R. 2000. *Rethinking writing*. London: The Athlone Press.

Heurley, L. 1994. Traitement de textes procéduraux. Etude de psycholinguistique cognitive des processus de production et de compréhension chez des adultes non experts. Ph. D. Thesis, Université de Bourgogne.

Holdcroft, D. 1994. Indirect speech acts and propositional content. In *Foundations of speech act theory. Philosophical and linguistic perspectives,* ed L. Tsohatzidis, 350–364. London: Routledge.

Hornsby, J. 2008. Speech acts and performatives. In *The Oxford handbook of philosophy of language,* ed. E. Lepore and B. C. Smith, 893–912. Oxford: Oxford University Press.

Jakobson, R. 1960. Linguistics and poetics. In *Style in Language,* ed. T. Sebeok, 350–377. Cambridge: MIT Press.

Kautz, H. 1990. A circumscription theory of plans recognition. In *Intentions in communication,* eds. Ph. Cohen, J. Morgan and M. Pollack, 105–133. Cambridge: MIT Press.

Kosseim, L., and G. Lapalme. 2000. Choosing rhetorical structures to plan instructional texts. *Computational Intelligence* 16 (3): 408–445.

Lemarié, J., R. F. Lorch, H. Eyrolle, and J. Virbel. 2008. SARA: A text-based and reader-based theory of signaling. *Educational Psychologist* 43 (1): 27–48.

Levinson, S. C. 1983. *Pragmatics*. Cambridge: Cambridge University Press.

Norman, D. 1981. Categorization of actions slips. *Psychological Review* 88 (1): 1–15.

Petrucci, A. 1986. *La scrittura. Ideologia e rappresentazione*. Torino: Einaudi.

Reason, J. 1990. *Human error*. Cambridge: Cambridge University Press.

Schaeffer, J-M. 1989. *Qu'est-ce qu'un genre littéraire?* . Paris: Editions du Seuil.

Searle, J. R. 1969. *Speech acts. An essay in the philosophy of language*. Cambridge: Cambridge University Press.

Searle, J. R. 1971. What is a speech act?. In *The philosophy of language,* ed. J. R. Searle, 39–53. London: Oxford University Press.

Searle, J. R. 1979. *Expression and meaning. Studies in the theory of speech acts*. Cambridge: Cambridge University Press.

Searle, J. R., and D. Vanderveken 1985. *Foundations of Illocutionary Logic*. Cambridge: Cambridge University Press.

Tonfoni, G. 1996. *Communication patterns and textual forms*. Bristol: Intellect.

Tsohatzidis, S. L. (ed.) 1994. *Foundations of speech act theory. Philosophical and linguistic perspectives*. London: Routledge.

Vanderveken, D. 1990. *Meaning and speech acts*. Volume I: *Principles of language use*; Volume II: *Formal semantics of success and satisfaction*. Cambridge: Cambridge University Press.

Vanderveken, D. 2001. Illocutionary logic and discourse typology. Special issue "Searle with his replies". *Revue Internationale de Philosophie* 216 (2): 243–255.

Vanderveken, D., and S. Kubo. 2002. *Essays in Speech act Theory*. Amsterdam: John Benjamins.

Virbel, J., and J.-L. Nespoulous. 2006. Des raisons de (ne pas) faire l'action. Approches logico-pragmatiques et perspectives (neuro)psycholinguistiques. *Psychologie de l'Interaction* 21–22: 211–236.

Vuillemin, J. 1984. Les formes fondamentales de la prédication: un essai de classification. *Langage et philososophie des sciences* 4: 9–30.

Zaraté, P., M. Munoz, J.-L. Soubie, and R. Houé. 2005. Knowledge management systems: A process oriented view. *Cybernetics and Systems Analysis* 41 (2): 274–277.

Chapter 3
The Issue of Textual Genres in the Medical Literature Produced in Late Imperial China

Florence Bretelle-Establet

Abstract A great abundance of Chinese medical texts have come down to us since the mid seventeenth century. This is the combined result of the large number of texts written in this period and the fact that the texts were better preserved than earlier. As a matter of fact, the large number of medical texts, coming from various social settings, set historians the quite daunting challenge of understanding what the texts really were and how they should be classified. The idea underlying this article was to go beyond the various modern generic classifications used thus far ("learned", "popular", and the like) and to highlight, instead, how medical texts themselves differ from each other. In this aim, I chose to compare a number of excerpts of medical texts written from the eighteenth century to the beginning of the twentieth century in different geographical and social settings. And to compare them, I decided to use some tools created by linguists who have been particularly interested in the issues of genre, notably those used by speech act theoreticians. In this article thus, I analyze these excerpts by paying close attention to the five levels of any discourse act, summarized in the well known formula "*Who (says) What (to) Whom (in) What Channel (with) What Effect*", following Harold D. Lasswell's classic communication paradigm. In other words, I try to shed light on how each of these texts differs from the others from the semantic, syntactic and emotional angles and whether these differences can be linked to the authors' social, geographical, chronological or intentional settings, and, finally, if we can speak of genres in Chinese medical literature.

A great abundance of Chinese medical texts have come down to us since the mid seventeenth century.[1] This is the combined result of the large number of texts written

---

[1] Only mentioning the medical books preserved today in the 113 biggest Chinese libraries, there are approximately 11,129 books written from 1600 onwards out of a total of 12,124 (Xue 1991). The catalogue of books written (and not always preserved) in all administrative units of the Chinese empire also shows a dramatic increase of titles from the seventeenth century (Guo 1987).

F. Bretelle-Establet (✉)
SPHERE (ex-REHSEIS, CNRS & University Paris Diderot), Université Paris 7, UMR 7219, 75205 Paris Cedex 13, France
e-mail: f.bretelle@wanadoo.fr

© Springer International Publishing Switzerland 2015
K. Chemla, J. Virbel (eds.), *Texts, Textual Acts and the History of Science*, Archimedes 42, DOI 10.1007/978-3-319-16444-1_3

in this period and the fact that the texts were well preserved. As it became more and more difficult to enter the more prestigious imperial bureaucracy, an increasing number of people turned to medicine, in late imperial times. In the resulting acute competition, and in an age of literacy spread, more and more medical practitioners wrote and strove to publish medical writings as proof of their scholarship and legitimacy. At the same time, the development of commercial printing centers, eager to print books that would sell well, allowed a better preservation of books than before.[2]

As a matter of fact, the large number of medical texts, coming from various social settings, set historians the quite daunting challenge of understanding what the texts really were and how they should be classified. As Charlotte Furth stressed, because of the lack of professionalization, the distinction between a "popular" and a "specialist" work was not great. However, it was possible, according to her, to distinguish between learned and a more popular literature, the latter eschewing "elaborate causal explanations and expositions of complex techniques".[3] Even if the "specialist" vs "popular" or "scholarly" vs "popular" classifications, made by historians and thus exogenous, are not satisfying, it is usually with these terms that historians, most often, qualify medical literature.

The idea underlying this article was to go beyond the various generic classifications used thus far and to highlight, instead, how medical texts themselves differ from each other. In this aim, I chose to compare a number of excerpts of medical texts written from the eighteenth century to the beginning of the twentieth century in different geographical and social settings. And to compare them, I decided to use some tools created by linguists who have been particularly interested in the issues of genre, notably those used by speech act theoreticians. In this article thus, I analyze these excerpts by paying close attention to the five levels of any discourse act, summarized in the well known formula "*Who (says) What (to) Whom (in) What Channel (with) What Effect*", following Harold D. Lasswell's (1948) classic communication paradigm. In other words, I try to shed light on how each of these texts differs from the others from the semantic, syntactic and emotional angles and whether these differences can be linked to the authors' social, geographical, chronological or intentional settings, and, finally, if, we can speak of genres in Chinese medical literature.[4]

---

[2] (Chao 2009) and (Bretelle-Establet 2002, 2009) provide statistical evidence of the increase in number of medical practitioners which led to the acute competition mentioned notably in (Leung 1987; Grant 2003; Furth 1999; Hanson 1997; Scheid 2007; Volkmar 2000). The commercial printing boom in China took place in the late Ming (1368–1644) and probably favored the better preservation of books written in late imperial period than earlier (Brokaw and Chow 2005; Brokaw 1996; Chia 1996 and 2002; McDermott 2006).

[3] (Furth 1987, pp. 10–11).

[4] The second chapter in particular of (Schaeffer 1989) "De l'identité textuelle à l'identité générique", the edited book of Genette (1986) and the numerous talks given by Jacques Virbel (IRIT, CNRS) in the "Histoire des Sciences, Histoire du texte" (REHSEIS/IRIT) seminar, notably: "La théorie des actes de langage" (13/03/2003; 24/04/2003; 12/10/2006), developed at length in his contribution to this volume, and "La question des genres textuels" (18/03/2004) provided me the major part of the tools I employed to analyze these texts.

3.1 Six Texts: Six Discourse Contexts

This article focuses on six excerpts taken from the medical literature produced in the far south of China, between the mid eighteenth and the beginning of the twentieth century. These texts do not belong to an academic corpus which for instance was endorsed by the official central institution of medicine, the 太醫院 (*taiyi yuan*) Medical Bureau of Medicine.[5] In fact, all these texts were produced outside the institution, just like the bulk of medical texts written in the period. Moreover, these six texts come from varied social circles that we can reconstruct from two types of source: biographies of the authors, written by local scholars and recorded in local official gazetteers and the books' prefaces. The information included in the biographies as well as in the very make up of the biographies—their length for instance—,[6] the information included in the prefaces and in the identity of the preface writers[7] allow us to deduce the status of an author, his degree of involvement in the practice of medicine, his links with other medical experts and local elites. Last but not least, all this information gives an understanding of the author's intention and his expected audience. Without getting into a detailed presentation of the life and works of each author of these six excerpts, let us highlight a few points about the status of these texts. They will be of paramount concern for analyzing whether we can find any correlations between an author's social status, his intention, his target audience and his ways of writing.

The 醫碥 (*yibian*, *Stepping-Stone for Medicine* hereafter, *The Stepping-Stone*) was written in 1751 by He Mengyao 何夢瑤 (1693–1764), a native of Nanhai 南海, in the province of Guangdong. The several prefaces attached to his books show him as a polymath who had met the great intellectuals of his time, like Hui Shiqi (1671–1741), an eminent scholar not only in classical culture but also in astronomy and mathematics. He Mengyao subsequently not only mastered classical scholarship but also mathematics, notably spherical trigonometry that had been brought to China by the Jesuits but was then mastered and widely discussed by Chinese mathematicians.[8] He Mengyao passed the imperial examinations, he was a 進士 (*jinshi* "presented scholar", the highest degree taken in metropolitan examinations) and held several posts as a magistrate in the provinces of Guangxi and Guangdong. Despite the prestige associated with officialdom, He Mengyao eventually abandoned his official functions to devote himself to medicine. During his career that thus combined officialdom, teaching and medicine, he wrote 20 or so books in

---

[5] On this central institution in Chinese history, which from the fifth century, was involved in medical matters, such as defining the content of the official learning of medicine, and, at times, commissioned medical texts, see (Gong 1983).

[6] See (Bretelle-Establet, 2002, pp. 70–77; 2009).

[7] The identity and status of the people chosen to write prefaces to these medical texts allow assumptions to be made about the status of the author himself, for whom, biographical data is often scarce, if it exists at all (Bretelle-Establet 2011).

[8] Catherine Jami, seminar Sciences en Asie, REHSEIS, 08/01/2008.

very different fields—poetry, official gazetteers, mathematics and medicine—.[9] According to He Mengyao's own preface of *The Stepping-Stone*, one of his six medical texts, his book has three objectives: firstly, to participate in the knowledge of the Way (Dao 道) that includes medicine; secondly, to enable people to study medicine: "Thus, in writing this book, I brought together the sayings of the physicians that I had learnt by heart when young, I removed the complex and obscure points, I clarified the passages that were not clear and I added my own considerations. We can use it as a stepping-stone to start learning. I don't dare say that it is a medium that can lead to the Way (dao). I wish only that those who make use of it can rely on it and progress, in the way one uses a stepping-stone. This is why I called my book "stepping-stone"."; and thirdly, to publicize a criticism of the medical fashion launched by Zhang Jingyue (1563–1640) who underlined the importance of replenishing the Yang in the body and advocated using warming and tonifying drugs.[10]

The 醫學精要 (*Yixue jingyao, The Essentials of Medicine, The Essentials*, hereafter) was written in 1800, by Huang Yan 黄岩 (1751–1830), a native of Jiaying 嘉應, in Guangdong province. While this man is described by his preface writers as embodying all the distinctive traits of the scholar—good at poetry, literary composition, and classical Chinese—,[11] he does not hold an imperial degree. Moreover, his biography recorded in one gazetteer is very short, suggesting that he did not belong to the local elites.[12] However, his book *The Essentials* received the preface of a scholar of the Hanlin academy, an imperial institution that brought together the most eminent scholars of the time. Like He Mengyao, but to a lesser extent, Huang Yan is a polymath: in addition to his medical book, (*The Essentials* and the 眼科纂要 *Yanke zuanyao, The Essentials of Ophthalmology*, 1867), he wrote several literary books. The first point Huang Yan addresses in his reading guidelines (*dufa* 讀法) gives an idea of his book's main objective: "This book was written to teach disciples. This is why it is a compilation of the best of the sages of the past and of what I have obtained from my own work. The language is simple but the intention is completeness. The notions are subtle but the words are clear. Characters and sentences have been chosen after reflection. Rhymes are clear and sound harmonious. One can chant or recite them by rote. He who wants to master this doctrine must read it entirely, and he must be familiar with it".[13]

The 暑症指南 (*Shuzheng zhinan, Guide for Summer-Heat diseases, The Guide*, hereafter) was written in 1843 by another native of Guangdong, Wang Xueyuan 王學淵, about whom we have no information as no biography was recorded in

---

[9] See He Mengyao's biographies in (Shi 1879, chap.128), in (Gui 1911, chap.39) and in Xin Changwu's preface to (He [1751]1994).

[10] (He [1751] 1994, p. 47). The author plays with the homophony 'bian' of the three characters 碥 "stepping-stone", 編 "book" and 砭 "critics", to give his book a pedagogical and critical tone. As underlined by (Leung 2003) a lot of medical textbooks have titles like "Stepping-Stone".

[11] (Huang [1800] 1918: Li Guangzhao's preface, 1831).

[12] See the very short biography of Huang Yan in (Wen 1898, Chap. 29).

[13] (Huang [1800] 1918, p. 1).

gazetteers and nobody other than the author wrote a preface to his book.[14] In his own preface, however, the author explains why he wrote the book: he wanted to correct the countless errors made about these types of disease he had heard throughout his life, a common topos used by authors, since antiquity, to write a new text: "My heart was full of regrets on hearing so many mistakes. This is why I wanted to gather all the summer-heat diseases into one single book in order to help their study, but I was confused, ignorant and superficial. I did not have good clinical experience until 1830 when a summer drought happened and the diseases then encountered all had the same cause, the summer-heat. I treated them following the techniques of Master Ye Tianshi (1667–1746) and everyone recovered [...] In his time, his doctrine was widespread but he had no time to write books, only his *Guide with clinical cases* is still extant, which is excellent for those who read it. His clinical cases about summer-heat diseases are particularly detailed and clear. I don't add my own opinion, but, in addition I also collected the good points from the other masters and wrote a book untitled *Guide for Summer-heat diseases*. It will allow students to know that summer-heat diseases are always caused by hot evil *qi* and prevent them from believing that some are Yin heat diseases and others Yang."[15]

The 評琴書屋醫略 (*Pingqin shuwu yilue, Short guide to medicine of the Pingqin shuwu* hereafter, *The Short guide*) was written in 1865 by Pan Mingxiong 潘明熊 (1807–1886). Like the previous authors, he is a native of Guangdong. He passed the first level of the imperial examinations but he did not hold posts in officialdom. As far we can ascertain from his biographies and prefaces, his main employment was medicine even if some preface writers depict him as a good musician as well.[16] Some 10 years after his *The Short Guide*, in 1873, he wrote another medical book where he gathered together the medical cases of the aforementioned Ye Tianshi. In his *Short Guide*, Pan Mingxiong aimed to bring together a series of health guidelines for his son and nephews who had left their hometown to study in the provincial capital, Guangzhou. Pan Mingxiong was afraid, he explained, that they would get sick if they were not careful about their health. Hence, he started to write down some basic rules for them. Then, he enlarged the content of his book and his target audience, having in mind the idea that his book could also help people to learn how to cure themselves, thereby avoiding "quacks'" mistakes: "I propose to print (this book) [...] in order that people who don't know anything about medicine can treat their diseases themselves and not be the victims of quacks."[17]

The 不知醫必要 (*Bu zhi yi biyao, What Someone Ignorant in Medicine Should Know*, hereafter, *What Someone Ignorant*) is a book written in 1881 by Liang Lianfu 梁廉夫 (1810–1894), a native of Chengxiang 城廂, in the province of Guangxi, a rural province at that time. Liang Lianfu succeeded in the imperial examinations,

---

[14] Only the title of the book and the name of its author are mentioned in the bibliographical section of the local gazetteer (Chen 1890, Chap. 52).

[15] (Wang 1843, author's preface 1838).

[16] See Pan Mingxiong's biography (Guo 1987, pp. 1989–1990).

[17] (Pan [1865] 1868, pp. 1–2).

he was a 附貢生 (*fugongsheng,* 2nd list of the "recommended man" intermediate degree, taken in provincial examinations). He obtained several posts in officialdom linked to teaching and civil service examinations. He began learning medicine at the middle of his life, at the age of 36, and, afterwards, practiced medicine in addition to his public career.[18] He only wrote one medical book, with the same objective as Pan Mingxiong's: to help people learn how to self-medicate and avoid "quacks": "If you are not a physician, you can have this book, the clinical symptoms are clear to read, it will allow you to avoid quacks' mistakes."[19]

Finally, the last text, the 經驗良方 (*Jingyan liangfang, Tried and Effective Recipes*), is an anonymous text, copied in a quasi-similar way in two Guangxi gazetteers, published in 1936 and 1946 respectively: in that of Rongxian and in that of Sanjiang, in the north Guangxi mainly inhabited by the Dong ethnic tribe. It is difficult to date this text. Its title is very common in the history of Chinese medicine. However the fact that this text mentions smallpox vaccination and gives a theoretical overview of hereditary syphilis suggests that this text belongs to the beginning of the twentieth century or, at least, that it was regularly updated. While the origin of this text remains unknown, a short introduction allows us to know why it was copied in these two official sources: to fill the gap of physicians in these remote and poor areas of Guangxi: "In the villages and cantons of our district, people believe in sorcerers and they don't turn to physicians; some would agree to turn to them, but towns are too far and physicians too few, it is not convenient. Here is a collection of recipes, they have been collected without following any order".[20]

This brief overview firstly points to the fact that these six texts, written over a period of two centuries, come from distinct social milieus, from the top of society, as in the case of *The Stepping-Stone,* written by a scholar who had passed the highest imperial examinations and who received the prefaces of eminent scholars to the quasi-anonymous *The Guide* which received no allographic preface, and whose author did not deserve an official biography. Secondly, this brief presentation highlights that these texts were written either in Guangdong or Guangxi, that is, in rich and integrated parts of the empire as well as in poor, rural and remote areas. Finally, and maybe even more importantly for the purpose of this analysis, these texts were produced by authors who claimed different statuses as writers: that of teacher for He Mengyao (*The Stepping-Stone*), Huang Yan (*The Essentials),* and Wang Xueyuan (*The Guide)*; philanthropic connoisseur for Liang Lianfu (*What Someone Ignorant)* and Pan Mingxiong (*The Short Guide),* and physician substitute in the case of the anonymous *Tried and Effective Recipes.* These texts were written for different recipients as well: apprentice medical practitioners, erudite families, and a populace without physicians. Their commonality lies in their directive function, explicitly announced in the prefaces. According to speech act theory, speech acts or illocutionary acts can be described as follows: "We tell someone how things are (assertive), we try to make someone do something (directive), we engage ourselves in doing

---

[18] See Liang Lianfu's biography in (Liang 1935, Chap. 4).

[19] (Liang [1881] 1936, pp. 2–3).

[20] Respectively in (Huang 1936; Wei 1946).

something (promissive), we express our feelings (expressive) and we change the world by saying something (declarative)".[21] This standard taxonomy of illocutionary acts is usually used at a sentence level. I take the liberty of applying this taxonomy to a text level in order to identify the types of texts I am dealing with as a whole. And, as stressed above, the medical texts under consideration explicitly aim to lead their readers to do something: either to learn how to treat others, in the case of *The Stepping-Stone, The Essentials* and *The Guide*, or to learn how to treat oneself in the case of *The Short Guide, What Someone Ignorant* and *Tried and Effective Recipes*.

What then do these authors, who come from diverse social, geographical and cultural settings, do to achieve their goal? This is what I will now try to establish by taking sample excerpts from each of these six books—their discourse on *huoluan* 霍亂 etymologically meaning "Sudden Disorder".[22] Now that I have identified the "who" and the "whom", I will shed light on the semantic (part 3.2) and syntactic (part 3.3) features of these texts (the "what" and the "channel" of the famous formula mentioned above) that mainly deal with the content and the form of a message and, in a fourth part, I will complete my work by analyzing the communicational properties of these texts (the "with what effect" of the same formula).

3.2 The Sudden Disorders *huoluan* Analyzed by Six Authors

All these excerpts thus speak of the same thing: *huoluan*. Today, in the biomedical taxonomy of diseases, this term translates cholera, a disease caused by a specific microbe, the Vibrio cholerae (Kommabacillus) or Koch's vibrio, discovered in 1883. Yet, the use of the term *huoluan* does not start with the end of the nineteenth century. Already used in Ban Gu's *History of the Han* (1[er] CE) and in the first transmitted medical text *The Yellow Emperor's Inner Canon* (2nd c. BCE-8th c. CE), this term has a long history. The etymology of the two characters that, once combined, form the term *huoluan* (*huo*=quick, sudden) and (*luan*=disorder) evokes sudden disorders. Since the beginning of the term's use in a medical context, these sudden disorders were located in the abdomen. Historians who analyzed the different clinical symptoms usually associated with this category of disease in the classical medical literature formulated the first hypothesis that, until the beginning of the nineteenth century, this term designated a set of pathological states that included all types of gastroenteritis and alimentary intoxication. As new signs—high contagiousness and high mortality notably—appeared in the medical literature in the

---

[21] (Schaeffer 1989, p. 102). Denis Vernant proposed to slightly modify this standard taxonomy, adding the metadiscursive acts for citation (Vernant 2005).

[22] This disease is treated in the third chapter of *The Stepping-Stone*, of *The Essentials*, in the second chapter of *What Someone Ignorant* and of *The Short Guide*. It is the first disease addressed by *The Guide* which has no chapter division. It is also dealt with in the beginning of *Tried and Effective Recipes*.

early nineteenth century for this category of disease, historians formulated a second hypothesis that the set of pathological states referred to by this term at that time had changed. The heretofore unseen contagiousness and mortality attached to *huolu-an* suggests that true cholera, caused by Vibrio cholerae, had then reached China, an hypothesis which seems plausible as what became known as the 'first Asiatic cholera pandemic' broke out in the 1820s.[23] Since the beginning of the nineteenth century, the term *huoluan*, according to the classic phenomenon of semantic shift, continued to designate all the ancient pathological states usually associated with this term as well as true cholera.[24] The texts under discussion, written between 1751 and 1936, were thus written in this period of transition. However, while these six never-before-analyzed texts produced in the far south of China are likely to provide insight into disease history, the object of this article is not to contribute to the history of cholera in China. Therefore, in order to avoid confusion or ill-considered parallels between complex nosological realities that are not always equivalent, I will translate *huoluan* as "Sudden disorders". Similarly, one must bear in mind that words like "spleen", "stomach", "blood", and the like do not always refer to the substances or organs of the same name in modern anatomy.

The confrontation of these six excerpts (their semantic features are summarized in Tables 3.1 and 3.2) reveals that with respect to their content, all these texts differ on two principal levels. Firstly, they do not provide the readers with the same types of information. Secondly, the actual information given can differ from one text to the other. *The Stepping-Stone,* with its 2173 characters, is the furthest from *Tried and Effective Recipes,* having only 149 characters. The longest text is made up of definitions, clinical descriptions, theoretical explanations, citations, therapeutic recipes and specific diets, clinical cases, advice on how to behave with patients. *Tried and Effective Recipes* only provides definitions of the disease, short clinical descriptions and therapeutic recipes. Between these two texts that are the most radically different from each other, and which are also the farthest apart chronologically, the four other texts either do not discuss certain points addressed in *The Stepping-Stone*, or address all its points but much more concisely. But they do not mention anything that was not addressed in the longest text. Thus, between the two texts that differ the most we see variations of a greater or lesser extent in the content, that I will now detail.

3.2.1   Semantic Features Shared by all the Excerpts

Some semantic features are shared by all the excerpts. Not surprisingly due to the nature of the books, all of them provide definitions of the disease including clinical descriptions and therapeutic instructions. Nevertheless, this information is not given in the same way. The clinical descriptions, first, forming the basis of diagnostics

---

[23] See (Yu 1943).

[24] On these classic semantic shifts, see the example of lepra addressed by (Grmeck 1983).

Table 3.1 Semantic features shared by all the excerpts

	The Stepping-Stone ... 1751	The Essentials... 1801	The Guide... 1838	The Short Guide... 1865	What someone ignorant... 1881	Tried and Effective Recipes 1936
Author's place, status, and target audience	Guangdong	Guangdong	Guangdong	Guangdong	Guangxi	Guangxi
	Highest degree Officialdom & physician linked to the highest elites of his time	No degree Physician Not part of officialdom Patronage from local elites	No degree Physician Not part of officialdom No relation with local elites	Lowest degree Physician Not part of official-dom Patronage from local elites	2nd degree Officialdom Patronage from local elites	Anonymous
	Speaking to students (2173 characters)	Speaking to students (759 characters)	Speaking to students (301 characters)	Speaking to families (330 characters)	Speaking to families (845 characters)	Speaking to population without physicians (138 characters)
Clinical signs	Abdomen and epigastric pain, vomiting, diarrhea, painful muscular cramps in the feet, abdomen or in the whole body, retraction of the tongue, death	Vomiting, physical and spiritual agitation, abdominal pain, diarrhea, cold-hot, cooling of the four extremities, cyanosis	Vomiting, diarrhea, abdominal pain, muscular cramps, short urination and dark urine, cooling of extremities, white tongue, cya-nosis, blue fingers, retraction of the tongue and genital organs, blotches on the upper part of the body	Abdominal pain, vomiting, diarrhea,	Vomiting, diarr-hea, muscular cramps, pain, retraction of genital organs, abdominal distension and pain, sweats and chills, cooling of the extremities	Vomiting, diarrhea, muscular cramps, black spots on the back
Subdivision of the disease	- 霍亂, huoluan, Sudden Disorder -幹霍亂, gan huoluan, Dry Sudden Disorder = 攪腸痧 jiaochang sha disease of the blocked intestines	-霍亂 huoluan Sudden Disorder -攪腸痧 jiaochang sha, Disease of the blocked Intestines = 幹霍亂gan huoluan Dry Sudden Disorder, -痧症shazheng -濕霍亂Shi huoluan Damp Sudden Disorder -霍亂轉筋 huo-luan zhuanjin Sud-den Disorder with muscular cramps	-霍亂 huoluan -幹霍亂 gan huo-luan Dry Sudden Disorder, =攪腸痧 jiaochang sha disease of the blo-cked intestines = 烏沙腸 wu sha chang Sudden Disorder with cyanosis	-霍亂 huoluan Sudden Disorder	-霍亂 huoluan Sudden disorder -轉筋霍亂 zhu-anjin huoluan Sudden disorder with muscular cramps -幹霍亂 gan huoluan Dry Sudden Disorder=絞腸痧 jiaochang sha Disease of the blocked intestines	-霍亂 huoluan Sudden Disorder -霍亂轉筋 huo-luan zhuanjin Sudden disorder with muscular cramps -幹霍亂 gan huo-luan Dry Sudden disorder
Therapeutic prescriptions	Hot and cold Decoctions (+/−40) Acupuncture using moxa Cutaneous stimula-tion (guasha) Forbidden foods: warming drugs and particular foods	Acupuncture Family pill Decoctions (+/−6)	Powder, decoctions (+/−10) Forbidden foods: Warming drugs and particular foods	Decoction (1 with variations) Forbidden foods	Decoctions (8) Powder (10) Forbidden foods	Decoctions (3) Cutaneous stimu-lation (tiaopo) Forbidden foods

are not identical from one text to another and do not require the same competences from the reader: a first group of texts (*The Essentials, The Short Guide, What Some-one Ignorant, Tried and Effective Recipes*) mentions only signs that can be detected through visual or auditory observation of the patients –vomiting, diarrhea, stomach pain, and the like as diagnostic elements; a second group (*The Stepping-Stone, The Guide)* in addition to these first signs, mentions the state of the pulse—deep and slow, quick, imperceptible, etc.—and mentions the seasonality of the disease.

Table 3.2 Semantic features specific to some texts

Author's place, status, and target audience	The Stepping-Stone ...1751	The Essentials...1801	The Guide... 1838	The Short Guide... 1865	What someone ignorant 1881	Tried and Effective Recipes 1936
	Guangdong	Guangdong	Guangdong	Guangdong	Guangxi	Guangxi
	Highest degree Officialdom & physician linked to the highest elites of his time	No degree Physician Not part of officialdom Patronage from local elites	No degree Physician Not part of officialdom No relation with local elites	Lowest degree Physician Not part of officialdom Patronage of local elites	2nd degree Officialdom & physician Patronage from local elites	Anonymous
	Speaking to students (2173 characters)	Speaking to students (759 characters)	Speaking to students (301 characters)	Speaking to families (330 characters)	Speaking to families (845 characters)	Speaking to population without physicians (138 characters)
causes	Evil *xie* (too much cold food and liquids, with cold wind attacks) blocked at the medium burner	Evil *xie* (cold wind/ unsuitable food/excess of yin and dampness) in spleen and stomach. Separation of Yin and Yang	Mix of evil wind, cold, summer-heat, food and water in intestines and stomach	No indication	Cold wind, Abnormal food Abnormal climate conditions Spleen injured by a damp cold blood and *qi* injured in foot yang-brilliance *Zuyangming* and foot-ceasing-*yin* *Zujueyin* conduits	No

Table 3.2 (continued)

Disease's processes	Yin and Yang obstructions and separation; exhaustion of bodily fluids; condensation of fire in the sinews/muscles	Trouble in the foot-major-yin conduit *Zutaiyin*/ Exhaustion of all in relation with Earth phase	Mix of Pure and impure in intestines and stomach	No indication	Obstruction of food, separation of Yin and yang, Blockage of the *qi*	No
Quotations or citations	Zhu Danxi (1280–1368), Liu Hejian (1120–1200), Wang Kentang (1549–1613), Zhang Zihe (1151–1231), Wang Haizang (1200–1264), Cheng Wuji (1066–1156), Luo Qianfu (thirteenth, fourteenth century), Dai Fu'an	*Huangdi Neijing* Dai Fu'an	Zhang Changsha (150–219) and Liu Hejian (1120–1200)'s therapeutic recipes	Wang Kentang (1549–1613) The Ancients	No	No
Type of disease	Exploding disease belonging to Fire Disease of spleen and stomach	*Shazheng* choleric disease	Implicit: Summer-heat disease	Exploding disease linked to the liver conduit and liver	Disease where the spleen is injured by a dampness cold	No

Table 3.2 (continued)

Expert advice	Avoid interpretation errors Pay attention to pulse states Be careful to distinguish different patho-conditions Question the patient about food taken and emotions Do not give ill-considered drugs	Remain calm in face of frightening symptoms	No	Ignore the pulse and just observe clinical signs	No	No
Clinical cases	Two	No	No	No	No	No
Pulse states	Vast and quick pulse….	No	Slippery and fluid pulse	No	No	No
Season	Between summer and autumn	No indication	Summer	No indication	No indication	In winter as well as in summer

The number of clinical signs detected by visual and oral observation of a potential patient varies with the texts: around thirty in *The Stepping-Stone*, fifteen in *The Essentials*, twenty or so in *The Guide*, six in *The Short Guide,* around twenty in *What Someone Ignorant in medicine* and six in *Tried and Effective Recipes*. Moreover, this fairly refined inventory of the clinical signs leads to a reasonably detailed diagnosis of the disease: with the exception of *The Short Guide*, all the texts agree to put two to four forms of the disease under the general heading of *huoluan*—乾霍亂 *gan huoluan* "Dry Sudden Disorders", 濕霍亂 *shi huoluan* "Moist Sudden Disorders", 霍亂轉筋 *huoluan zhuanjin* "Sudden Disorders with cramps", 烏沙腸 *wushachang* "Sudden disorders with cyanosis". The most serious of these, Dry Sudden Disorders, is moreover designated by its "popular" name, 攪腸痧 or 絞腸痧 *jiaochangsha* "disease of the blocked intestines".

All the texts, then, provide therapeutic advice. Note that the place this therapeutic advice occupies in these excerpts as well as the degree of its specificity differ from one text to the other. The number of prescriptions varies from 1 to 40 and they draw on different therapeutic fields. While all these texts promote hot or cold decoctions made of dried or fresh medicinal herbs or minerals, certain therapies are only mentioned in some of them. Acupuncture, for instance, sometimes combining moxa is only written about in *The Stepping-Stone* and *The Essentials*. *Tried and Effective Recipes* and *The Stepping-Stone* mention another therapeutic method: 刮痧 *guasha* 挑破 *tiaopo*, a technique which involves palpation and cutaneous stimulation where the skin is pressed, in strokes, by a round-edged instrument or needle. Note also that the authors have chosen different formats in which to present their therapeutic advice: either only mentioning the title of the formulas and referring the reader to another part of the book for more information or detailing the composition of the formulas and the ways to make them in the main text. I will come back to this point later which leads to different reading practices.

While all the texts provide readers with clinical and therapeutic information, they do not all contain the same information (Table 3.2).

3.2.2   Content Features Specific to Certain Texts

Firstly, the texts do not all provide theoretical explanations on the etiology or physiopathology of the diseases. Nothing of that kind can be found in *Tried and Effective Recipes* neither in *The Short Guide,* which only recalls the link made by the Ancients between this disease and a wider set of diseases. The other texts include theoretical explanations and strive to classify the Sudden Disorders in a more general theoretical framework: spleen and stomach disease, summer-heat disease, liver disease, spleen disease. The theoretical explanations that are all in keeping with the framework of Yin/Yang and the Five Phase doctrine of systematic correspondence are important to a greater or lesser degree and require a more or a less well refined knowledge of it.[25] *The Stepping-Stone* reserves a lot of space to the presentation of the different causes and the diverse developmental processes of the diseases including obstruction of the medium burner because of something evil 邪*xie*, separation of Yin and Yang, troubled circulation of the liquids necessary to human activity, condensation of fire in the muscles provoked by the exhaustion of bodily liquids leading sometimes to muscular cramps. *The Essentials* recalls what the classic *The Yellow Emperor's Inner Canon* had said about the development of these diseases, and evokes the dysfunction of the foot-major-yin conduit (*Zutaiyin*) of the spleen. It also mentions the causes of the diseases: the spleen and stomach being attacked by different types of evil element: abnormal food, abnormal wind or cold, overabundant yin and dampness. The separation of Yin and Yang and the impossibility for the *qi* to circulate properly in the body are invoked to explain the most serious forms of the diseases. *The Guide* only evokes the effect of some of the six classic external pathogenic factors: wind, cold, summer-heat, dampness. *What Someone Ignorant* evokes some of these factors as well—cold wind—, and also abnormal climate, abnormal eating and evokes troubles along the foot-yang-brilliance (*Zuyangming*) of stomach and foot-ceasing-yin (*Zujueyin*) of liver conduits.

Secondly some of the texts use citations while others do not: *What Someone Ignorant* and *Tried and Effective Recipes* do not cite anyone; the other texts mobilize their peers but not always in the same way and for the same function: *The Guide* in fact alludes to two masters to highlight the lineage of two of its therapeutic prescriptions (it is the recipe of x and y); *The Essentials* cites T*he Yellow Emperor's Inner Canon* to recall some theoretical elements; *The Short Guide* cites Wang Kentang, a medical author from the sixteenth century, in proposing a therapeutic recipe and he cites the Ancients, to recall what these "sages" used to say about the diseases that happened suddenly and that included, according to him, the Sudden Disorders. In *The Stepping-Stone,* He Mengyao not only resorts to his peers to teach the reader the different theories developed by some of his forebears but also to include in his text, in two clinical cases cured by Luo Qianfu, another physician, some individualized flesh and blood patients.[26] I will develop later what functions the citation can meet but at this stage just note that not all the texts make use of citation.

Thirdly, not all the texts provide advice on how to behave with patients, for making a good diagnosis and adopting an adequate therapeutic strategy. *The Stepping-Stone* warns its readers against possible misinterpretation, such as, for example,

---

[25] In the third century BCE, the Yin and Yang and the Five Phases (*Wuxing*) were combined into a single doctrine that explained all the changes in the universe, be they natural or political. This doctrine was also adopted in the medical literature of the Han dynasty to explain bodily changes and became one of the most important tools for conceptualizing health and disease as the Han medical literature was adopted as canonical medical literature (Unschuld 1985, 1986, 2003).

[26] Even though some authors do not cite anyone, they often recopy earlier texts verbatim. In the translations, I indicated these passages by a footnote. In fact, as (Volkmar 2000) has shown, plagiarism was extremely frequent in Chinese medicine.

believing that a patient is affected by a cold disease because the cramps are only in his feet. In the clinical encounter, it also recommends to be very cautious in differentiating between hot, cold and damp patho-conditions, and to take care in pulse taking to avoid confusion and to prescribe the suitable treatment. It also advises the reader to question patients in order to get precise information on the food eaten and on the types of emotions felt. We find a piece of advice of this sort in *The Short Guide* which recalls that in sudden disorders, and contrary to what *The Stepping-Stone* promotes, pulse diagnosis is not important. *The Essentials* also advises readers to be calm when facing a certain number of signs that at first could seem serious but that are in fact necessary for the patient's recovery.

Finally, the extent of the specification or the explanation of the semantic content of these excerpts is not equal in all these texts. The authors assume that their readership already has, to a greater or lesser extent, the capacity to understand these texts. *The Stepping-Stone* assumes its readers have a command of pharmaceutical knowledge, to the extent of knowing what "antagonist drugs" and "astringent drugs" are, and of being able to elaborate a decoction by only reading the title "Decoction to stop thirst" or "Decoction of the seven *qi*." It also supposes the reader knows the acupuncture points, *guasha* methods and how to read the pulse. *The Guide* does not assume its readers have deep theoretical medical knowledge; however, it does assume that its readers are able to understand the pulse and have a command of pharmaceutical science to the extent of being able to identify the "warming drugs". *The Essentials* assumes that its readership has mastered the elementary theoretical elements (the conduits and the five phases) but has no knowledge of the pulse; it assumes its readership knows a little about therapeutics, as it recommends prescriptions whose ingredients and their utilization are not always explained. By contrast, the authors of the three other texts do not require capabilities other than literacy: *Tried and Effective Recipes* does not require any knowledge of medical theory, of pulses or pharmaceutical science. In fact, anyone who has mastered literacy can understand the recipes and put them into practice. *The Short Guide* requires a very basic knowledge of theory, but no capabilities in pulse interpretation or in pharmaceutics, since all the recommended recipes are presented with a description of the quantities of their ingredients. Even someone ignorant in pharmacology could prepare the prescriptions. *What Someone Ignorant* assumes its readership to have a basic knowledge of the conduits but not necessarily any knowledge in pulse states or in pharmaceutics as, here again, the author always describes, the ingredients, their quantities and how to make each recipe.

To summarize on this first point, the comparison of the semantic features of these excerpts reveals that while all the texts speak about the same thing, they don't speak about it in exactly the same terms. In fact, we find significant variations on two levels that we must differentiate clearly. On the first level, the type of information the authors provide to their readership can vary. All give descriptions of the disease and therapeutic treatments, but some texts do more than that. Some provide theoretical explanations about the causes and the physiopathology of the disease; some equip the reader with advice on how to be a good medical expert; some give proofs by echoing ancient theories or practices. On a second level, we find significant variations within the information itself: while all these authors agree upon classifying various troubles that all of them found to be quite similar under the category "Sudden Disorder", their understanding of the causes and development of the disease, their methods for identifying them properly and for treating them, differed substantially. These second-level variations are not surprising for any historian working with Chinese medical texts and well aware that medicine in China has never been a coherent and unified tradition. These variations only attest to the fact that, during the two centuries under consideration and within the different milieus that studied the body and its ailments, there was not a consensus; but rather, that observing patients, understanding the causes and the development of diseases and treating them, were grounded in medical fashions that varied according to either the epoch, the geographical region or social milieu. Whatever the importance of these second-level semantic variations for the history of medicine, they are not, however, of primary concern for the present project which attempts to characterize the different texts and thus pays more attention to the first-level variations (the different types of information provided, or not, by all these texts) and try to understand to what these variations are linked. In Table 3.3, I tried to highlight how the first-level semantic variations (theory, quotations, clinical cases, etc.) are distributed among these texts.

According to this table, the author's social milieu (deduced from having obtained a high/medium/low degree, holding a post in the bureaucracy, having contact with local elites) does not seem to be a fundamental criterion for explaining the variations in content of these texts: the texts that are the most similar in semantic content— *The Stepping-Stone, The Essentials, The Guide* on one hand and *The Short Guide, What Someone Ignorant,* and *Tried and Effective Recipes* on the other—come in fact from distinct social milieus. Conversely, the texts that come from similar social circles do not have much commonality: *The Stepping-Stone* and *What Someone Ignorant,* both coming from high degree-holders, involved in prestigious officialdom, have only one content feature in common. The commonly-used distinction between "scholarly" and "popular" literature thus does not hold. However, the chronology as well as the declared function of the texts (teaching or self-medication) seem to impact on the content of these texts: the texts prior to 1843 have at least three common semantic features: they all include theoretical elements, quotations or citations, and assume the readership to have mastered the rudiments of medicine in order to understand the texts. *The Short Guide, What Someone Ignorant,* and *Tried and Effective Recipes* written between 20 and 50 years apart and written for families or patients without physicians, have two common negative features though: they don't present clinical cases and they don't expect their readership to have prerequisite knowledge of medicine in order to understand the text.

It is not easy to see, at this point, if it is the chronology or the declared function of the texts that impact on the content of the texts. The books written before 1843 are also those written to teach medicine and thus aim at a specific audience, apprentice medical practitioners; the texts written afterwards and that appear to be the most

Table 3.3 Content variations within the 6 excerpts

Texts/Date	The Stepping-Stone ...1751	The Essentials...1801	The Guide... 1838	The Short Guide... 1865	What someone ignorant 1881	Tried and Effective Recipes 1936
Author's place, status, and target audience	Guangdong Highest degree Officialdom & physician linked to the highest elites of his time	Guangdong No degree Physician Not part of officialdom patronage from local elites	Guangdong No degree Physician Not part of officialdom No relation with local elites	Guangdong Lowest degree Physician patronage of local elites Not part of officialdom	Guangxi 2nd degree Officialdom & physician Patronage from local elites	Guangxi Anonymous
	Speaking to students (2173 characters)	Speaking to students (759 characters)	Speaking to students (301 characters)	Speaking to families (330 characters)	Speaking to families (845 characters)	Speaking to population without physicians (138 characters)
Theory	+	+	+	–	+	–
Expertise advices in face of patients	+	+	–	+	–	–
Citation or quotations	+	+	+	+	–	–
Clinical cases	+	–	–		–	–
Prerequisite to understanding	+	+	+	–	–	–
Total +/–	5+/0–	4+/1–	3+/2–	2+/3–	1+/4–	5–

similar were written for self-medication and aimed at families. However, from this first analysis, the factors that seem to distinguish these texts the most acutely are not the social divide between "low " or "high" circles, but more chronological or functional. I will now turn to what Schaeffer, in the footsteps of the logicians, calls the syntactic structure of the texts,[27] that is, the set of elements that encode the message and investigate whether the proximity found between some of these texts on the content level is confirmed or not at the syntactic level.

3.3   The "Channel" or the Wide Range of Elements that Encode the Message

Comparison of these six texts reveals that the formal elements that were chosen by the authors to encode their messages are not identical. Let us now highlight these differences.

3.3.1   Long and Short Texts: The Vade-mecum versus Reference Books?

When we consider these six different excerpts that speak about the same thing, the first obvious difference lies in their different lengths: *The Stepping-Stone* is the longest, with 2173 characters, while *Tried and Effective Recipes* with only 138 characters is the shortest. Between these two extremes, we have two groups of quite similar texts: *What Someone Ignorant* and *The Essentials* are three times shorter than *The Stepping-Stone*; *The Guide* and *The Short Guide* are seven times shorter. The different lengths that we can observe is in part linked to the various levels of semantic richness of the texts, as we have seen previously. *The Stepping-Stone* addresses many more different themes than *Tried and Effective Recipes* that only gives a few clinical descriptions and therapeutic recipes. However, as Tables 3.1 and 3.2 show clearly, some very short texts, such as *The Guide,* address nearly all the themes developed by *The Stepping-Stone*, but do so more concisely. The exhaustiveness of the content thus may not be the only factor that plays a part in the length of the texts.

3.3.2   Simple Language that Uses Textual Connectors and Discourse Markers Frequently or Sparingly

At first glance, all these texts use simple language that does not seem to differ radically from one text to another. But, in fact, a consistent difference does differentiate

---

[27] (Schaeffer 1989, p. 112).

the language used in these texts: the frequency of the use of textual connectors and discourse markers—that is, particles in the language of sinologists. Among these particles, it is important to distinguish between, on the one hand, the characters that are used to clarify, show cause/result, indicate time, sequence ideas, add information, illustrate, predict and so on and so forth and that are usually called "textual connectors" or "text connectives" and whose function is to contribute to the cohesion of the text (Table 3.5); and on the other hand, the particles that do not change the meaning of an utterance and have a somewhat empty meaning (equivalent to the English "oh", "well", "you know", "I mean", "right?") and that are usually called "discourse markers". For instance, the particles *ye* 也, *yi* 矣, *hu* 乎 ("it's sure", "that's it", "but why then") can be labeled "discourse markers", while *you* 又, *gu* 故, *ze* 則 ("also", "it is because", "then") can be called "textual connectors". As is often explained by sinologists, one of the functions of these "discourse markers" was to segment the texts which, before the introduction of modern punctuation in the beginning of the twentieth century, were often written continuously.[28] It is true that in the sample of texts under consideration, the text that makes the greatest use of these discourse markers was not punctuated. The unpunctuated *Stepping-Stone* actually used 1 discourse marker every 57 characters, while the punctuated texts rarely used them. However, the use of these discourse markers is not only motivated by the lack of punctuation and a concern of readability. *Tried and Effective Recipes* does not use punctuation and nor does it use discourse markers. In fact, these particles do more than punctuate and allow writers/speakers to add something other than simple readability to their discourse. But before developing this particular point, let us go back to the question of the different uses of textual connectors and discourse markers in these excerpts.

As Table 3.4 shows (as well as the boxes/circles in figs 3.1, 3.2, 3.3, 3.4, 3.5, and 3.6 and in translations, at the end of this chapter), not all the texts use these textual connectors and discourse markers in the same frequency. On the one hand, *The Stepping-Stone* (1 textual connector and discourse markers every 10 characters), *The Short Guide* (1 every 9 characters) use textual connectors and discourse markers frequently. *The Essentials* (1 every 20 characters) and *What Someone Ignorant* (1 every 25 characters) rarely use them. Table 3.4 suggests that chronology does not play a significant role in the differing use of textual connectors and discourse markers: the medical writing before the end of the eighteenth century does not seem to use more textual connectors and discourse markers than the same language at the end of the nineteenth century and beginning of the twentieth century.

---

[28] On the history of punctuation in Chinese, see (Wilkinson 1998). In fact, before the introduction of modern punctuation in the twentieth century, authors often used some punctuation markers -circle, mid-comma, for indicating breaks or to underline some passages. In the texts under consideration, we find the traces of these ancient punctuation markers. On the distinction between textual connectors and discourse markers, see notably (Schiffrin 1988).

Table 3.4 Different uses of textual connectors (t.c.) and discourse markers (d.m.)

Title/Date	The Stepping-Stone ...1751	The Essentials...1801	The Guide...1838	The Short Guide... 1865	What someone ignorant...1881	Tried and Effective Recipes 1936
Author's place, status, and target audience	Guangdong Highest degree Officialdom & physician linked to the highest elites of his time	Guangdong No degree Physician Not part of officialdom Patronage from local elites	Guangdong No degree Physician Not part of officialdom No relation with local elites	Guangdong Lowest degree Physician Patronage of local elites Not part of officialdom	Guangxi 2nd degree Officialdom & physician Patronage from local elites	Guangxi Anonymous
	Speaking to students	Speaking to students	Speaking to students	Speaking to families	Speaking to families	Speaking to population without physicians
Number of characters	2,173	759	301	330	845	138
Number of t.c. and d.m.	228	44	22	36	34	9
1 t.c. or d.m. every n characters	10	17	14	9	25	15
Number of different t.c and d.m	37	17	12	19	23	6

Fig. 3.1 Excerpt from He Mengyao 何夢瑤 (ca. 1692–1764), 醫碥 *Yibian, The Stepping-Stone for Medicine* [1751] second edition, date unknown. Chap. 3. 霍亂 *Huoluan*, Sudden Disorders. Note the unpunctuated writing, small-size characters for the author's comments and the frequent use of textual connectors and discourse markers (*circles*)

As this first analysis pinpoints, the use of discourse markers and textual connectors is not indispensable. We can thus argue that to use or to avoid them is the result of the authors' choice. The analysis of not only the quantity but also the variety of textual connectors and discourse markers used in all these texts suggests that one reason that may explain this differential use of these characters may be found in the authors' endeavors to use more or less sophisticated language. *The Stepping-Stone* not only uses a lot of textual connectors and discourse markers but also resorts to the widest range of this type of words: 37 different often synonymous textual connectors or discourse markers (see Table 3.5). Its author thus seems particularly careful to avoid repetition and uses all the nuances of his language. As I mentioned above, this text was written by a polymath trained in classical studies who moreover wrote a great number of books which received prefaces from members of the top of the elites. The frequent use of textual connectors and discourse markers may thus be linked to the author's endeavor to adopt a good writing style. On the contrary, sparing particles could be linked to another target: to avoid unnecessary words to have shorter, handier books, easy to carry around when making medical rounds, for instance. And in fact, handiness (便 *bian*) is a concern expressed in two texts: in his

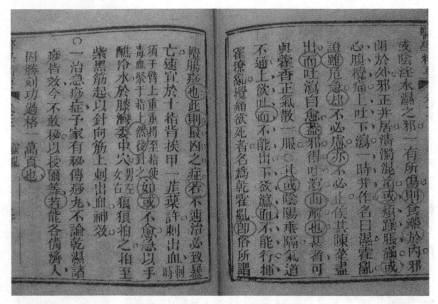

Fig. 3.2 Excerpt from Huang Yan 黄岩 (1751–1830), 醫學精要 (*Yixue jingyao, The Essentials in Medicine* [1800] 1867, Chap. 3, 霍亂 *Huoluan*, Sudden Disorders. Note the wide range of punctuation markers and the less frequent use of textual connectors and discourse markers (*circles*)

introduction, the author of *What Someone Ignorant* informs us that his book will be convenient for people who move frequently, merchants and bureaucrats.[29] The short introduction to *Tried and Effective Recipes* also underlines that in Guangxi's rural areas, it was neither easy nor convenient to find a physician. While some authors were keen to use a more sophisticated language, even if it meant having a large cumbersome volume that could only be examined in a medical office (the modern edition of *The Stepping-Stone* has 797 pages), others wanted to write short, handy, easy-to-carry texts. One way to be more concise was to avoid unnecessary words, such as textual connectors and discourse markers.

The various uses of these textual connectors and discourse markers may finally be linked to what the authors were trying to do when they wrote: convincing by using explanation (causal assertives) and/or prediction (predictive assertives) that need connectives such as "if… then", "This is why", "because"; or convincing by using descriptive assertives and directives like in this imaginary utterance "Things are like that. Do this and that" which need few connectives; or convincing by interacting with the readership, through the use of discourse markers such as "right?", "that's it", "for sure". It is a point which I will come back to later when we compare the different types of speech acts chosen by authors to meet their communicational target. Let us just say, for the moment, that there are different communicational properties behind the frequent or rare use of textual connectors and discourse markers.

---

[29] (Liang [1881]1936, pp. 3–4).

Fig. 3.3 Excerpt from Wang
Xueyuan 王學淵 (active
in 1830), 暑症指南 (1838)
*Shuzheng zhinan, The Guide
for Summer-Heat Diseases*
(no dividing chapter) 霍亂
Huoluan, Sudden Disorders.
Note the punctuation markers
on the right of the characters
and the less frequent use of
textual connectors and discourse markers

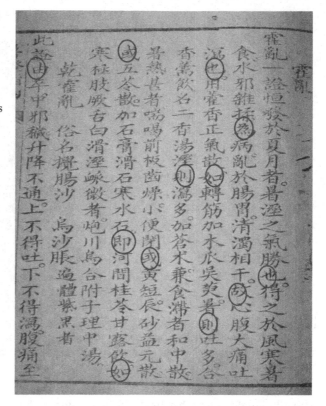

3.3.3   Texts in Prose and Texts in Verse

Another stylistic feature distinguishes these texts: the use or not of metric constraints. In fact, two texts aimed at different readerships—apprentice medical practitioners for *The Essentials* and families for *What Someone Ignorant*—were written
under phonetic and metric constraints. The author of *The Essentials* writes in blank
verse and sometimes in rhyme in different places in his text: in the first symptomatic
description of Dry Sudden Disorder, and when he enumerates the different etiolo-
gies of the disease. *What Someone Ignorant* is written under even more constraints.
Not only does the author choose an identical format to present his message in the
first three parts of the first chapter—a symptomatic description of a particular form
of the disease, followed by an explanation of the causes, then, an explanation of
physiopathology—but also, within this first part, the explanation of the causes is
written in five characters, the explanations of the physiopathology that follow are
all in seven characters (with the exception of the last two), and all these segments
finish with the same sentence: "and then disease occurs". These constraints give the
texts a particular rhythm that surely aimed at facilitating memorization. Memorization has always been highly valued in classical studies in China and this learning

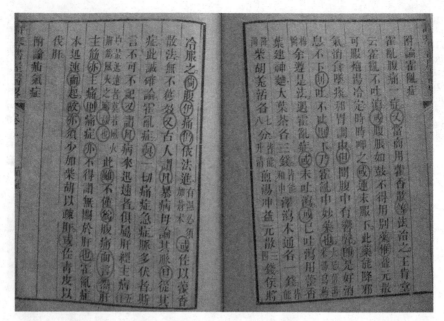

Fig. 3.4 Excerpt from Pan Mingxiong 潘明熊 (1807–1886), 評琴書屋醫略 (1865) *Pinqin shuwu yilue, Short Guide to Medicine of the Pingqin Shuwu* (Chap. 2) 霍亂 *Fulun huoluan, Added Doctrine on Sudden disorders*. Note the punctuation markers and the frequent use of textual connectors and discourse markers

method migrated into other fields of knowledge such as medicine.[30] Versification facilitated memorization particularly for those who started learning after childhood, as was often the case for those who had decided to learn medicine after successive failures in the imperial examinations.[31] In his short introduction, the author of *The Essentials* explains the reasons for his stylistic choice in the following terms: it would help pupils to rote chant his text and learn it by heart.

3.3.4 A Different Order in the Presentation of Information

The order of presentation of the information and arguments—definitions, etiological and physiopathological explanations, recipes, quotations, etc...—is far from being the same in all these texts and this is worth some consideration. In a field very far from medicine, N. Goodman and J. Virbel have clearly shown that the simple

---

[30] In Chinese classical education, huge numbers of characters were memorized, by repetition and copying, and even if memorization as a didactic tool was at times condemned in the history, massive memorization played a crucial part in the classical education until the end of the empire, in 1911. See notably (Mizayaki 1976; Elman 2000; Elman and Woodside 1994; Gernet 2003).

[31] See (Leung 2003, vol. 2, pp. 89–114).

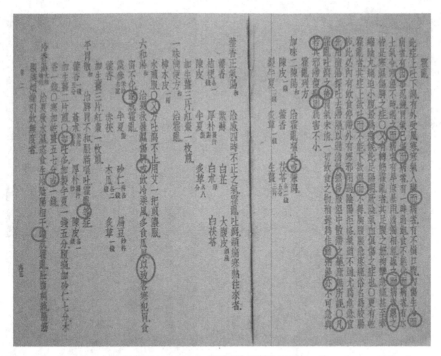

Fig. 3.5 Excerpt from 梁廉夫 (1810-1894), 不知醫必要(1881) *Bu zhi yi biyao, What Someone Ignorant in Medicine Should Know* (Chap. 2) 霍亂 *Huoluan, Sudden Disorders*. Note the punctuation markers and the very rare use of textual connectors and discourse markers

permutation of the episodes in a narrative gives the narrative very different tones and textual statuses.[32] Let us now see how all these authors chose to present information.

While *What Someone Ignorant, The Stepping-Stone* and *The Guide,* with some nuances though, first describe and define the disease, explain its possible causes and development, reserving the therapeutic section for the end of the chapter, the *Short Guide* does the contrary: in its first section, it describes the disease and gives therapeutic advice, then, at the end of the chapter, it recalls, in a somewhat accessory way, what the Ancients used to say about these types of diseases that appeared very suddenly. *The Essentials* chooses another order: it begins with the description of the most serious form of the disease, whose popular name is immediately recalled, and with the therapeutic strategy to adopt urgently in this particular form of the disease. Afterwards, the author recalls some theoretical elements and describes the clinical signs of the common form of the disease. Finally, the author comes back to the most serious form, its clinical signs, its popular name, and the therapeutics suited for acute diseases in general and for which, he says, he has a pill, a family secret,

---

[32] See (Virbel 1997, pp. 251–272).

Fig. 3.6 Excerpt of an anonymous text 經驗良方 (published in 1936 and 1946) *Jingyan liangfang, Tried and Effective Recipes*, 治霍亂方 *Zhi huoluan fang, Recipes for Curing Sudden Disorders*. Note the absence of punctuation and the very rare use of textual connectors and discourse markers

that he agrees to detail. *Tried and Effective Recipes* first describes three forms of the disease and then gives recipes.

The order in which the information is presented in these texts suggests the idea of a hierarchy of information—what is the most significant information to give my reader first—which, at the same time, suggests that these texts were written for different purposes: the texts which did not leave room for theoretical explanations or that put the theories at the end of the chapter seem to have been written to be consulted in an emergency, in the acute phase of the disease (*The Essentials*, *Tried and Effective Recipes*, *The Short Guide*); the others aim at a readership which is not under the constraint of emergency and can spend time on first reading about the causes and the development of the disease (*The Stepping-Stone*, *What Someone Ignorant* and *The Guide*).

The order in which the therapeutics are presented also varies. Some texts (*The Short Guide* and *The Guide*) chose to include their prescription in the general discussion of the disease; others (*What Someone Ignorant*, *The Stepping-Stone*, *Tried and Effective Recipes* and, to a lesser extent, *The Essentials)* have a clearly distinguished chapter for the theoretical explanations or descriptions of the disease and a different chapter for the therapeutics. Another distinction in the ways of presenting therapeutic advice or recipes can be noted. As already mentioned, while some au-

Table 3.5 Frequency and diversity of textual connectors and discourse markers

	The Stepping-Stone	The Essentials	The Guide	The Short Guide	What someone Ignorant	Tried and Effective Recipes
Discourse markers (equivalent of "well", "right?", "anyway"…	26 也 *ye*, it's sure 1 耳 *er*, so 1 哉 *zai*, isn't it? 1 乎 *hu*, but why then 8 矣 *yi*, that's it! 1 勿 *wu* don't…	4 也 *ye* it's sure 1 无疑 *wu yi* undoubtedly	2 也 *ye* it's sure 1 矣 that's it!	3 也 *ye* it's sure	1 也 *ye* it's sure	
Textual connectors indicating how the text is developing. They are used for clarifying, indicating time, sequencing ideas, adding information, expressing condition or concession, emphasizing, summarizing etc….	3 如 *ru* if 16 或 *huo* or 33 则 *ze* then 20 而 *er* but, and, 13 若 *ruo* if 5 即 *ji* then 7 故 *gu* this is why 5 又 *you* also 13 亦 *yi* also 2 於是 *yushi* then 4 然 *ran* then, but 5 因 *yin* because of 5 盖 *gai* in fact 3 且 *qie* moreover 2 大抵 *dadi* generally speaking 1 是為 *shiwei* due to 5 更 *geng* and also, 3 但 *dan* but 4 仍 *reng* and 8 後 *hou* after 2 按 *an* as (cause) 1 雖 *sui* even 1 惟…後 *wei…hou* only after 1 是 *shi* this is because 2 再 *zai* again 1 上 *shang* as mentioned above 5 先 *xian* first 5 乃 *nai* and, then 5 并 *bing* equally 4 兼 *jian* equally 3 凡 *fan* whatever	6 而 *er* but, and 2 即 *ji* then 1 為 *wei* due to 9 或 *huo* or 4 则 *ze* then 2 又 *you* also 1 雖… 却 *sui… que* in spite of.. 2 亦 *yi* also 1 盖 *gai* in fact 3 若 *ruo* if 1 然後 *ranhou* after 1 如 *ru* if 1 因 *yin* because of 2 及 *ji* and 2 并 *bing* equally	1 故 *gu* this is why 3 如 *ru* if 1 若 *ruo* if 3 则 *ze* then 1 兼 *jian* equally 4 或 *huo* or 1 即 *ji* then 3 亦 *yi* also 1 又 *you* also 1 和 *he* and	3 又 *you* also, again 1 等 *deng* notably 6 或 *huo* or 1 惟 *wei* only 2 但 *dan* but 1 即 *ji* then 3 则 *ze* then 1 乃 *nai* then 2 仍 *reng* again 2 凡 *fan* whatever 1 與 *yu*! 2 故 *gu* this is why 1 雖 *sui* however 1 為…而言 *er… wei yan* to speak of 1 然 *ran* in fact 3 亦 *yi* also 1 而 *er* and	8 而 *er* and then 1 因 *yin* because of 1 总之 *zongzhi* to summarize 2 又 *you* in addition 1 更 *geng* moreover 1 先 *xian* first 1 然後 *ranhou* then 1 若 *ruo* if 1 凡 *fan* whatever 1 後 *hou* after 1 雖 *sui* even 1 亦 *yi* also 1 若 *ruo* if 1 则 *ze* then 3 或 *huo* or 1 等 *deng* for instance 1 如 *ru* if 1 於…时 *Yu…shi* when 1 兼 *jian* and also 1 便 *bian* in that case 1 已致 *jizhi* until 2 逐 *zhu* progressively	3 又 *you* also 2 亦 *yi* also 1 并 *bing* equally 1 如 *ru* if 1 既 *ji* immediately 1 而 *er* but
Total	228	44	22	36	34	9
No. different t.c. and d.m. used	37	17	12	19	23	6

thors chose to include all the details about the prescribed formulas in the main text, others prefer to send the reader to other parts of the book. *What Someone Ignorant, The Essentials* and *The Short Guide,* by indicating the name of the recipes, possibly the type of therapeutics they belong to (invigorating, cooling, etc…), the kind of patho-conditions they cure, the quantities of ingredients and the ways of preparing them in the same place, immediately gives the reader all the indications allowing him to prepare these recipes and to understand them. On the other hand, *The Stepping-Stone* and *The Guide* prefer another format: they only mention the name of the recipe and send the reader to another part of the book. What these variations are linked to is not clear. What is sure is that these two different ways of presentation lead to different ways of reading or consulting the texts.

The question of the pagination, of paragraph indents, and of typographical changes would be very interesting to develop in this analysis, even if it is clear that these choices were often in the hands of the editors rather than the authors.[33] Moreover, not all the documents used here are first editions, which, for some of the texts, are no longer available. The editions available to me still show clear contrasts (see Figs. 3.1–3.6). Some authors/editors have chosen to add some paratextual elements in order to make the book easier to read: while *The Stepping-Stone*, in its second edition, has no punctuation and uses continuous writing, the other texts have clear indents and punctuation markers. *The Essentials*, in an 1867 edition, uses small and big circles to signal the beginning of paragraphs. It uses rows of circles or mid-commas to the right of the characters in order to emphasize one passage and it uses indents. *The Guide* and *The Short Guide* use small circles to mark sentence breaks. *What Someone Ignorant* uses small and big circles and indents to introduce new chapters or new recipes.

Comparison of these excerpts shows that the authors do not resort to the same formal elements to encode their messages. As summarized in Table 3.6, we have: concise texts *versus* diluted texts; texts which use connectives widely *versus* texts that use connectives sparingly; punctuated texts *versus* unpunctuated texts; texts in verse *versus* texts in prose; texts that first focus on how to see and do *versus* texts that first focus on how to see and think; texts that place all the information in one single place *versus* texts that interrupt the message and send the reader to other parts of the book. Are these different ways of writing linked to chronology, geographical setting, social milieu or the declared function of the book?

These differences *a priori* do not seem to be particularly linked to the chronology, nor to the author's status, nor to the target audience. They do, however, suggest very different types of reading or consultation: a quick pragmatic reading *versus* a slower reading for comprehension. On the one hand, we have the highly portable vade-mecum, for those who have no time for etiological or physiopathological explanations, but who must be able to quickly identify the disease and prepare recipes,

---

[33] On the role of editors for making a book easier to read and sell better, see (Widmer 1996 and Chia 2002).

Table 3.6 Different encoding of the six excerpts

Text/date	The Stepping-Stone ...1751	The Essentials...1801	The Guide...1838	The Short Guide... 1865	What someone ignorant 1881	Tried and Effective Recipes 1936
Author's place, status, and target audience	Guangdong Highest degree Officialdom & physician linked to the highest elites of his time	Guangdong No degree Physician Not part of officialdom Patronage from local elites	Guangdong No degree Physician Not part of officialdom No relation with local elites	Guangdong Lowest degree Physician not part of officialdom Patronage of local elites	Guangxi 2nd degree Officialdom & physician Patronage from local elites	Guangxi Anonymous
	Speaking to students (2173 characters)	Speaking to students (759 characters)	Speaking to students (301 characters)	Speaking to families (330 characters)	Speaking to families (845 characters)	Speaking to population without physicians (138 characters)
Length	++	+	–	–	+	–
Number of t.c and d.m	++	+	+	++	–	+
Metric Constraints	–	+	–		+	–
Cross-reference	+	+/-	+	–	+	–
Make understanding first	+	–	+	–	+	–
Make doing first	–	+	–	+	–	+

without wasting time searching in other parts of the book for other information (*Tried and Effective Recipes, The Short Guide* but also, because of its clear presentation, *What Someone Ignorant*). On the other hand, we have the reference work, not handy, that first gives the reader all the explanations about the disease, maybe different theories expressed by earlier physicians, before indicating the name of a few formulas that the reader is then required to find in another chapter in order to learn how to prepare. It is the case of *The Stepping-Stone*, which moreover requires first of all that the reader punctuates and segments the text to get its points.

The analysis of the channels used by the authors to encode their message traces a line between texts written for doing before understanding and texts written for understanding before acting. Let us now finally see what the authors do to guide their readership to follow their first intentions.

3.4 Communicational Targets: Authors' Intention and Communicational Properties in the Six Texts

As I mentioned at the beginning, the intention behind these six texts, as explained by the authors themselves, was to teach the reader how to treat people or to self-medicate. From these explicit intentions, we can draw two conclusions about the nature of the texts: firstly, all the texts can be considered as written discourses aimed at a readership (we are thus in a dialogical and not in a note-taking situation, involving a speaker and a hearer); and secondly, the authors did not intend their text to narrate something but rather to engage people do something.

The tools elaborated by speech act theoreticians seemed quite useful in identifying what all the authors did to achieve their communicational target. For these theoreticians, language is not only used to describe or to give some statements; to speak is to act with or against someone to transform our worlds.[34] Thus, and always in a comparative perspective, I strove to identify the different types of illocutionary acts (assertive, directive, promissive, etc.), the different illocutionary force markers used in all these texts. A few problems arose from the attempt to apply a theory elaborated within the framework of modern spoken English to ancient written Chinese.

The first problem lies in the difficulty of segmenting pre-modern Chinese texts into utterances and thus into individual illocutionary acts. This problem is linked to the lack of either punctuation or punctuation markers that raise issues. As already mentioned, the first and second editions of *The Stepping-Stone* was unpunctuated. Punctuation markers were added in later editions. In this case, the punctuation, even though it can vary from one edition to another, corresponds to modern punctuation,

---

[34] John Austin and John Searle were at the origins of this theory, which, since then, has been developed or transformed by other linguists or philosophers of language (Austin 1962; Searle 1969).

a point indicating for instance the end of a sentence.[35] Thus segmenting the text into individual sentences is easy for this text. However, the other texts have been printed with punctuation markers that do not always correspond to the modern values of the same markers. For instance, in *What Someone Ignorant*, the small circle can break a sentence, isolating on the one hand the nominal group, and on the other, the verbal group. As noted earlier, this text was written under heavy metric and rhythmic constraints, and the punctuation in this text is less to indicate semantic or syntactic breaks or pauses, than to give a special rhythm when the text is read aloud. The author strives to write segments of the same length, counting the same number of characters, in order to give a rhythm to facilitate memorization. In order to identify a meaningful sentence in this particular text, I have therefore been obliged to disregard the author's punctuation.

A second problem arose when I tried to identify the illocutionary force markers and notably those used for translating directives. While the imperative mode is a syntactic force marker easy to locate in French or in English, it is not the same in Chinese because Chinese does not conjugate verbs and does not need personal pronouns, even in what would be the indicative mood (in written Chinese "*kan shu* 看書" can be interpreted as "I read book" or "Read book!"). We can sometimes find a particle at the end of the sentence that indicates an order. But with the exception of *The Stepping-Stone* that once uses this type of particle, such particles are rare in these texts. However, the reading of these excerpts shows that the Chinese language uses other resources for translating orders, as we will soon see.

A third problem, not linked specifically to the Chinese language, is that a unique sentence can contain illocutionary acts of different natures, as in this imagined single sentence which contains 3 types of illocutionary act: "If you feel cold you probably have fever (assertive), take rest (directive) and you will feel better (assertive), I promise (promissive)". In the translations provided at the end of this chapter, I chose to indicate, in front of the translations, the different types of illocutionary acts made in each sentence.

Finally, a speech act can perform different illocutionary acts, according to the context in which it is performed. Take as an example, the sentence "Dinner is served at 8 pm" pronounced in a hotel. It can be assertive in the context of a report on the hotel; it can be promissive if it comes from the mouth of the hotel chef to the hotel manager or clients; it can be directive if it comes from the hotel manager to the chef or to the clients.[36] Language is thus not univoqual, and the same sentence, *i.e* without syntactic or semantic change, can have different communicational values, such as describing, ordering or promising. The second part of the fifth text, *What Someone Ignorant,* devoted to recipes, in which the author lists and quantifies the

---

[35] For Chiu et al. 2004: "Chinese does not provide sufficient morphological indicators to syntactic structures, thereby making punctuation a relatively free reference to syntactic structures as well as to semantic domains".

[36] Example taken from Jacques Virbel's presentations in REHSEIS.

different ingredients necessary for a particular decoction illustrates this issue. Are these sentences assertive acts—they describe the composition of the recipe—or directive acts—they implicitly tell the reader what he has to do to prepare the recipe? This question is difficult to settle.

Despite all the difficulties attached to this theory and to its application to Chinese, I found it interesting to use, expecting that it would highlight the communicational properties of these texts and thus give an additional angle of approach for this attempt of systematic comparison.

3.4.1 Speakers and Hearers: Varying Levels of Personal Authority in the Discourse

All these texts have a common point, they include definition: "If there are x and y signs, it is the Sudden disorder" or "the Sudden disorder is a syndrome where the stomach is painful". The act of defining introduces a hierarchical distance between the speaker and the hearer: it is the speaker who knows and has the authority to define and the hearer who is unknowing and therefore accepts the definition. We can even think that if this kind of utterance was pronounced in a medical office by the physician with his patient "You have signs x and y, you thus have the disease z", the illocutionary act would be declarative. Just as a judge who, by his utterance "I sentence you to twenty years imprisonment", transforms the social situation of the hearer, from free to convicted, the physician transforms the social status of the patient, who, from the simple state of having varied signs is, after the physician's declaration, a diagnosed sufferer of a disease (with all the social changes, exclusions in particular, that can result in a person's life from being identified as affected by a particular disease, think of epilepsy or glaucoma for airline pilots, HIV, etc...).

All these texts thus introduce a hierarchical imbalance between speakers and hearers, however, the presence of the speakers is noted differently: *The Essentials* is the text where the speaker's presence is the most palpable through the use of different markers. Firstly, the speaker introduces his text by citing himself ("Nai'an says"); secondly, he uses the personal pronoun "I" *wo* 我 and *yu* 予 which is not necessary in written Chinese; thirdly, he distinguishes himself from his hearers by addressing them directly: "I am teaching it to you (disciples) *yi shou ni deng* 以授尔等"; finally, he uses many typographical markers, either punctuation markers that are intended to emphasize some particular points, or small size characters for his own comments.[37] In the *Short Guide*, Pan Mingxiong also uses the personal pronoun "I" *yu* 予 and typographical changes for his commentaries. In *The Stepping-Stone,* He Mengyao does not use personal pronouns, however, he resorts very

---

[37] According to (Blanks 2005, pp. 40–43) typographical change constitutes a linguistic marker of the author's presence, to the same extent as the use of a personal pronoun.

frequently to small size characters to add personal comments. Moreover, as already noted and as will be developed below, this author often resorts to discourse markers. These markers, in addition to helping the reader to segment unpunctuated texts, also provide metalinguistic indications about the unfolding discourse, such as translating the speaker's emotion about what he is saying. Interestingly, these three texts that thus bear the presence of the speakers to a higher extent than the others are also the texts that cite other famous masters. Contrastingly, we have some completely impersonal texts. The authors of *What Someone Ignorant, The Guide* and, not surprisingly, the anonymous *Tried and Effective Recipes* never use personal pronouns or typographical changes to introduce personal comments. If Liang Lianfu, author of *What Someone Ignorant,* uses small size characters in the second part of his text devoted to recipes, it is only to specify the general properties and the quantities of the required ingredients, and not to introduce personal comments. Furthermore these writers resort to neither citation nor quotation, that is, do not inscribe their message in a precise individual lineage.

The first impression that arises from the reading of these texts that pay attention to the position of the actors—speaker and hearer—and their ways of including themselves, or not, in the discourse, is that while all these speakers present themselves as more authoritative than their hearers, their endeavor to establish their authority is expressed differently. While some authors ground their authority in individual experience—their own and that of other physicians, by way of quotations or citations—, others prefer to enunciate a discourse where their own presence and that of the famous masters they might refer to remain hidden for the benefit of a general truth, a scientific objectivity. It is also interesting to note, as we will now see, that these latter texts, in comparison to the others, prefer to use descriptive assertives rather than causal or predictive assertives, giving the text an implicit general authority that we could illustrate in this way: things are like that, it is not me who is going to explain to you why.

3.4.2   Directive Texts Versus Explaining and Predicting Texts

The application of J. Searle's standard taxonomy on this sample of texts shows that all the texts are made up of directives and assertives that is, of speech acts aimed at leading the reader to do something, and of speech acts aimed at either defining (assertive of definition), describing (descriptive assertive), explaining (causal assertive), predicting (predictive assertive) the state of the world, and, in the present case, the Sudden Disorder. However, if we take into account D. Vernant's modifications to this standard taxonomy, that notably reintroduced Austin's idea that some speech acts had metalinguistic functions ("I demonstrate", "I show" which speak about the speaking activity of the speaker) and should be identified as such, we can say that while all these texts are made up of assertives and directives, only a few of them contain metadiscursive acts, that is, they quote or cite other texts or other sayings.[38]

---

[38] See (Vernant 2005).

Before indicating in which proportions these texts are made up of such illocutionary acts, I shall first underline how the Chinese language translates directives.

As mentioned above, the Chinese language does not conjugate verbs, and there are no morphological changes to indicate tense or mood like the imperative or the infinitive. It does not mean that the Chinese language cannot express injunctions and these texts clearly show it. Firstly, the context makes clear that some verbs could not be translated differently but by the use of an infinitive or imperative ("add this and that", "refer to the section of the cold damage disease"), which introduce direct directives or conditional directives if the directive speech act is preceded by a conditional proposition introduced by a conditional particle ("if this happens, do that"). Secondly, we have lexical markers including the modal auxiliary verbs "must", "can", that introduce indirect directives[39]. Finally, the authors can express their injunctions without resorting to syntactic markers nor to lexical markers, but by using an implicit formula whose illocutionary force is even more directive, like in this example: "You perspire anytime and in great quantity: Decoction of four colds!".

As Table 3.7 shows, the proportion of assertives and directives is not equal in all these texts. *Tried and Effective Recipes* is the only text that gives its readers more orders than descriptions or explanations. In this respect, it is the text that seems the closest to a notice mainly devoted to action. On the contrary, *The Essentials* and *What Someone Ignorant* are made up of more assertives than directives. The other texts are made up of nearly half assertives and half directives. All these texts, therefore, can be considered, to some extent, as texts for action.

Another feature which seems to distinguish these texts is the types of assertives performed. Assertives, indeed, do not all have the same function: they can describe, define, explain, and predict the state of the world. Behind these different functions, the speaker thus endows himself with different capabilities. As noted above, the act of defining, which can be considered as an assertive (and in some contexts as a declarative) introduces an imbalanced hierarchy between the speaker and the hearer. We can assume that the act of explaining or predicting the unfolding world, that is making causal or predictive assertives, gives the speaker a particularly enlightened intermediate status between the world and the hearer: he presents himself as being able to explain why things happen and how they will happen. Interestingly, these texts do not all refer to the same types of assertives to the same extent. In addition to directing the reader to do something, *Tried and Effective Recipes* contents itself to defining and describing the disease but not to explaining or to predicting its development. Conversely, He Mengyao, the author of *The Stepping-Stone*, explains, predicts, and describes the disease to the same extent. Finally, between these two contrasting texts, the other books—*The Essentials, What Someone Ignorant, The Short Guide* and *The Guide*—give a larger space to description, reserving only one fourth of their assertions to explaining or predicting the disease.

---

[39] We find approximately the same verbs having directive illocutionary force in all these texts, with a lower or higher frequency.

Verbs with directive illocutionary force	The Stepping-Stone...	The Essentials...	The Guide...	The Short Guide...	What Someone Ignorant...	Tried and Effective Recipes
	3 當 one must 7 不可 one can't, one mustn't 8 宜 one must, it is suitable 2 必 one must 7 須 one must 1 忌 Abstain from	2 不必 one mustn't 2 可 one must 2 宜 one must, it is suitable 1 須 one must	1 可 one can, one must	2 不得 one mustn't 1 可 one must, one can 1 必須 one must 1 不可 one can't, mustn't 1 須 one must 1 大忌 Abstain absolutely from	1 宜 one must, it is suitable 1 不可 one can't, one mustn't	3 忌 abstain from

Table 3.7 Approximate counting and identification of illocutionary acts in these excerpts

Types of illocutionary acts	The Stepping-Stone	The Essentials	The Guide	The Short Guide	What Someone Ignorant	Tried and Effective Recipes
Assertive of definition	7	3	1	2	1	3
Descriptive Assertives	20	17	14	6	36	1
Causal Assertive	10	3	2	1	8	0
Predictive Assertives	11	2	5	3	1	0
Total of Assertives	48 (42%)	25 (58%)	22 (59%)	13 (52%)	46 (67%)	4 (44%)
Conditional Directives	11	5	5	2	1	1
Direct directives	15	6	9	3	8	0
Indirect directives	21	3	1	4	4	4
Implicit Directives	13	1	0	1	10	0
Total of Directives	60 (53%)	15 (35%)	15 (41%)	9 (36%)	23 (33%)	5 (56%)
Metadiscursive citative	5 (5%)	2	0	3 (12%)	0	0
Metadiscursive expositive	0	1	0	0	0	0
TOTAL	113	43	37	25	69	9

The table reads: Of the illocutionary acts identified in *the Stepping-Stone*'s excerpt, 48 (or 42%) are assertive, 60 (or 53%) are directive, 5 (or 5%) are meta-discursive.

The comparative analysis of the illocutionary acts carried out in these texts reveals three very different communicational targets: the texts that aim to lead the reader to do something (*Tried and Effective Recipes*), those that want to lead the reader to identify the disease properly and to act (*The Essentials*, *What Someone Ignorant*, *The Short Guide* and *The Guide*) and one text that wants its readers to be able to identify, to understand, to predict and to act (*The Stepping-Stone*).

3.4.3   How do the Speakers Convince Their Readership to Do Something?

Another significant distinction in these texts is the way the authors commit themselves to the results of their advice or instructions Indeed, if these texts are devoted to action, the authors need to convince their readers to do what they say or to think as they want them to do. As we have already noted, the ways these authors grounded their authority is different: some grounded their authority in personal experience while others preferred to refer to a general truth. If we now look at how these authors commit themselves to the results of their advice, we note a difference (see Table 3.8). *What Someone Ignorant*, which is impersonal, differs from all the others: while all the other texts use sentences like "using this or that is wonderfully efficient *shenxiao* 神效", "all these have great effectiveness *xie you qi gong* 皆有奇功", formulas that carry in themselves the mark of a personal value judgment, *What Someone Ignorant* prefers a style that refers to an objective value: instead of saying "this recipe is effective, is excellent, and so on", the author introduces all his therapeutic recipes with the same utterance: "This recipe cures *zhi* 治 x, y".

In fact, the speakers use different tools to convince their readers to follow their instructions or ways of thinking, or, in other words, to be sure that the perlocutionary effect, which refers to what is commonly called the effect on the receiver, is satisfied. One is quotation or citation. Citation meets very different functions. One of them is to gain authority, as Bhatia clearly underlined: "in order to become acceptable to the community…, one must relate his/her knowledge claims to the accumulated knowledge of others in the discipline".[40] Therefore, by referring to prestigious peers, by saying "x and y said that", the speaker not only relates his knowledge to the prestigious scholars of the past, but he also increases the chances that the perlocutionary effect be satisfied, that is, the reader believes him immediately and acts accordingly. *The Stepping-Stone*, *The Essentials* (The Classic says: …), and *The Short Guide* (Wang Kentang says:…), by using citation and relating

---

[40] (Bhatia 2004, p. 190; Vernant 2005).

Table 3.8 How the authors convince their readers to obey their injunctions

Commitment formulas	The Stepping-Stone…	The Essentials…	The Guide…	The Short Guide…	What the Ignorant…	Tried and Effective Recipes
	愈 *yu* s/he recovers 神效 *shenxiao* marvelously effective 效 *xiao* effective 最效 *zui xiao* extremely effective 立愈 *li yu* s/he recovers immediately 2而愈 *er yu* and s/he recovers 極妙 *ji miao* extremely marvelous 妙 *miao* marvelous	神效 *shenxiao* marvelously effective 皆效 *xie xiao* effective for all 都效 *dou xiao* all are effective 皆有奇功 *xie you qi gong* all are surprisingly effective 2治 *zhi* it cures	最妙 *zuimiao* most marvelous 愈 *yu* s/he recovers	此藥能 *ci yao neng* this drug can… 妙藥 *miaoyao* marvelous drug 無不效 *wu bu xiao* always effective 3皆能 *xie neng* they all can…	10治 *zhi* it cures	立刻見效 *li ke jian xiao* effectiveness immediately visible 愈 *yu* s/he recovers 神效 *shenxiao* marvelously effective 效 *xiao* effective

their own work to their forebears, use this tool. Only *What Someone Ignorant* and *Tried and Effective Recipes* do not use citations.

Another tool, only used by *The Stepping-Stone* to convince its readership to act and think according to what is written, is the use of discourse markers which have a strong illocutionary force. We had previously noted that this text included a lot of textual connectors and discourse markers, whose function is usually thought to be that of replacing punctuation markers when texts were unpunctuated. But in fact, the author might well have used so many discourse markers for another target. By littering his argument with a lot of discourse markers such as "isn't it", "it is sure", "that's it", "I ask you", the author of *The Stepping-Stone* interacts with his readership, expecting thereby to strengthen the perlocutionary effects of his writing. By using discourse markers having a strong illocutionary force and other locutions such as "some people believe that… but how is it possible to be so mistaken ….?", "How is it possible to make such mistakes!", "to state that is really impossible", the author's goal is to produce a complete adhesion to the speaker's arguments and annihilate any opposition. Interestingly, this text, which uses quotations and interactional discourse markers, was written not only for teaching but also to convince people to stop following a particular medical fashion.

To summarize on this last point, applying different tools from the speech act theories helps to highlight that the communicational properties of these texts are very different (see Table 3.9). We have texts where the speaker's presence (his ways of thinking, his commitment to results) is highly marked and texts where the speaker and other individualized people are hidden for the benefit of an objective discourse; we have texts that look like simple notices for use and texts that combine instructions for use, explanations, and descriptions.

3.5   Conclusion

The issue underlying this analysis was to highlight how medical texts differed from each other. Thus I submitted a small sample of texts dealing with a same issue to a semantic, syntactic, and communicational comparative analysis. This systematic comparison first allows to see that these texts do not share the writing conventions that are usually shared by people who belong to a well-defined professional community.[41]

This analysis then allows us to confirm two points highlighted by linguists specializing in the question of genre: the literary act is a complex semiotic act, and, as we multiply the analytical approaches, "we are more likely to find increasing flexibility, fluidity and tentativeness in our understanding of generic integrity".[42] Indeed, when we analyze these six excerpts from the single perspective of their semantic content, we find that the texts are informative in varying degrees. When

---

[41] On the conventions of writing within professions, see (Bhatia 2004, p. 136).

[42] See (Schaeffer 1989; Bhatia 2004, p. 181).

Table 3.9 Communicational properties of these excerpts

Titles/dates	The Stepping-Stone ...1751	The Essentials...1801	The Guide... 1838	The Short Guide... 1865	What the ignorant 1881	Tried and Effective Recipes 1936
Author's place, status, and target audience	Guangdong Highest degree Officialdom & physician linked to the highest elites of his time	Guangdong No degree Physician Not part of officialdom Patronage from local elites	Guangdong No degree Physician Not part of officialdom No relation with local elites	Guangdong Physician Lowest degree Patronage of local elites Not part of officialdom	Guangxi 2nd degree Officialdom & physician Patronage from local elites	Guangxi Anonymous
	Speaking to students (2173 characters)	Speaking to students (759 characters)	Speaking to students (301 characters)	Speaking to families (330 characters)	Speaking to families (845 characters)	Speaking to population without physicians (138 characters)
Directive/assertive	=	-	=	=	-	+
Personalized texts	+	+	+	+	-	-
Perlocutary force (citations, discourse markers)	++	+	+	+	-	-

we consider them from the point of view of their encoding, we distinguish ver-
bose *versus* concise texts, texts that invite action first *versus* texts that first invite
comprehension, texts of differing levels of readability, texts to be learned by heart
versus others to be consulted and meditated upon. Finally, if we look at the texts
from their communicational properties, we distinguish texts that aim at making the
reader follow a series of instructions without explanations *versus* texts that have
the same purpose but by providing the reader with explanations and predictions;
we distinguish highly personalized texts *versus* texts where individual opinions or
judgment are hidden for the benefit of a certain objectivity; texts which, in order to
convince the readers, interact with them *versus* texts that do not invite interaction.
Some texts seem to be closer to others on the semantic or syntactic or emotional
levels. But as soon as we look at these texts from all three perspectives, it becomes
quite difficult to establish a particular typology that would distinguish one genre
from another. The identification and the delimitation of genres, in whatever field,
thus raise difficult problems.[43]

If firm conclusions cannot be drawn yet, this attempt to make a systematic comparison has led to the establishment of an analysis grid that stresses firstly that
texts can be characterized by far more features than simply being "theoretical",
"practical", "popular" or "scholarly". Of course, as other historians have already
underlined, we can distinguish texts for "doing" from texts for "thinking". But what
this multi-approach analysis shows is that this distinction does not only rely on
the presence or lack of theoretical elements but rather on a wider set of criteria:
the frequency of directives; the order of presentation of the argument that notably
induces very different reading modalities. What this attempt to describe texts from
many perspectives stresses, next, and already outlined by linguists working in the
field of literature, is the fact that when an author claims, in the paratext (notably in
its title and/or preface), generic determination for his text, it does not necessarily
imply that the rules he follows are reducible to the rules that are often associated
with the generic name chosen.[44] Let us take as an example the three texts that claim
to be pedagogical books: *The Stepping-Stone, The Essentials* and *The Guide*. It is
true that these three texts share the highest number of semantic features—clinical
signs, prescriptions, explanations of causes and the development of the disease and
citations. However, if we look at them from the other points of view developed
here, we see significant variations: some texts are in prose while others are in verse;
some try to be concise and handy while others do not; some ground their authority
in individualized opinions while others ground their authority in a neutral objectivity. Note that *What Someone Ignorant,* announced as a book on self-medication is,
to some extents—verse and rhythm constraints, order of information presentation
(explaining before acting), clear pagination—closer to some textbooks than to the
Short Guide which was also announced as a book on self-medication.

---

[43] (Genette et al. 1986, p. 163).

[44] (Schaeffer 1989, p. 128).

   This sample of six excerpts is of course very small. Applying this analysis grid to a larger number of medical texts produced in late imperial China will perhaps help to better highlight the different ways of writing medical texts at that time in China and perhaps distinguish more clearly different types of medical communities.

3.6   Translations

In order to bring out the discourse characteristics in each text—frequency of textual connectors and of discourse markers, illocutionary force markers, concise or expansive writing styles, etc—I chose to provide a very close translation to the original Chinese. This appeared to be a very difficult challenge which furthermore did not always produce particularly nice results, in terms of readability and style. When I found it really necessary, I have added connectors to the translation, but these connectors are not marked with boxes. I capitalized the names of the five phases (Wood, Fire, etc…) but used small size letters for terms such as "cold", "warm", "heat" even if one must bear in mind that these terms, beside corresponding to feelings and to external conditions, are also used as general terms to classify the phenomena and the elements of the universe.
Legend:

- Small characters: comments or indications inserted into the main text in smaller characters by the authors
- Boxes: textual connectors and discourse markers
- Abbreviations for the different types of illocutionary act:

DA	Descriptive Assertive
CA	Causal Assertive
PA	Predictive Assertive
AoD	Assertive of Definition
CD	Conditional Directive
DD	Direct Directive
ID	Indirect Directive
ImD	Implicit Directive
CM	"Citative" Metadiscursive
EM	"Expositive" Metadiscursive

1st text: *The Stepping-Stone for Medicine (Yi bian),* by He Mengyao, 1751

Sudden Disorder	AoD
Huo signifies that feet and hands are moved by an agitation which is sudden, it is sure, *luan* signifies that inside the evil *xie* and the correct *zheng* are in disorder, it is sure.	
Its symptoms are: the epigastrium and the abdomen are painful, there is either vomiting or diarrhea, or vomiting and diarrhea at the same time, in serious cases then feet have muscle cramps, (when) muscle cramps are serious then from the feet they reach the abdomen, or all the muscles/sinews of the body are spasmodic, the tongue retracts enrolled and death occurs.	AoD
If there is no vomiting nor diarrhea, but only muscle cramps and abdominal pain, we call it Dry Sudden Disorder, and this is exactly what we commonly call the Disease of the Blocked Intestines, it's sure.	AoD
The disease's origin is that something evil *xieqi* has collected and obstructs the middle burner, isolating the upper and the lower (parts of the body), the correct qi no longer circulates.	CA
The *yang* qi of the upper (part of the body) cannot flow downward nor communicate with the *yin*, therefore it causes an obstruction in the upper body, this is why there is vomiting.	CA
The *yin* qi of the lower (part of the body) cannot flow upwards nor communicate with the *yang*, therefore it causes downward pressureoppression in the lower part this is why there is diarrhea, it is sure.	CA
It is what we call the separation of *yang* in the upper (part of the body) and *yin* in the lower (part of the body), it's sure, and also to have the *yin* in the upper (part of the body) and the *yang* in the lower (part of the body) or also that *yin* and *yang* are irregular, that the pure and impure, the correct and the evil mix together, that cold and hot fight against each other, being opposed to each other without communicating, as a result the circulation of blood, qi, camp qi, and defensive qi is disturbed, here it goes with the current, there it is against the current, they fight altogether in the middle (part of the body), that's it.	AoD
Up, vomiting, down, diarrhea, then the evil located in the middle burner succeeds in dispersing and disappears, some recover even without drugs.	PA
If the evil qi is extremely hot, vomiting and diarrhea go so far as to exhaust the regular qi, the evil condenses and as a result doesn't move, that's it.	PA
The origin of the cramps lies in the condensation of evil Fire in the muscles/sinews, caused by the long-term exhaustion of blood and bodily fluids, and thus it is not something sudden, it is sure.	CA
If because of Sudden Disorder vomiting and diarrhea, you loose the bodily fluids suddenly, then the exhausted blood and fluids are even more exhausted.	PA
But if it is not because of vomiting or diarrhea but because you caught a cold wind from outside, the skin structures become blocked, the hot qi remains enclosed and increases and accordingly the heat of the evil Fire increases even more —it is what happens for the muscle cramps of Dry Sudden Disorder as well as for ordinary muscle cramps- In that case, the contractions are highly painful, and it is not possible to relieve/stop them, that's it.	PA and DA
To judge by its shocking appearance, it belongs to Fire, there is no doubt.	DA
Sometimes people believe that it is because of cold, they state that cold provokes muscle spasms, but don't they know in fact that the muscle spasms are just a simple forced contraction, and that they have nothing in common with the unusual pain of the muscle cramps, how is it possible that they are so mistaken?	DA
Or they suspect that as the muscle cramps are mainly in the feet, it clearly belongs to cold and *yin* (diseases).	DA
Truly, the feet are in the lower part (of the body), they belong to *yin* and cold, if in addition you catch a cold wind, you will have violent muscle spasms.	DA
It doesn't matter whether the Sudden Disorder shows cold or hot symptoms, all have something evil collected in the middle burner, the *yang* qi cannot flow downward, the two feet must have cold, this is why they have muscle cramps.	CA
If it is because of the Fire heat, then all the muscles/sinews should have cramps isn't it, so why do they only affect the feet I ask you?	CA
To say that it is because of cold that only the feet have muscle cramps, is also right, but to say that it is only because of cold, and that there is no evil Fire in it, this however is not possible, for sure!	
If there was no evil Fire, then the *yin* would condense and would remain immobile, it goes against the principle of violent cramps.	PA
Furthermore Sudden Disorder is a disease of the spleen and the stomach, this is why only the feet have muscle cramps, we can't say that it is necessarily because of cold, for sure.	CA
Danxi, to cure the muscle cramps of the whole body that enter into the abdomen, makes a very salty decoction that he puts into a vessel and he warms it gently, and he uses it to irrigate the dryness the blood is hot right?, once cooled it becomes strong and violent, we can also use it to soften hard masses.	DA
In order to cure Sudden Disorder, you always must examine carefully to identify which evil is the cause.	ID
Hejian says that it is caused by excessive heat, a very violent Fire, it's sure.	CM
Whatever the violent diseases or violent deaths, they all belong to Fire, it is sure.	DA

The *Zhunsheng*[45] says that its origin lies in a damp spleen, damp is abundant in the spleen, consequently, there is obstruction and heat production at the same time, that's it. — CM

For Zhang Zihe, it is the combination of the three qi of wind, damp and heat that thus produces the evil; Indeed, the damp-Earth of the spleen is controlled by the wind-Wood of liver the clear qi of the Wood is blocked by dampness, it can't go upward, if it can't go up then dampness increases and does not circulate, and if it receives the Wood it is dominated, naturally it condenses and consequently produces heat that develops suddenly, and in response to this development, the Fire of the heart goes up in smoke, this is why there is vomiting, the dampness of the spleen flows downward, this is why there is diarrhea, it's sure. — CM

Wang Haizang says equally that it is the combination of wind, damp, heat and food that produces the evil. — CM

As for the *Mingli lun*[46] it says that for the most part Sudden Disorder is caused by troubles coming from food and drink, like in Sudden Disorder caused by cold, where an evil qi enters into the middle burner, the qi of the stomach is not at peace, accordingly the *yin* and *yang* eventually become separated. — CM

One must combine the theories of these masters, one cannot lean toward one of them, otherwise this would lead to making errors. — ID

Generally speaking, this disease is the most prevalent between summer and autumn. — DA

Indeed during summer months people eat cold food and drink cold water, their cold and damp qi fight against the very hot qi in the stomach, this is why cold and heat are not balanced, then nothing can't stop the obstructions, and a little time after the disease occurs. — DA

If from the outside you catch a cold wind, or from the inside you are harmed by ingesting food, then there is obstruction, it develops suddenly, it's sure. — PA

The development of the disease always begins with abdominal pain, if the pain goes up and gets close to the heart then vomiting happens first, if the pain goes down and gets close to the navel then diarrhea happens first. — PA

As a result of a wind attack, then there is aversion toward wind and perspiration, as a result of a cold attack then there is aversion toward cold and no perspiration, as a result of a damp attack then the body is heavy, as a result of a heat attack the heart and spirit are troubled. — PA

To treat it, ask which kind of food has been eaten, which of the seven emotions have been activated, and next distinguish whether it is a cold, hot, repletion or depletion (patho-condition). — DD

The one who has shortness of breath, white lips, cold flesh, muscle contractions in the four limbs, aversion toward cold and preference for heat, deep and slow pulse, clear mind, has a cold (patho-condition), for sure. — AoD

Sudden Disorder being a brutal disease has Fire for origin, but Fire is contained by cold, once it develops Fire flows away and cold remains alone, that's it. — CA

Decoction to regulate the middle burner — ImD
See cold attacks — DD

Irregular pulse: decoction for promoting circulation in the vessels and lighten cold in the limbs — ImD
See cold limb syndrome — DD

Unceasing vomiting, original qi dispersed, enormous spread of yang towards the outside, or thirst for cold drinks but incapability of drinking the water asked for, or fever and agitation, desire to remove clothes, you must not make the mistake of believing that it is a hot (patho-condition), — ID

Decoction to regulate the middle burner is suitable; in serious cases, then, decoction with aconite to regulate the middle burner — DD
see again cold attacks — DD

If it is not effective, then decoction for the four cold limbs — ImD
see cold limb syndrome — DD

In addition one must absolutely abstain from ingesting cold things. — ID

For those who have muscle cramps, remove atractylodes macrocephalae from the decoction to regulate the middle burner, add a stem of fresh aconite. Or add 1 qian of iced gelatin cut and browned to the decoction to regulate the middle burner. — DD

Because of cold the blood became solidified, it is like iced gelatin, this is why you must add this to the treatment, and also you use it to moisten the dryness. — ID

Or make a thick decoction with salt and then firmly hold legs and calves so that they do not fold up into the abdomen. — DD

---

[45] Abbreviation for the 證治準繩 (*Zheng zhi zhunsheng, Rules for diagnosis and treatment)*, 1602, by Wang Kentang (1549–1613).

[46] Abbreviation for the 伤寒明理論 (*shanghan ming lilun, Clear discussion on cold damage)* by Cheng Wuji, eleventh–twelfth centuries.

Or burn 27 moxa cones on the *chengshan* point, it is highly effective.   DD
One method: males pull their genital organs with their hands, females press their breasts. [47]   ID
The one who has a hot body, thirst, thick breath, dry mouth, preference for cold, aversion to heat,   AoD
confused heart and mind, deep, quick pulse, it is a hot (patho-condition), for sure.

If furthermore the four limbs are heavy and articulations are painful, it is combined with dampness, for   AoD
sure.

In Sudden Disorder caused by Summer-heat, it is suitable to take a cool decoction of herba elscholtziae,   ID
see Summer-heat attacks. Drink it in one go and very cold as if it was coming from the bottom of a well; the
powder of cinammomun, digitalis and atractylodes macrocephalae is also marvelous.

Those whose dampness is abundant: Decoction to eliminate dampness, see dampness attacks, decoction of   ImD
fructus chebulae.

Those whose Summer-heat and dampness are mixed together: Powder of the two perfumes.   ImD
Those who are very hot and drink a lot: powder of the five drugs containing poria cocos see dampness   ImD
troubles.

Those who have muscle cramps: cold juice of the dry fruits of chaenomeles speciosa, or it is possible   ImD
also to take a juice of herba escholtziae, or grill and grind twenty gardenia stems into powder, make an
infusion with hot water.

For those who are thirsty and agitated, whose fluids, after vomiting and diarrhea, have been exhausted,   ImD
for sure: Decoction to stop the thirst, cold decoction to reduce the spleen, decoction of poria cocos and
alismatis rhizomes, decoction of wheat and asparagus.

After Sudden Disorder, if diarrhea does not stop and the abdomen is painful, if we fear that it is turning   CD
into dysentery, pills of coptis roots are suitable.

After Sudden Disorder, if you see blood in the diarrhea: decoction to stop the blood and decoction of   CD
silicate of aluminum.

Since all these hot and cold patho-conditions are similar and difficult to determinate, you must observe   ID
the pulse.

Because in Sudden Disorder the qi is blocked, the pulse is so deep and hidden that it is imperceptible, or   DA
it is choppy, blocked as if it was stopped.

You must distinguish, thank to its sound, whether it has strength or not.   ID
In fact this is also difficult to differentiate, for sure.   DA
You must not give medicine without due consideration, but begin by determining (the disease). Yin Yang   ID
Water and the *shagua* method (extracting the *sha*) are very effective and suit Dry Sudden Disorder
particularly well.

This method consists in using a bowl in thin porcelain, warmed with hot decoction, a perfumed oil is put   DA
on the borders to make it slippery, without wounding the flesh, you scrap the *yuxue* point on both sides
of the vertebral column, and you extract the evil from depot and palace organs, you also scrap the soles
of the feet and palms of the hands', to extract it from everywhere, when the hot blood succeeds in going
out, red or crimson petechiae appear, red is mild, crimson is serious, black is even more serious.

To be more precise the qi clots and then the blood thickens, the blood thickens and then the qi is even   CA
more blocked, if the blood disperses and the qi circulates, then recovery is immediate, for sure.   and
                                                                                                           PA

For those who are overwhelmed with the seven emotions: Decoction of the seven qi.   ImD
Those who have pain under the ribs, is (because) the Wood destroys the Earth, that's it: Decoction of   ImD
radix bupleuri and chaenomeles speciosa.

Moreover those who are affected by an evil wind from outside: Decoction of the six harmonies, see   ImD
Summer-heat attacks. Double herba agastachis, and boil with storax pills. See diverse attacks.   et DD

Generally speaking, storax pills and agastachis powder to regulate Qi, see wind attacks, are the most   DA
appropriate to calm the spirit and regulate the qi, they must be used, for sure.   and
                                                                                                        ID

---

[47] From the 千金藥方 (*Qianjin yaofang*, Essential Prescriptions Worth a Thousand Pieces of Gold),
by Sun Simiao (6th–7th) Chap. 20, recipe 14.

Sudden Disorder is always caused by something evil, as soon as something evil is present it immediately   DA
causes repletion; even if people are depleted, one must not even give them supplements, but only after   an ID
vomiting and diarrhea have occurred, one must observe carefully.

If everything has been expulsed, the person is overwhelmed with tiredness, his pulse is thin and weak, in   PA
that case the evil thing has already left and it is a real depletion, for sure.

If vomiting and diarrhea are not finished, the pulse is successively choppy, blocking and strong, the   PA
person is agitated and not at peace, in that case the evil is not yet exhausted, for sure, one must make the   and
distinction.   ID

Those who don't stop vomiting and have diarrhea, who get dizzy and have fits of dizziness, whose limbs   CD
are cold with muscle cramps, who seem incurable: Decoction of fructus evodiae.

Those who continuously perspire until having cold limbs: Decoction against cold in the four limbs, see   ImD
cold limbs syndrome

Those who, once vomiting and diarrhea have stopped, urinate and perspire normally and who, suddenly,   ID
become hot inside and outside, must also be warmed.

Those who, once vomiting and diarrhea have stopped, don't urinate or don't have motions, feel pain in   DD
the stomach, have an obstructing mass: Four noble drugs, see qi, and add 1 liang of rherum officinalis.

Once vomiting and diarrhea are finished, the diaphragm rises, the obstructing mass is close to breaking   CD
down: Decoction to regulate the middle burner, see cold attacks, add citrus frusca, and poria cocos.

When vomiting and diarrhea have expulsed everything but vomiting and diarrhea still continue, the   CD
abdomen is still painful, it is suitable to take the decoction of one ingredient with leaves of Pisum   And
sativum, it is particularly recommended for Dry Sudden Disorder. [48]   ID

Luo Qianfu [49] cured a man who was harmed by rotten meat and milk, he had a Sudden Disorder with   DA
vomiting and diarrhea, his pulse was deep and quick heat hadn't yet left, surely, wasn't strong at the
palpation, the things that had harmed him had already left. He took half a bowl of freshly drawn water that he
mixed with cinamommum powder, atractylodes macrocephalae rhizome, he made him take it slowly.

He was a little better, moreover to strengthen the yin, he (Luo Qianfu) dug the soil two feet deep, he   DA
introduced freshly drawn water that he agitated, clarified, we call it Earth Juice, he drank it again and
then he was cured.

To strengthen the yin Earth, we must emphasize yin surely; [50] in case of extreme dryness and Summer-   ID
heat, there is no other solution. [51]

He also cured a man of 80 years old, who had been affected by a Summer-heat Sudden Disorder, with   DA
vomiting, diarrhea, dizziness, Fire hot head, cold feet.

He used a cold decoction of cinamomum, digitalis and honeydew to eliminate heat and strengthen qi,   DA
lower the superficial Fire, bring peace to the mind, he added poria cocos to separate the yin and the yang,
he mixed this with fresh water and made him drink it and he recovered.

Moreover, he used a decoction with ginseng and atractylodes macrocephalae to regulate the middle   DA
burner and he recovered. [52]

In post-partum Sudden Disorder, there is no method other than those mentioned above however one   ID
must be very careful with the baby and avoid depletion.

In Dry Sudden Disorder, one can neither vomit nor have diarrhea, the evil consequently gathers in the   CD
middle burner, if one uses warming and heating (drugs) death occurs immediately.

Brown salt and put it in freshly drawn water that you heat, drink a lot of it to provoke vomiting;   DD

If there is no vomiting, keep on drinking it; if vomiting occurs keep on drinking it, and stop after   CD
vomiting three times, Generally speaking stop immediately when the evil is exhausted, don't be obstinate!

This method is very effective, be it for those who want to vomit but who can't or for those who can't   DA
vomit all they should, it is suitable for both cases.

It must be very salty and then it is marvelous.   ID

---

[48] Same sentence in the 秘傳證治要訣及類方 (*Mi zhuan zheng zhi yao jue ji lei fang, Secrets for diagnosis and curing with recipes*) by Dai Sigong (公元1368-1644年), Chap. 1.

[49] A physician who lived during the thirteenth and fourteenth centuries.

[50] This passage is slightly different in another edition.

[51] Inspired by (copied from?) the名醫類案 (*Mingyi lei'an, Classified Case Records of Celebrated Physicians*), by Jiang Guan (1503-1565), Chap. 2, Shu暑.

[52] idem.

After vomiting, if the epigastrium is painful, that you want to go to the toilets, that it does not work even   ImD
in forcing, agastachis powder to regulate the qi see wind attacks, add 1 qian of fructus aurantii, when fresh it is
even quicker.

If a lot of things flow down: fudan pill[53]. See Summer-heat attacks, if there is diarrhea then it is not possible to use this   CD
pill

If it is not effective, you must use shenbao pills see harm by food and drink, but this pill must reach the large   CD
intestine to work.

If it stagnates in the upper part, then it is not suitable, you must take laifu pills crushed into fine powder,   CD
mix with a decoction, take 100 small yangzhengdan grains see qi, and, hopefully, it can lead the drugs
taken earlier to flow down.[54]

Dai Fu'an's method[55]: first take a thick salty decoction to provoke vomiting, then mix storax and laifu   DD
pills, and successively introduce agastachis powder to regulate the qi to which you add radix
aucklandiae, fructus aurantii seu ponciri, decoction of magnolia officinalis, powder to save life,
decoction of fructus malvae verticillata.

Those who have red blotches that develop quickly between the nape of the neck and the heart, take green   CD
artemisia vulgaris mixed with water, drink it and you will recover.

Or needle the *weizhong* point, and at the same time you bleed the top of the ten fingers, it is also   ID
marvelous.

When Fire is extreme, drugs must be antagonistic.   ID

An ancient recipe: make a salty decoction boil and introduce the urine of a child, take it, it will lower   DD
down (Fire), and allow *yin* to be reached, that's sure, and simultaneously it allows blood to circulate.

Whatever (kind of) Sudden Disorder you mustn't give puree.   ID

Indeed, the obstruction of the evil has not yet been transformed, once the cereal is eaten, the obstruction   CA
will increase, leading often to death.

Abstain also from alcohol, ginger decoctions, garlic, black plum puree, hot beverages, and all astringent   DD
warming drugs.

---

[53] Pill promoted in the太平惠民和劑局 (*Taiping Huimin he ji ju fang, Formulas of the Bureau of People's Welfare Pharmacy*), Chap. 5. 引杜先生方.

[54] Entirely similar in 秘傳證治要訣及類方 (*Mizhuan zhengzhi yaojue ji leifang*) by Dai Sigong (1368–1644), chap.1, itself copied from the 證治準繩 (*Zheng zhi zhunsheng, Rules for diagnosis and treatment*)of Wang Kentang.

[55] A physician whose name appears in the張氏醫通 (*Zhangshi yitong, Zhang's Medical Compendium*) (1695) by Zhang Lu.

2nd text: *The Essentials of Medicine, (Yixue jingyao),* by Huang Yan, (ca. 1751–1830), 1800, edition of 1867 (Chap. 2, pp.15–16)

Sudden Disorder	EM
Nai'an says: up, one wants to vomit but nothing comes out. Down, one wants to evacuate diarrhea but it doesn't work. Agitation, depression, and confusion happen simultaneously, they are precisely the symptoms of the Blocked Intestines.	AoD
Turn quickly to the *weizhong* point, refer to the foot map in the second chapter of convulsive symptoms, and needle until blood appears.	DD
Or needle the ten fingers altogether.	DD
Thanks to this intervention, blood circulates and the evil is dispersed.	CA
I have a pill that can stop choleric diseases.	DA
The Classic says: a troubled qi from the foot-major-yin conduit flows upward inappropriately, and then it is Sudden Disorder.[56]	CM
It says also: in the bursting out of the suppressed Earth energy[57], there is vomiting and Sudden Disorder.	CM
Therefore the disease of Sudden Disorder happens undoubtedly because the stomach and the spleen have received something evil, that's all.	CA
This evil is not one and only one.	DA
First there is the evil carried by food or drink.	DA
Second the evil comes from cold wind.	DA
Third the evil comes from an excess of *yin* from dampness.	DA
Each harms in the same way.	DA
That is food is obstructed inside, the evil is blocked from outside.	DA
The evil and the correct are mixed together.	DA
The pure and the impure are mixed together.	DA
Either there is agitation and distension or the heart/spirit and the abdomen are disturbed and painful. Up, vomiting, down, diarrhea, both can happen at the same time, we call it Damp Sudden Disorder.	AoD
Even if this patho-condition seems frightening, one must not worry however, nor make it stop.	ID
Wait until all that had gathered goes out entirely.	DD
And then vomiting and diarrhea will stop on their own.	PA
Indeed the evil must be vomited and evacuated by diarrhea so that it is eliminated, it's sure.	CA
In serious cases, one can take agastachis powder once to regulate the qi.	CD
If someone has *yin* and *yang* that get separated, has qi that does not circulate, who, up, wants to vomit but has nothing that is expulsed, down, wants to have a motion but has nothing that is expulsed, who is restless, and has such a pain that he wants to die, we call it Dry Sudden Disorder, it is precisely what we commonly call the Choleric Disease of the Blocked Intestines, it's sure.	AoD
This is in fact the most serious disease.	DA
If we don't treat it quickly, it definitely leads quickly to death.	PA
One must quickly press some chives on the back of the ten fingers, needle and be sure to bleed them.	ID
When needling, one must press very strongly from the top of the arm to the fingers so that the poisoned blood collects in the fingers and then needle them.	ID
If he does not recover, quickly put some cold water on the *weizhong* point in the back of the knee, left for the boys, right for the girls, and massage violently with the hand palm.	CD
Massage until the sinews become purple.	DD
With needles pierce the sinews so it bleeds.	DD
It is prodigiously effective.	DA
To treat acute choleric diseases, my family possesses a pill which was handed down secretly, it is effective for both Dry and Damp Sudden Disorder, and today I don't dare keep it secret.	DA
I transmit it to you my pupils, if you can have it to save people, it will be much better than spending a lot of time searching in thousands and thousands of pages, it is sure.	DA
There are a lot of (types of) choleric diseases, some people because they eat raw vegetables or fatty things, feel in a split second exhaustion in their feet and hands, their abdomen oppressed, a slight pain; some, in a split second, have their face turn from white to black, their hands and feet become cold and they fall unconscious; some have their abdomen disturbed and painful, and at the same time feel cold then hot; the secret familial pill against choleric diseases is effective for all these.	DA

---

[56] Excerpted from the *Lingshu, Huangdi neijing*, Chap. 10.
[57] Translation of Wu Liansheng, Wu Qi, p. 416.

Those who are basically puny and who frequently develop choleric diseases, there is also nothing better   CD
than the schefflera venulosa and radix vicitis quinatae, boil it until you have a thick soup, before taking
the small intestine of a pig, make it brown with vinegar and introduce it into the soup and boil the
intestine, take this 4 to 5 times, it can strike at the root of the evil.

Agastachis powder to regulate qi   DA
It cures diseases contracted by external cold wind, the inner harm caused by food and drink, Sudden
Disorder vomiting and diarrhea and also all the troubles caused by the irregular qi of the miasmas from
mountain fog, you can also use it modified.

Agastachis herba to regulate the Qi in the epigastrium   DA
Fresh perilla frutescens, mild platycodon grandiflorum, orange peel, poria cocos, atractylodis
macrocephalae rhizome, magnolia officinalis, in the same proportions. Pinellia ternata rhizome, massa
fermentata medicinalis, radix angelicae dahuricae, ginger (*Zingiber officinale*) rhizome, jujubes (*Ziziphus
zizyphus*), can also cure the troubles caused by external, internal (factors) and mountain fog miasmas.

For the muscle cramps of Sudden Disorder, add chaenomeles speciosa.   CD
For those who have been seriously affected by food, add a drug that promotes digestion.   CD

Secret family pill against choleric diseases.   DA
It cures the Sudden Disorder vomiting and diarrhea, the different heart and abdomen pains and all the
clinical signs of choleric diseases, and also it has a surprising effectiveness for the troubles contracted by
an external evil cold, for the harm caused internally by raw vegetables, for oppression of the heart and
abdomen, for untimely vomiting and diarrhea, cold qi, blocked qi and abdominal pain.

Alpinia chinensis rosc 4 liang, notopterygii rhizome 1 liang 5 qian, magnolia officinalis 1 liang 5 qian, orange   ImD
peel 1 liang 5 qian, atractylodes, saposhnikovia divaricata 1 liang 5 qian, critrus fusca 2 liang, herba agastachis 2
liang, yazao roast it slightly and remove the husk 5 qian, aasarum sieboldi 1 liang, shenggan 1 liang bujingzi and fruits
of cinnamomum camphora, for these two ingredients, use a slightly salt wine cook them 3 or 5 times, these two ingredients
are the ruler, take four, five or six liang

Mix until you have a puree and make pills of the size of sterculia seeds, use a talc coating or you can also   DD
use a preparation made of isatis and cinnabar. Adults take some 3 to 5 qian with cold water. For children,
it depends on their weight. If the disease explodes violently, quickly crush some pills, mix them with
water and make (the patient) drink, all will survive.

3rd text: *The Guide for Summer-heat Diseases, (Shuzheng zhinan),* by Wang Xueyuan, 1843

Sudden Disorder	DA
The disease of Sudden Disorder usually happens during the summer.	
The Summer-heat and damp qi prevails, it's sure.	DA
We get it from wind, cold, Summer-heat, food and water; all these evil mix together and cause the disease.	DA
The disorder is located in the intestines and in the stomach.	DA
The pure and the impure are mixed together.	DA
This is why the epigastrium and the abdomen are very painful and there is vomiting and diarrhea, it's sure.	CA
Use agastachis powder to regulate the qi.	DD
If there are muscle cramps, add chaenomeles speciosa and fructus evodiae	CD
If it is because of Summer-heat, then vomiting is frequent.	PA
Add elscholtzia powder usually called the decoction of two perfumes.	DD
If it is because of dampness, diarrhea is abundant.	PA
Add atractylodes.	DD
Those who at the same time have a blocked digestion, powder to harmonize the stomach.	CD
Those whose Summer-heat is acute, who pant, whose front's teeth are dry, whose urine is blocked or ☐ yellow and short, Changsha's powder to tonify the essential energy. [58]	CD
Or five drug powder containing poria cocos. Add gypsum fibrosum, talc and calcite.	DD
It is precisely Hejian's cinnamom, poria cocos and liquorice hydrolat.	DA
If someone is very cold, with cold extremities, white slippery and moist tongue, thin pulse, make a decoction to regulate the middle burner with prepared aconite root and aconite.	CD
Dry Sudden Disorder - that we usually call Disease of the Blocked Intestines or Distension with Cyanosis <small>all the body is deep purple</small>	DA
This disease's origin is apoplexy.	CA
The evil and fetid things can't go up or down.	DA
Up, one can't vomit, down, one can't evacuate diarrhea.	DA
Abdomen is extremely painful.	DA
Fingers become dark black.	DA
The symptoms are those of a *yin* disease.	DA
They are the most serious signs.	DA
If one uses warming drugs death will occur immediately.	PA
Eating a thick puree provokes death as well.	PA
If the tongue has a cramp and retracts, if testicles or women's genital organs retract and enter into the abdomen, then it is difficult to cure, it is sure.	PA
Quickly use browned salt with Yin Yang Water.	DD
Warm it and drink abundantly.	DD
It is the most effective to provoke vomiting.	DA
Or it is also possible to make a decoction of lemon.	ID
If blotches appear between the nape of the neck and the heart, cool immediately.	CD
Drink the juice of green artemisia vulgaris and water and recovery will occur.	DD
There is also the case where the entire body is deep purple, it is what we usually call the Distension with Cyanosis.	AoD
Again use browned salt and Yin Yang Water to provoke vomiting.	DD
For the preparation of moxas and recipes, look at the classified recipes, hereafter.	DD

---

[58] Entirely copied from a commentary of the 醫宗金鑒 (*Yizong jinjian, Golden Mirror of Medicine)*, compiled by Wu Qian in 1742

4th text: *The Short Guide of Medicine of Pingqin Room (Pingqin shuwu yilue),* by Pan Mingxiong, 1865

Appended discussion on the disease of Sudden Disorder Sudden Disorder is a disease where the abdomen is painful.	AoD
It is also the disease which responds to agastachis powder, among others, used by the merchants.	AoD
Wang Kentang says that against Sudden Disorder without vomiting and diarrhea, or when the abdomen is swollen like a drum, you must not take other drugs, at a pinch you can take the powder to tonify the Essential Energy, you prepare it in a decoction that you cool and drink slowly and continuously.	CM
If you don't stop taking this drug, it can triumph over the evil qi, bring down food and mucus, harmonize the stomach and balance the middle burner.	PA
However if you hear noises inside the abdomen, then it is a good thing.	PA
If nothing goes out downward, then there is vomiting, if there is not vomiting, then things go out downward.	PA
Consequently this is a wonderful drug against Sudden Disorder, it is sure.	DA
You must absolutely abstain from ginger decoction, thick purees, and black plum puree.	ID
For my part, I respect this method.	DA
When I (or you) face Sudden Disorder where either vomiting and diarrhea have not yet occurred, or where vomiting and diarrhea have already happened, I use (use) the leaves of herba agastachis, fermented with large tea leaves, 3 qian of each all can harmonize the middle burner alismatis rhizome, caulis akebiae, 1 qian each all can bring down the impure radix bupleuri, angelica 7 or 8 fen each all can bring up the pure I boil (boil) and infuse with the powder to invigorate essential energy for 3 or 4 qian.	ImD
I wait (or, wait) until it is cold and drink.	DD
If the abdomen is still painful, I still keep following this method.	CD
If there is dampness, you must add some atractylodes rhizome.	CD
Or use herba agastachis powder as an aid.	DD
This method is always effective.[59]	DA
Moreover the Ancients used to say that for whatever violent disease, we should not discuss the pulse but observe its clinical signs[60].	CM
This doctrine is particularly true for Sudden Disorder, as well as for all acute and painful diseases, where the pulse is quick and deep.	DA
We must always remember this statement.	ID
Furthermore they used to say: whatever diseases that happen suddenly, all are diseases linked to the liver conduit.	CM
This is because among the Five Phases, there is no quicker phase than Blowing Fire (or wind and Fire), and the liver is the Depot of Blowing Fire (wind and Fire), that's it[61]	CA
And this is not only true for abdominal pain.	DA
In fact, the liver rules sinews and also pain.	DA
Consequently it is not possible to say that diseases with pain as well have no link with the liver, it is sure.	ID
As Sudden Disorder is a disease that develops quickly, this is why we must also add a little bupleuri in order to disperse all that has been collected in the liver.	ID
Or use mandarin peel as an assistant in order to inhibit the hyperactivity of the liver.	DD

---

[59] In the passage that follows "For my part, I respect this method", there is an ambiguity, due to the fact that Chinese verbs do not conjugate and one does not need personal pronouns in written Chinese either. Therefore, this passage can be understood as a description of what the author does when he faces cases of Sudden Disorder, but also as an injunction aimed at the reader. In both cases, however, the illocutionary target is directive: either the author wants the reader to do as he himself does (I do that, implicitly, do as I do), or the author gives him a series of instructions.
[60] According to the belief that in acute forms of any disease, and unlike the other clinical signs, the states of the pulse become confused and cannot provide reliable information about the disease.
[61] This sentence sounds strange as the liver is associated with Wood and by extension with wind.

5th text: *What Someone Ignorant in Medicine Should Know (Bu zhi yi biyao),* by Liang Lianfu, 1881

Sudden Disorder (With) this disease, up, there is vomiting, down, there is diarrhea.	DA
Some receive cold wind from outside.	DA
The cold qi enters into the Depots and then disease occurs.	CA
Some do not take care of what they eat.	DA
Raw food harms the insides and then disease occurs.	CA
Some because of misfortune starve.	DA
The qi in their stomach is harmed and then disease occurs.	CA
Some have once eaten too much.	DA
Food cannot be digested and then disease occurs.	CA
In some places, the qi is disturbed. cold and warm are out of season.	DA
One finds it and then disease occurs.	CA
Some droughts alternate with lashing rain. The pure and the impure mix together.	DA
It affects you and then disease occurs.	CA
Briefly speaking all are diseases where the spleen is harmed by a damp cold.	DA
Furthermore some have Sudden Disorder with cramps.	DA
The sinews from the feet to the abdomen contract with cramps and acute pain.	DA
In acute cases, the genital organs are reduced into small balls.	DA
Pain presses the hypogastrium.	DA
It is the most serious sign.	DA
It is the disease where the qi and the blood are harmed together in the foot-yang-brilliance *Zuyangming* and foot-ceasing-*yin Zujueyin* conduits, it is sure.	DA
In addition there is Dry Sudden Disorder.	DA
(With) this disease, up, you want to vomit but you can't.	DA
Down, you want to evacuate diarrhea but you can't.	DA
The thorax and abdomen are both swollen and are acutely painful.	DA
We call it commonly the Disease of the Blocked Intestines.	AoD
It must be that inside there is a stagnation of food and drink.	CA
And that a evil cold from outside is blocked.	CA
Yin and *yang* are opposed,	DA
The qi no longer circulates.	DA
It is very dangerous.	DA
First you must take a salty soup to provoke vomiting and evacuate the obstructing things, in order to let the pure qi circulate.	ID
Then, you must take drugs that warm the middle burner and disperse the obstructions.	ID
There is never a mistake.	DA
After the vomiting and diarrhea of the Sudden Disorder, the qi of the stomach is not yet pure.	DA
It is recommended to eat and drink food and beverage little by little and in small quantities.	ID
However, it is also not recommended to give thick purees to eat too quickly.	ID
If the evil obstructs and again causes blockages, then there will be much damage.	PA
Classified recipes (against) Sudden Disorder	DA
Decoction of mandarin peel and of two more ingredients harmonizing cures Sudden Disorder vomiting or the associated diarrhea	DA
Mandarin peel, 1 qian Herba agastachis Poria cocos each 2 qian and 5 fen Pinelliae rhizome cut 2 qian Roasted herbs 1 qian Fresh ginger 3 slices	ImD
Decoction of agastachis to regulate the qi harmonizing Cures those who have been affected by a seasonally abnormal qi, who have Sudden Disorder vomiting and diarrhea, who have headache and feel hot and cold alternatively.	DA
Herba agastachis Perilla frutescens Radix angelicae dahuricae Pericarpium arecae washed with alcohol Platycodon grandiflorum 1 qian each Magnolia officinalis cut with juice of ginger Atractylodes rhizome dried white poria cocos mandarin peel pinelliae rhizome cut roasted herbs 8 fen each	ImD
Add three pieces of fresh ginger, one jujube and boil.	DD

Simple and practical decoction of one ingredient harmonizing Cures Sudden Disorder	DA
Bark of camphor tree 1 liang.	ImD
Boil with water.	DD
Moreover when vomiting and diarrhea don't stop, use a bunch of artemisia vulgaris boil and take it thickened.	CD
Decoction of the six harmonies harmonizing It cures the harm to the spleen caused by excessive heat and dampness in summer and autumn. If you eat too much cold (food), too much melon, until having blocked cold in the stomach. The food stagnates and is not digested. Progressively, it becomes the Sudden Disorder.	DA
Angelica root fried with rice its bark removed Pinellia ternata rhizome cut Fructus amomi crushed 1 qian dolichos crushed and browned Herba agastachis red poria cocos Chaenomeles speciosa 2 qian each roasted herbs 1 qian	ImD
Add three slices of ginger, one jujube and boil.	DD
Powder for pacifying the stomach harmonizing It cures disharmonies of the spleen and stomach, abdominal distension, vomiting, Sudden Disorder, and the like.	DA
Herba agastachis 1 qian 5 fen atractylodes washed in rice water peel of magnolia officinalis browned in ginger juice orange peel 1 qian each	ImD
Add two slices of ginger and boil. If vomiting is frequent, add 1 qian and 5 fen of Pinellia ternata rhizome. Painful abdomen: add 7 Fructi amomi and 1 qian of radix aucklandiae. cold: add 5 to 7 fen or 1 qian of dried ginger.	DD
Decoction of lengxiang very warming prescription It cures Sudden Disorder progressively developed because of mutual attacks on yin and yang, linked to the desire to eat raw and cold things, in damp summer and autumn. Abdomen is fiercely painful, sides are big and full, physical and psychological agitation leading to excessive eating.	DA
Dried grains of Alpinia katsumadai washed, roasted and crushed aconite cut Alpinia chinensis rosc 1 qian each Flos caryophylli 7 fen Radix symplocoris paniculatae roasted herbs 1 qian each	ImD
Once boiled wait it becomes cold. It is when vomiting happens that it must be taken.	DD
Decoction of aconite and of late rice very warming prescription It cures Sudden Disorder with the four extremities cold, with frequent nausea and little vomiting.	DA
Dried ginger 1 qian slightly browned aconite prepared Pinellia ternata rhizome cut in pieces of 1 qian 5 fen roasted herbs 1 qian	ImD
Add a pinch of late rice, two big jujubes and boil.	DD
Classified Recipes against Sudden Disorder with cramps	DA
Decoction of Chaenomeles speciosa warm It cures unstoppable vomiting and diarrhea, troubled with cramps	DA
Fructus evodiae 1 qian soaked fennel 1 qian browned Chaenomeles speciosa 3 qian roasted herbs 1 qian 5 fen	ImD
Add three slices of fresh ginger, ten leaves of perilla and boil.	DD
Sishun fuzi tang very warming prescription It cures Sudden Disorder with muscle cramps, vomiting and diarrhea, cold extremities, aphasia due to exhausted qi, cold perspiration.	DA
Aconite a piece of 1 qian 5 fen dried ginger browned roasted herbs 1 qian each	ImD
Classified recipes against Dry Sudden Disorder	DA
Alum powder slightly cooling It cures (the cases) where one wants to vomit but can't, one wants to evacuate diarrhea but can't, with associated abdominal pain, and that we commonly call the Disease of the Blocked Intestines.	DA
Alum 1 qian	ImD
Crush into powder, mix it with Yin Yang Water and take. Mix the same quantity of cold water and boiling water, it is precisely the Yin Yang Water.	DD

6th text: *Tried and Effective Recipes (jiyan liangfang),* anonymous, 1936 and 1946

Recipes for curing Sudden Disorder	AoD
Up, vomiting, down, diarrhea, we call it Sudden Disorder.	
With muscle cramps, we call it the Sudden Disorder with muscle cramps.	AoD
Vomiting but no diarrhea, we call it Dry Sudden Disorder.	AoD
The disease exists from winter to summer.	DA
The recipes use camphor tree, Cunninghamia lanceolata, dried leaves of maple, an old rope, the bristles of a broom, the soil from a stove, perilla leaves, in equal quantities, add a pinch of fresh salt, brown and roast some nails and some cotton threads, boil together until you obtain a thick soup, drink it like tea, it is effective immediately.	ID
Furthermore you carefully observe the sick person's back, if there are black spots, you pierce with a needle and bleed, he will recover immediately.	CD
There is also powder of alum combined with Yin Yang Water. You take 1 to 2 qian, it is miraculously effective.	ID
You can also take a pinch of salt, put it on the knife blade that you heat until it is red, you infuse it in Yin Yang Water, it is also very effective.	ID
For this disease, refrain from eating rice; refrain from eating rice soup as well; a fortiori refrain from ginger.	ID

References

Anonymous. 1936. 經驗良方 (Jingyan liang fang, Tried and Effective Recipes). In 融縣志 (Rong xianzhi, Gazetteer of Rong). Huang Zhixun (Compilator).

Austin, John L. 1962. *How to do things with words*. Oxford: Oxford University Press.

Bhatia, Vijay K. 2004. *Worlds of written discourse. A genre-based view*. London: Continuum International Publishing Group Ltd.

Blanks, David. 2005. *Les marqueurs linguistiques de la présence de l'auteur*. Paris: L'Harmattan.

Bretelle-Establet, Florence. 2002. *La santé en Chine du Sud, (1898–1928)*. Paris: CNRS, Asie Orientale.

Bretelle-Establet, Florence. 2009. Chinese biographies of experts in medicine: What uses can we make of them? *East Asian Science, Technology, and Society* 3 (4): 421–451.

Bretelle-Establet, Florence. 2011. The construction of the medical writer's authority and legitimacy in late imperial China through authorial and allographic prefaces. *NTM Zeitschrift für Geschichte der Wissenschaften, Technik und Medizin* 19:349–390.

Brokaw, Cynthia J. 1996. Commercial publishing in late imperial China: The Zou and the Ma family business of Sibao, Fujian. *Late Imperial China* 17 (1): 49–92.

Brokaw, Cynthia J., and Kai-Wing Chow. 2005. *Printing and book culture in late imperial China*. Berkeley: University of California Press.

Chao, Yuanling. 2009. *Medicine and society in late imperial China. A study of the physicians in Suzhou, 1600–1850. [= Asian thought and culture, 61]*. New York: Peter Lang.

Chen, Lanbin 陳蘭彬. 1890. 高州府志 (*Gaozhou fuzhi, Gazetteer of Gaozhou*). Compilator. Taibei: Taiwan xuesheng shuju.

Chia, Lucille. 1996. The development of the Jianyang book trade, Song-Yuan. *Late Imperial China* 17 (1): 10–48.

Chia, Lucille. 2002. *Printing for profit: The commercial publishers of Jianyang, Fujian, 11th-17th century*. Harvard-Yenching Institute Monograph series. Cambridge: Harvard University Asian Center.

Chiu, Yu Tseng. 2004. Speech prosody: Issues, approaches and implications. In *From traditional phonology to modern speech processing*, Eds. Gunnar Fant, Hiroya Fujisaki, Jianfen Cao, and Yu Xi. Beijing: Foreign Language Teaching and Research Press.

Elman, Benjamin. 2000. *A cultural history of civil examinations in late imperial China*. Berkeley: University of California Press.

Elman, Benjamin, and Alexander Woodside. 1994. *Education and society in late imperial China.* Berkeley: University of California Press.

Furth, Charlotte. 1987. Concepts of pregnancy, childbirth and infancy in Ch'ing dynasty China. *Journal of Asian Studies* 46 (1): 8–35.

Furth, Charlotte. 1999. *A flourishing Yin: Gender in China's medical history, 960–1665.* Berkeley: University of California Press.

Genette, Gérard, et al. 1986. *Théorie des genres.* Paris: Seuil.

Gernet, Jacques. 2003. L'éducation des premières années (du 11ᵉ au 17ᵉ siècles). In *Education et Instruction en Chine,* eds. Christine Nguyen Tri and Catherine Despeux, vol. 1, 7–60. Louvain: Peeters.

Gong, Chun 龔春. 1983. 中國歷代衛生組織及醫學教育 (*Zhongguo lidai weisheng zuzhi ji yixue jiaoyu. Medical teaching and Public Health Organization in Chinese History*). Beijing: Weisheng bu kejiao.

Grant, Joanna. 2003. *A Chinese physician. Wang Ji and the 'Stone Mountain Medical Cases History'.* London: RoutledgeCurzon.

Grmeck, Mirko. 1983. *Les maladies à l'aube de la civilisation occidentale.* Paris: Payot.

Gui, Dian 桂坫. 1911. 南海縣志 (*Nanhai xianzhi Gazetteer of Nanhai*). Taibei: Taiwan xuesheng shuju.

Guo, Aichun 郭靄春. 1987. 中國分省醫籍考 (*Zhongguo fensheng yiji kao. Reference of medical books in each province of China*). Tianjin: Kexue jishu chubanshe.

Hanson, Marta. 1997. *Inventing a tradition in Chinese medicine: From universal canon to local medical knowledge in South China, the seventeenth to the nineteenth century.* PhD diss., University of Pennsylvania.

He, Mengyao 何梦瑶. [1751] 1994. 醫碥 (*Yibian. The stepping-stone for medicine*). Beijing: Renmin weisheng chubanshe.

Huang, Yan 黄岩. [1800] 1918. 醫學精要 (*Yixue jingyao, Essentials in medicine*). Shanghai: Cuiying shuju yinhang.

Huang, Zhixun 黄志勛. 1936. 融縣志 (*Rong xianzhi, Gazetteer of rong*). Compilator. Taibei: Taiwan xuesheng shuju.

Huangdi neijing 黄帝内經. 1997. Trans. Liansheng Wu, Qi Wu. Beijing: China Science and Technology Press.

Leung, Ki Che Angela. 1987. Organized medicine in Ming-Qing China: State and private medical institutions in the lower Yangzi region. *Late Imperial China* 8 (1): 134–166.

Leung, Ki Che Angela. 2003. L'instruction médicale et sa vulgarisation dans la Chine des Ming et des Qing. In *Education et Instruction en Chine,* eds. Christine Nguyen Tri and Catherine Despeux, vol. 2, 89–114. Louvain: Peeters.

Liang, Chongding 梁崇鼎. 1935. 貴縣志 (*Gui xianzhi, Gazetteer of Gui*). Compilator.

Liang, Lianfu 梁廉夫. [1881] 1936. 不知醫必要 (*Bu zhi yi biyao. What someone ignorant in medicine should know*). Shanghai: Kexue jishu chubanshe.

McDermott, Joseph. 2006. *A social history of the Chinese book. Books and literati culture in late imperial China.* Hongkong: Hong Kong University Press.

Mizayaki, Ichisada. 1976. *China's examination hell. The civil service examinations of imperial China.* New Haven: Yale University Press.

Pan, Mingxiong 潘明熊. [1865] 1868. 評琴書屋醫略 (*Pingqin shuwu yilue. The short guide of medicine of the room Pingqin*). Guangzhou: Guangzhou keben.

Schaeffer, Jean-Marie. 1989. *Qu'est-ce qu'un genre littéraire?* Paris: Seuil.

Scheid, Volker. 2007. *Currents of tradition in Chinese medicine 1626–2006.* Seattle: Eastland Press.

Schiffrin, Deborah. 1988. *Discourse Markers, (Studies in Interactional Sociolinguistics).* Cambridge: Cambridge University Press.

Searle, John R. 1969. *Speech acts: An essay in the philosophy of language.* Cambridge: Cambridge University Press.

Shi, Zheng 史證. 1879. 廣州府志 (*Guangzhou fuzhi, Gazetteer of Fuzhou*). Compilator.

Unschuld, Paul U. 1985. *Medicine in China: A history of ideas*. Berkeley: University of California Press.

Unschuld, Paul U. 1986. *Nan-ching, The classic of difficult issues*. Berkeley: University of California Press.

Unschuld, Paul U. 2003. *Huang Di Nei Jing Su Wen, nature, kowledge, imagery in an ancient Chinese medical text*. Berkeley: University of California Press.

Vernant, Denis. 2005. Pour une analyse de l'acte de citer. In *Citer l'autre,* eds. M.-D. Popelard and A. J. Wall, 179–194. Paris: Presses Sorbonne Nouvelle.

Virbel, Jacques. 1997. Aspects du contrôle des structures textuelles. In *Perception auditive et compréhension du langage,* eds. J. Lambert and J.-L. Nespoulous, 251–272. Marseille: Solal.

Volkmar, Barbara. 2000. The physician and the plagiarist. The fate of the legacy of Wan Quan. *East Asian Library Journal* IX (1): 1–77.

Wang, Xueyuan 王学渊. 1843. 暑症指南 (*Shuzheng zhinan. Guide for summer-heat diseases*). n.p: n.p.

Wei, Renzhong 魏任重. 1946. 三江縣志 (*Sanjiang xianzhi, Gazetteer of Sanjiang*). Compilator. Taibei: Taiwan xuesheng shuju.

Wen, Zhonghe 溫仲和. 1898. 嘉應州志 (*Jiaying zhouzhi, Gazetteer of Jiaying*). Compilator.

Widmer, Ellen. 1996. The Huanduzhai of Hangzhou and Suzhou: A study in seventeenth-century publishing. *Harvard Journal of Asiatic Studies* 56 (1): 77–122.

Wilkinson, Endymion. 1998. *Chinese history: A manual*. Cambridge: Harvard University Press.

Xue, Qinglu 薛清录. 1991. 全國中醫圖書聯合目錄 (*Quanguo zhongyi tushu lianhe mulu, Catalogue of medical books preserved in the Chinese libraries*). Beijing: Zhongyi guji chubanshe.

Yu, Yunxiu 余云岫. 1943. 流行性霍亂與中國舊醫學 (*Liuxing xing huoluan yu zhongguo jiu yixue. The cholera epidemic and the ancient medicine of China*). *Zhonghua yixue zazhi* 6:273–285.

Chapter 4
Zoological Nomenclature and Speech Act Theory

Yves Cambefort

Abstract To know natural objects, it is necessary to give them names. This has always been done, from antiquity up to modern times. Today, the nomenclature system invented by Linnaeus in the eighteenth century is still in use, even if the philosophical principles underlying it have changed. Naming living objects still means giving them a sort of existence, since without a name they cannot be referred to, just as if they did not exist. Therefore, naming a living object is a process close to creating it. Naming is performed by means of a particular kind of text: original description written by specialists, and more often accompanied by other, ancillary texts whose purpose is to gain the acceptance and support of fellow zoologists. It is noteworthy that the actions performed by these texts are called "nomenclatural acts". These texts and acts, together with related scientific and social relationships, are examined here in the frame of speech act theory.

In his seminal work *How to Do Things With Words*, the philosopher and linguist John L. Austin (1962) argued that the aim of language was not just to express ideas, but "to do things". In fact, he mostly used the verb "to perform" (hence the adjective "performative" to qualify this role of language), and he did not seem to mean that speech actually could "make" "things", i.e. create objects, but words and concepts. This is precisely what a particular sort of text purports to do: "nomenclatural acts", or "acts of biological nomenclature". These acts are twofold processes performing two actions at once: giving names to still unknown organisms, while at the same time conceptualizing them into discrete kinds called "taxa" (singular "taxon": species, genus, family, etc.). Making something new (even if virtual) is the most obvious characteristic of this sort of text; but nomenclatural acts also involve other kinds of speech act, especially enumeration and instruction. Therefore, even if their primary aim seems to be an adjustment of [scientific] words to the world, they also have normative aspects, i.e. they also have the aim of making an adjustment of the world to words. This ambiguity will be studied in the frame of the speech act theory ("SAT", as developed particularly by Austin, op. cit., and Searle 1969, 1975, 1979,

Y. Cambefort (✉)
SPHERE (ex-REHSEIS; CNRS & University Paris Diderot), Université Paris 7, UMR 7219
Paris, France

© Springer International Publishing Switzerland 2015   143
K. Chemla, J. Virbel (eds.), *Texts, Textual Acts and the History of Science*,
Archimedes 42, DOI 10.1007/978-3-319-16444-1_4

etc.). In conclusion, some difficulties of both nomenclatural acts and of speech acts in general will be outlined, especially the question: if something is made (created), what is it?[1]

4.1 From the Bible to Linnaeus to Austin

The naming and classification of plants and animals has long been one of the major aims of biology and even more so since Linnaeus's ground-breaking work in the eighteenth century [2]. Linnaeus has been portrayed as being a second Adam, giving names to all the animals he knew, just as the first Adam named the animals God had just created[3]. Although this portrayal was intended to be ironic criticism (and was attributed to Linnaeus's fellow botanist Albrecht Haller as such[4]), it can also be seen as praise. Indeed, Linnaeus was especially proud of his invention of a binomial system for both naming and classifying plants and animals. He introduced his system in 1735, in the first edition of his *Systema Naturae*, a work which was reissued many times during his life. As he was a botanist first and foremost, he went into his system in more detail in theoretical botanical texts, e. g. in his "Philosophy of Botany" (*Philosophia Botanica*, 1751):

> § 151. Botany has two fundamental aims: Arrangement & Naming. (…)
> § 210. When arrangement is done, naming–the second aim of Botany–foremost imposes names.[5]

Linnaeus's system of classifying and naming plants and animals is still in existence under the form of nomenclatural texts.

4.1.1 Nomenclatural Texts and Speech Acts

If *Genesis* is too far away from us, we might compare processes of naming and identification of plants and animals with other "speech acts". In Biology, the most important speech acts are "nomenclatural acts" (an expression used by taxonomists themselves), that have the aim of naming and describing "new" taxa. To some

---

[1] Ontological and other theoretical problems about the "nature", essence, or existence of animal taxa, especially species, will not be examined (see Lherminier and Solignac 2005 for extensive discussion); for general philosophical problems about "natural objects" and their names, see e.g. Putnam (1973), Schwartz (1977), Kripke (1980), Li (1993), Recanati (2008), etc.

[2] Although nomenclatural texts exist both in zoology and botany, the present paper will be devoted mostly to zoological texts (contrary to Daston (2004), a paper which partly covers this one, but from a different perspective).

[3] Genesis 2: 19.

[4] Blunt (1971).

[5] "151. FUNDAMENTUM Botanices duplex est: Dispositio & Denominatio. (…) 210. DENOMINATIO alterum Botanices fundamentum, facta dispositione, nomina primum imponat." (my translation).

extent, nomenclatural acts can be compared to speech acts such as ship naming: "I name this ship *Normandy*." These acts are strictly human, but they are nevertheless not inconsequential, for example the ship in question will bear her name for ever (or at least for an indefinite period of time). In an analogous way, taxonomists give names to organisms (plants and animals), under certain circumstances and following certain procedures, and these organisms bear these names for an indefinite period of time. But important differences immediately appear between ship naming and nomenclatural acts. Let us make a preliminary list of four of these differences:

1. whether a given organism already has or does not have a name is established in advance; the person who intends to give the organism a name must formally declare that this particular organism is still unnamed, unknown in a sense, and so in a way "new", thus this naming process could also be identified as a sort of creation;
2. nomenclatural acts are not spoken (uttered) but written: even if many of them have long been presented orally in academic meetings, they are now included in texts which must be published; their possible effects are delayed in relation to their production, i.e. they do not achieve success (become "valid") until they are published;
3. nomenclatural acts are not indisputable proclamations (like ship naming), but proposals which can be cancelled at any time (any cancellation also being a proposal, and not valid until it is published); the nomenclator, who is assumed to be competent to perform such acts, must in addition make every effort to ensure that not only will they achieve success (i.e. that the new taxa will be recognized as valid), but also that this success will last (i.e. that the new taxa will continue to be recognized as valid) for an indefinite period of time; in a way, nomenclatural acts are also similar to decisions in justice ("I pronounce you guilty"), that can be appealed before another (usually higher) court, except that any person who can perform nomenclatural acts (introduction of new names and taxa) can also perform their cancellation;
4. a nomenclatural act does not refer to an individual entity (as in the case of a ship or culprit), but to a group of organisms called a "taxon" (plural "taxa"), most often a species; these acts introduce and individuate taxa at the same time as they create names to designate them; taxa individuated by their names are the formal units of animal and plant classification.

All these points will be examined and discussed in the present paper. At this moment, let us insist again that nomenclatural acts are twofold processes, performing two actions at one and the same time: (1) naming taxa, as well as—in a way—(2) creating them, or at least giving them formal existence. This entails an interesting consequence: name creation almost always succeeds (once created, a zoological name will subsist almost indefinitely), whereas the introduction (creation?) of taxa often fails (a taxon considered as new might, in fact, already have been named, which therefore nullifies the 'new' taxon). Be that as it may, these four characteristics of nomenclatural acts lead us to question the nature of such acts within an SAT framework: this question will be examined in the course of and at the end of the present paper. Before that, the rest of this section will recall some basic principles of zoological nomenclature. A second section will be devoted to the study of examples of nomenclatural acts: primary or "original" descriptions (of species and of higher taxa), and secondary nomenclatural acts, the latter being aimed at questioning the validity of the former.

4.1.2 Living Organisms and Their Names

4.1.2.1 Naming Plants and Animals

For a long time, the primary aim of Natural History was to acquire knowledge of the entire living world, through the organisms that comprise it. An inventory of all the world's living objects (beings) involves both their description and naming: a description enables a particular living object to be recognized; but, when it is known (or in order for it to be known), it must receive a name; for if a particular object does not have a name, how is it possible to refer to this object? This basic principle is expressed by Linnaeus in a famous aphorism: "*Nomina si nescis, perit & cognitio rerum*" (If you don't know the names, then your knowledge of the things also perishes)[6]. It might be possible to refer to an object without naming it, provided that this object is well known; but is it possible when one deals with a poorly known or even unknown object? Clearly, naming is as necessary as describing, and it became highly convenient to do both actions at the same time, in the same "act", in order to attach a name to a description, and the pair (name + description) to a particular object (a specimen), or rather to a concept (the taxon to which this specimen belongs). The following figure schematizes the relationships between these three elements[7]:

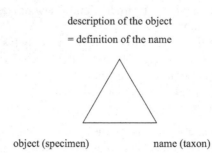

description of the object

= definition of the name

object (specimen)   name (taxon)

But another question arises: which objects are to be named? Should every living being receive a name, in the same way as humans and pets? Obviously, such a procedure runs up against insurmountable difficulties: how to name each of the billions of billions of insects, crustaceans, and other innumerable creatures actually

[6] Linnaeus 1751, p. 158.

[7] See a comparable but different triangular figure *in* Recanati (2008), p. 155.

Table 4.1 Example of Linnaean classification of a particular species: the domestic dog (*Canis familiaris*)

Species *familiaris* (the dog in the most restricted meaning)
Genus *Canis* (the dog, the wolf, and related species)
[Family Canidae (dogs *sensu lato*, foxes, and related genera)]
Order Carnivora (dogs, cats, and related families)
Class Mammalia (Mammals)
[Phylum Vertebrata (Vertebrates)]
Kingdom Animalia (Animals)

living on earth and seas? Therefore, naturalists have chosen to sort living beings into discrete kinds, called "taxa" (species, genera, etc.), and only name taxa. The most numerous of all taxa are species, which represent the basic unit of classification, immediately above the individual level. Although it has proven impossible to give a universally accepted definition of species, everybody agrees that species is/ are indispensable for a coherent understanding of the living world[8]. In the Linnaean system, species are clustered into genera, genera into families, etc., according to a hierarchy which Darwin called "ranking" and which is shown in Table 4.1 (the example of the domestic dog, *Canis familiaris* Linnaeus).

The categories created by Linnaeus are: Kingdom, Class, Order, Genus, and Species. The Family was introduced in the nineteenth century by Latreille, and the Phylum by Cuvier. Intermediate categories (superfamily, subfamily, subgenus, etc.) are useful in large and complex groups, especially arthropods. In Linnaean nomenclature, each species is designated by a pair of names which together make a "binomen", hence the designation of "binomial" nomenclature. The first name in the pair designates the genus (here *Canis*, the dog in a broad sense), into which the relevant species is placed (classified): *familiaris*, the dog in a strict sense, the "domestic" dog (there are additional species in the genus *Canis*). In the Linnaean system, a "specific" epithet (e.g. *familiaris*) has no meaning if it is separated from its genus; for this reason, it must always be quoted in association with its genus: *Canis familiaris*.

An important nomenclatural consequence follows from the existence, at least theoretical, of the "species" category: any new name Gs [Genus + species] created by an author in order to designate a particular animal specimen—let us say σ—can also be applied to the species Σ of which σ is an example, and to any other individual belonging to this species. By convention, the specimen σ will be considered as the type of the species named Gs, or rather of the name "Gs", this point having shifted during the history of taxonomy. Strictly speaking, only one type can exist for each species or name: hence the modern term of "holotype" (from the Greek *holos*, "whole", "entire"), which means that the unique specimen on which a description

---

[8] See again Lherminier and Solignac (2005).

is founded is the one and only basis for it. The concept of type dates back from the time of the "typological" concept of living species[9]. Today, this concept is rather obsolete: for modern biologists, a species is an aggregate of more or less variable populations, none of them—and *a fortiori* no individual—being more "typical" than the others. This difficulty is resolved by a more recent nominalistic conception: any holotype is no longer considered as the type of the species to which it belongs, but as the type of this species name. Other representatives of the same species have no particular relationship with the type: they are neither "tokens" nor "occurrences" of the latter; they are just other examples of this particular species (as is the type itself).

4.1.2.2   A Short History of Biological Nomenclature

The naming, describing, and classifying of living beings has been practiced since Antiquity. In the Middle Ages, it was mostly applied to medicinal plants, and was one of the causes of the quarrel between Nominalism vs. Realism. During the sixteenth and seventeenth centuries, naming, describing, and classifying were used to sort and arrange natural and artificial objects in "cabinets of curiosities", and later cabinets of natural history. Only during the eighteenth century did nomenclature take a scientific and systematic character, with Linnaeus's *Systema Naturae*. The first edition of this work dates back from 1735, but it is only in the tenth edition (1758) that Linnaeus developed the binomial naming of animals which is the main characteristic of his system[10]. For this reason, this tenth edition and its date have been chosen by convention as the starting point of animal nomenclature. We shall see below how important this book and its date are, since the basic principle of biological nomenclature is *priority*, according to which the oldest name (for animals: from 1758 onward) must take preference over any other.

Although the meanings of the two words are slightly different, "systematics" is sometimes referred to as "taxonomy"[11], from which the word "taxon" has been derived[12]. The species *Canis familiaris*, genus *Canis*, family Canidae, order Carnivora, class Mammalia, phylum Vertebrata are examples of taxa. Taxonomy deals more with taxa (objects of classifications), systematics with systems (classifications of objects).

Taxonomy was the main task of botanists and zoologists until the middle of the 1950's; they produced a profusion of publications: more than a million animal species (mostly arthropods) have been described and named, most of them being the subject of more than one paper (original description, corrections and emendations,

---

[9] Farber (1976).

[10] Linnaeus (1758). The binomial nomenclature of plants was introduced earlier (Linnaeus 1753).

[11] "*Taxonomie*" is the original spelling of the word as it was introduced in French by the Swiss botanist Augustin Pyrame de Candolle in 1813.

[12] The word *taxon*, derived from *taxonomie*, was coined in 1828.

re-descriptions, various additions, etc.). All of that makes a "grey literature", little read and flooding libraries, but indispensable when taking all living beings into account in a somewhat exhaustive way. Such a task might be considered as both impossible and useless: one might decide to abandon it and be content with a large scale description of the world, a solution which was advocated for example by the French scholar Réaumur, in the eighteenth century:

> As long as hundreds of species of flies and small butterflies do not offer us anything more remarkable than slight differences in the shape of their wings, or of their legs, or varieties of their color, or different distributions of the same colors, it seems to me they can be left mixed with each other.[13]

But such an attitude would be unsatisfactory, in the framework of modern scientific and philosophical concepts; moreover, it would not guarantee the precision and accurateness which are necessary when some practical use of living organisms is considered. For example, for a particular "noxious" animal species (often an insect), there is often another species (another insect) which is its exclusive parasite. If one wants to use the latter to struggle against the former, it is necessary to recognize and identify both species: this is the basic principle of "biological struggle".

4.1.2.3 Legal Status of Nomenclatural Texts

An important characteristic of nomenclatural acts and texts has to be stressed: although they are generally recognized as belonging to science, the creation of zoological (and botanical) names—and these names themselves—are ruled by "laws" and "codes", which give these acts and texts a sort of legal status[14]. In the same perspective, those types selected and described in nomenclatural acts appear as proof of these acts. This legal status may appear more important than purely scientific aspects: when most "up-to-date" papers are likely to be quickly forgotten, taxonomic papers will keep their importance and value as long as there are animals to manage, in one way or another, dead or alive, in fields or museums.

Despite this theoretical importance, most professional biologists had abandoned taxonomy by the middle of the 1950's, and it dramatically declined in the second half of the twentieth century. As a consequence, amateur taxonomists more or less replaced professionals in some taxonomic groups, especially insects, with the result that a relatively high proportion of authorial authority has been lost, although in other large groups, e.g. crustaceans, professional taxonomists still predominate. Be that as it may, the status of taxonomy as a scientific discipline declined in the last decades of the twentieth century. Recently, however, taxonomy has started to recover and improve, in part because of modern communication media: since the early 2000's, some journals devoted to animal taxonomy have been published

---

[13] Réaumur (1734), p. 3 (my translation).
[14] Minelli (2005).

in electronic form on the internet: New Zealand's *Zootaxa*, the Paris Muséum's *Zoosystema*, etc. They are also available on paper, in conventional printed form. *Zootaxa* is the world's leading journal of animal taxonomy, as far as the number of articles is concerned: between 2001 and 2012, it published some 250,000 pages, describing more than 25,000 taxa. Today, taxonomy and systematics are reaching high standards of excellence; they use modern methods and are being recognized once again as important scientific disciplines, especially in the context of the crisis in "biodiversity" and the extinction of many organisms, which probably will characterize the decades to come.

4.2   Primary Nomenclatural Acts: Naming and "Original Descriptions"

The basic nomenclatural texts are the "original descriptions", which have always been and still are produced in large number, and published in quite an array of journals and books, from older journals, like the academic societies' periodicals (for example, the Société Entomologique de France, founded in 1832, continues to publish issues of its two journals), to more modern ones, including the two mentioned above (*Zoosystema* and *Zootaxa*), which are available both in print and on the internet. Original descriptions are considered as, and called "primary nomenclatural acts" by taxonomists themselves. Most original descriptions deal with species, some with genera. Descriptions of families or higher categories are less frequent, except in poorly known groups, mostly fossils. We shall consider first the simplest case: description of species; description of genera will be considered in a second paragraph.

4.2.1   Species Description

As implied by its name, an "original description" is the first text to be published in order to make any still "unknown" (or at least unpublished) taxon known. In addition to giving its characters, an original description also introduces a new name for the relevant taxon, this name being written in the first line of the description. Original descriptions still have to be printed on paper and distributed in the form of journals or books: purely electronic media (internet, CD, DVD, etc.) is not accepted[15]. This difficulty has been solved in the following way: when new taxa are published in electronic form, traditional paper versions are also made available on special demand, and some copies are deposited in libraries. Original descriptions are accompanied by manuscript texts, especially labels, which contain important information on type specimens.

---

[15] See Huber (2007) for a discussion on non-availability of names published only electronically.

Example # 1 Ahyong (2006).

From the mass of taxonomic papers, I have selected almost at random an example which is both recent and short, in order to present the main features of species descriptions. It deals with a description of a sort of squillid (a group of crustaceans sometimes nicknamed "sea-grasshoppers" or "sea-mantes"). The following analysis is an enumeration of the successive parts which compose this sort of text.

(1). The Preliminary Sections

Title Articles of this sort, especially short notes, have explicit titles: the author's aims are clear from the first line of his text. Here, the new species is announced (but not yet named), established in its genus and other higher taxa, geographically situated, and a cladistic analysis is announced, which aims to establish the new species more clearly and firmly in its taxonomic context.

Author's Name and Address Their purpose is not mere identification, but authors wish to give guarantees on their competence and authority in order that the nomenclatural act they are performing will be accepted. Here, the author belongs to a respectable institution, a fact which gives his article (his new species) every guarantee of reliability.

Abstract and Résumé They are mere developments of the title: they introduce the new species' formal name, indicate its geographical distribution, and give its most important diagnostic characters. In the French translation (which is given since this is a French journal), the English title is translated as well.

Introduction This paragraph is strictly taxonomic and historical; it is rather long, due to the complexity of crustacean taxonomy and nomenclature. Situated at the beginning of the paper, it works as a sort of *captatio benevolentiae* in order to persuade a reader of the author's expertise, which is important since he is introducing a new hypothesis: three specimens studied by previous authors did not in fact belong to the species the authors proposed but to a new, still unrecognized one; hence the naming and description of this new species by the author under study. Cases like this are frequent, especially for smaller animals (insects, crustaceans, etc.): due to more elaborate study methods and deeper knowledge of taxa, almost as many new species are found in collections (museums, institutes, universities), as are freshly collected from nature. But more refined study methods, even with the latest tools, do not guarantee that the later opinion is better than the earlier one(s).

"Material and Methods" A traditional section, but showing here that nomenclatural activity calls upon new methods. Here we find a cladistic analysis, not only to explain the relationship of the new species to related taxa more precisely, but again also in an attempt to demonstrate that the species under study is really new.

Systematics The two first lines of this section are in fact the last of the preliminaries. They give the taxonomic and nomenclatural framework for the new species:

family and genus, which are formally designated by their name, author's name, and publishing date (these elements will be explained below).

(2). The "Original Description" s. str.

The new name All texts of this kind begin with a complete "binomen" [Genus + species]. Therefore, a binomen may be considered as the "title" of a description, the initial superscription giving the content of the following paragraph(s). In the case under study, which is typical for the description of a species, only the specific name (or epithet) is new, the genus name dating back from Manning, 1968, as explained in the line above; but the combination of an older genus name (*Carinosquilla*) and a new specific epithet (*mclaughlinae*) makes a new binomen. New zoological binomina are governed by various rules and constraints, detailed in a *Code of Zoological Nomenclature*, as has already been explained. In addition to the purely "legal" aspects, recent editions up to 1985 contained useful information on the "Transliteration and Latinization of Greek words," and other "Recommendations on the formation of names."[16] The basic rules are:

- a specific name must be a noun or an adjective, Latin or Latinized, in reasonable conformity to Latin morphology and syntax (see below, paragraph "Etymology");
- it must not convey an insult toward any person, group, institution, organization, state, etc.;
- it must not contain any explicitly political or religious meaning.

But the fundamental rule is that of **priority**: a specific epithet must never have been used before in the relevant genus, i.e. it is mandatory that the epithet *mclaughlinae* has never been used in the genus *Carinosquilla*. If it has already been used in this genus, it would be a case of "homonymy". Another case of nullity is "synonymy": when the supposed new taxon has already been given a name (by another or the same author). The rule is that more recent names become "homonym" or "synonym", respectively, of older names (see Sect. 4.3.1, below).

The mention "n. sp." Up to the 1850s, only the binomen [Genus + species] was written. Readers were supposed to understand whether it was new or not, with the help of the mention (or not) of previous quotations: if there were none, the binomen was assumed to be new. By the 1850s, descriptors started to mention explicitly that the name (the species) they were introducing was genuinely new. Since original descriptions were written in Latin, they added a mention in Latin, under various forms: "*nov. sp.*" (*nova species*), "*sp. nov.*" (*species nova*), etc. The short form "*n. sp.*" has the advantage of being readable in English as well: "new species." This mention, which follows the binomen here for the first and only time, indicates the beginning of the formal description of a new species (its naming having already been done).

---

[16] *International Code of Zoological Nomenclature* 1985, p. 182–229.

Let us pause here for a while. The whole nomenclatural (or speech) act, in its two parts, is accomplished in the mention "n. sp." Without it, the new binomen is not formally declared. Under a constative form, the mention "n. sp." actually means something like: "I—the author—consider that the taxon (the species) which has just been named is 'new', i.e. that it has never been described by a zoologist; I hereby give a formal description of it, in order that it can be recognized by readers whose expertise is similar to mine." Therefore, this nomenclatural act can also be considered as a speech act, of Austin's "illocutionary" kind, and the formula "new species" (under any form) might appear as a "marker of illocutionary force", a sort of label warning the reader that a nomenclatural act is being performed. When the formula "new species" was not in use, there was no such label and Austin's "uptake" was less likely to take place. Another role of the "new species" formula is to introduce the diagnosis and description; therefore, it can be considered as an "initializer" of both.

Remark: What Exactly Is New?   One may wonder about the "newness" of the species being described. Perhaps the word "new" is here considered as a synonym of "unknown". Such "creation" of a species could also be a metaphor, in the same way as many expressions used in biology. Another meaning of this novelty is that it takes into account the nature of taxon of the species under description. For Simpson and Mayr, a taxon possesses the twofold character of an object: both natural and artificial (cultural):

> A taxon is a group of real organisms recognized as a formal unit at any level of a hierarchic classification.[17]
> Two aspects must be stressed. A taxon always refers to concrete zoological objects. Thus, *the* species is not a taxon; but a given species is. Secondly, a taxon must be formally recognized by a taxonomist.[18]

In order to be recognized as extant, a given taxon (here: a particular species) must be formally identified. Therefore, an original description really and authentically "creates" a taxon: we are close to Adam's or even God's action in *Genesis*! Again, it should be remembered that such a "speech" or "discourse" is not a verbal utterance here, producing its effects instantaneously, but a published text, and actually when the 'creation', the new state in the world, takes place is open to question (see below).

Even if it can be accepted that a new taxon is (or will be) "created", the only new element of this text is the one the reader can see: the specific name (or epithet). If it turned out that the taxon in question (here the species *Carinosquilla mclaughlinae*) was not in fact "new" (i.e. the species already existed, it was already formally described and named), the performance (the illocutionary act) would fail, and the specific name (epithet) would "fall into synonymy"; but it would subsist, even if no longer in use. This is an obvious difference between names and taxa, and possibly also between nomenclatural acts and speech acts: the "newness" of a supposed new

---

[17] Simpson (1961), p. 19.

[18] Mayr (1969), p. 5.

taxon is not objective, not permanent; after its existence has been acknowledged by its descriptor(s), it is open to testing by fellow zoologists, who either accept or refuse it (see below). On the contrary, the creation of a new name is definitive and irrevocable, even if the taxon has already been introduced by an earlier author.[19] Any pair of "new" name and taxon created by the same act will remain attached as long as zoology subsists; but their link will be stronger if the nomenclatural act is "successful", i.e. if the name is accepted and considered "valid" after testing by fellow zoologists. On the contrary, if the nomenclatural act fails, the name will be invalid, for reasons of homonymy or synonymy (see below).

All the following sections of the paper under study may probably be considered not as the nomenclatural (or speech) act proper, but as additional texts used to increase its chances of success.

Iconography Illustrations are an inclusive part of the description and have a special status: in the case of the absence of type material, a figure can be selected as the type of a name[20].

Synonymy The brief lines under the mention "Appendix 1 and 2," which are not always present, indicate that the specimens under study had previously been examined by specialists, who concluded they belonged to a species described earlier by another specialist, Raoul Serène in 1954: *Carinosquilla carinata*. The mention "*Non* Serène 1954" means that the author under study (Ahyong) does not agree with his predecessor's opinion: for him, the specimens represent a new species; hence, his description of the latter.

Type Material According to latest editions of the *Code*, it is mandatory for original descriptions of new species to be accompanied by designation of actual specimens called "types"[21]. Ecological data are here given for the male "holotype", but also for the two available female "paratypes." Although the holotype is the one and only reference for an original description and "proof" of a new name, nomenclatural codes and rules have always considered that a new name applies to other specimens belonging to the same species as well, including "paratypes" (see below). Any name published without a holotype designation (as well as without a diagnosis/description) must be considered a *nomen nudum* (a "naked name"), i.e. without object, without meaning, without proof. From the point of view of nomenclature, a *nomen nudum* is null and void, and it is possible to reuse it (the one and only possibility of reusing a published name). Such a holotype designation is merely a descriptor's "decision", subject to his own appreciation, although it is of special importance, because the holotype will be the basis of the name as long as either subsist[22]. The

---

[19] The only exceptions are "naked names" (*nomina nuda*), see below.

[20] In fact, as explained in the *Code*, the type is not the illustration but the specimen illustrated, even if it is impossible to localize the latter.

[21] See e.g. *International Code of Zoological Nomenclature* 1999, articles 61–75. See also, as far as Botany is concerned, the fine paper by Daston (2004).

[22] In the case of destruction or loss of the holotype, it is possible to designate a substitute called "neotype".

holotype should belong to the most characteristic gender of this particular species. In our example, as the male specimen has been designated as the holotype, the two female specimens are designated as "paratypes" and are also described: they complement the description of the holotype, and give a more complete idea of the species it belongs to, i.e. a more complete definition of the new name. But designation and description of paratypes are not mandatory: an author can decide to designate and describe only the holotype, and leave other available specimens as mere "additional material"[23].

Etymology Sometimes entitled "*Derivatio nominis*", this short paragraph gives the meaning of the specific name (or "epithet"). Here, this epithet is made from a person's name, Latinized and declined in the genitive case of the feminine gender (since this person is a woman). Dedications of new species play important roles in communities of zoologists. Although sometimes presented to honor respected colleagues, or to acknowledge help received during the author's work, they are mostly aimed at gaining support from influential members of the author's community and always for the same purpose: to obtain the most guarantees that the nomenclatural act being performed will be successful.

Distribution A rather short summary, but geographical distribution can help to new species be better and more clearly defined.

Measurements The corresponding numerical data are more usually given in the description or diagnosis.

Diagnosis Although printed in a smaller font in comparison to the following paragraphs, this paragraph is very important since it gives the "diagnostic" characters, i.e. those characters that are necessary and sufficient to separate and identify the new species from all those species already known in the same genus. The syntax is identical to the "Description" section: the sentences have no verb, or the verb "to be" is implied everywhere.

Description Contrary to the diagnosis, the description includes a verbal depiction as complete and as detailed as possible of the new species, all of whose observable characters (diagnostic or not) are described. From a textual point of view, the description (like the diagnosis) takes the form of an enumeration. The section is introduced by the word "Description", which functions as an "initializer" of the enumeration, or perhaps a second initializer after the formula "n.sp." The description is organized into paragraphs which correspond to significant groups of characters following a rational order (from the front to the rear of the animal being described). Each paragraph is introduced by a "classifier": "Eyestalk", "Antennular peduncle", "Rostral plate", etc. Each group of characters is described in the form of a sentence, starting with a capital letter; but these sentences are not grammatically correct, since they have no verb (except that the auxiliary "to be" seems implied

---

[23] Some authors also use the concept and the word "allotype" to designate among paratypes one specimen of opposite gender to the holotype.

everywhere). An additional paragraph to the description "Color in alcohol" reminds the reader that the specimens under study are not true "natural objects": they are not present in a natural environment but in a museum collection; they have been preserved under artificial conditions, which give them the status of "artifacts".

It is not explicitly stated in this example to which specimen(s) the description and diagnosis refer. Strictly speaking, since the holotype is the one and only basis and proof for the new name, both the original description and the diagnosis should refer to the holotype alone. But the following "Remarks" (see below) imply that all the specimens available have been used in this case, which in turn implies that the author under study is confident that they all belong to the same taxon and can be designated by the same name, even if the holotype is the one and only proof of this name and exemplar of the taxon referred to by this name.

(3). The Annex or Additional Sections

At the end of the description, the whole nomenclatural act has been performed. All the following parts are sorts of annexes whose aim is to reinforce it as much as possible in order that it achieve success, i.e. that the new species is recognized as valid by the author's community, i.e. zoologists specialist in the animal group in question. Annexes give additional arguments, especially suitable in this particular case, where the author has expressed his own disagreement with previous specialists. In the present example, there are six annex parts.

Remarks The author insists on the unique characters which—in his opinion—separate specifically (i.e. at the species level) the three specimens he has just described as a new species from other species already described in the relevant genus. The style is emphatic: his new species "is unique in the genus" in such and such characters; but at the same time, it resembles some of those older species. The section ends with considerations on morphological characters attached to sex.

Key to Species This section is different from those both preceding and following it in that it is printed in the whole width of the page and not in two columns. The reasons are not clear and probably do not involve a decision by the author.

A "key" or "table" is a traditional *topos* in nomenclature. Historically, the tradition of dichotomous tables dates from Plato and the Academy, but they were strongly criticized and refused by Aristotle. Plato's tables were philosophical in nature, therefore different from the tables or "keys" used in taxonomy. The word "table" refers to the printed form of this sort of text. The word "key" is taken in its metaphorical meaning of a tool used to solve a problem, which in itself poses two questions: what exactly is the problem? and what is a "key" from an SAT point of view? Taking the second question first, the key is apparently a text of the genre "instruction" which is supposed to help a reader to identify any specimen in the crustacean genus *Carinosquilla*. As there is probably less than a dozen people, in the world, who know what a *Carinosquilla* is, and who probably do not need such a key, its true aim is clear (and answer our first question): demonstrating as clearly as possible the necessity of separating the three specimens under study into a new species, which possesses characters different from the other nine previously known species. This key is clever since it ends with the new species, although any other order would have been possible; another shrewdness lies in the fact that the older species with which the three specimens in question were identified by previous authors (*C. carinata*, see example, p. 308, col. 2, and above, Synonymy section), is situated in the key, but not immediately next to the new species, giving the impression that both species are not so closely related, after all!

Relationships (Cladistic Analysis) While a table or a key is an ancient tool, cladistic analysis is relatively recent, having been introduced in the 1950's by German entomologist Willi Hennig[24]. From this time on, cladistics has become to be considered as a very important tool in many aspects of biology, enabling scientists to reconstruct phylogenetic patterns (some say processes as well) in the groups they study. In the case under study, the author explains that the 10 species of the genus *Carinosquilla* can be divided into four groups, and that his new species does not belong to the group which comprises the older species with which his three specimens were previously identified. Again, this is another way of confirming the validity of his new species, as well as demonstrating his expertise.

Acknowledgements They are another way to express both social bonds in the "microcosm" of fellow specialists, as well as thanks for financial support (in the hope they will be continued).

References Nomenclatural and other taxonomic papers often use references older than those quoted in other modern papers, although the case is not so obvious here as it can be elsewhere, with the reference section quoting articles and books from the nineteenth, and even the eighteenth century, for example Linnaeus.

Appendix 1 and 2 These sections merely give data used to perform the cladistic analysis provided earlier in the paper.

As a conclusion to the additional sections, it may be possible to recognize appendices 1 and 2 as another sort of speech act, perhaps Austin's "expositive", whose aim would be to strengthen the readers' conviction before they left the page. For, as we shall see, the condition of success of nomenclatural acts depends on other specialists' adherence and support (even if silent).

4.2.2 Genus Description

(1). The Genus Name

The most important difference with the species description is the name: the genus name consists of one word which must be capitalized (i.e. its first letter must be in uppercase); more importantly, a new genus name must be original, i.e. it must be a combination of letters never yet used in (animal) nomenclature.[25] This condition is getting more and more difficult to comply with, even if documents and registers (including on the internet[26]) do exist, to check whether a particular name has already

---

[24] See especially Hennig (1966).

[25] It may have been used in plant nomenclature.

[26] http://uio.mbl.edu/NomenclatorZoologicus/

been used. The validity rules are the same as for the specific epithet: a genus name can fall into homonymy or synonymy (or both).

(2). The Mention

"*gen. n.*" A mention "*gen. n.*" follows the new name. It means *genus novus* ("new genus") in Latin and is homologous with the mention *sp. n.* ("new species") in the species description. As in the previous case, it means that a new kind of organism (a taxon) is being named and introduced, and that it is given a generic rank. It is another sort of nomenclatural act, as well as another sort of speech act, and the basic rules governing species names are applied here as well.

(3). The Type Species

The type of a genus is a species, often a new species which does not fit in any genus already known. But there are other cases where a new genus is created (and named) to accommodate already known species, for example on the occasion of a taxonomic revision.

(4). Diagnosis and Description

These two parts work in the same way as their homologues in species descriptions. The diagnosis gives those characters whose combination is characteristic (or diagnostic), and has never been found before in any other animal. Illustrations can be considered as parts of diagnosis. The description of the species type is more complete. Both diagnosis and description attempt to mention only those characters which are supposed to have a generic value.

4.3   Secondary Nomenclatural Acts: Testing Names and Taxa

In the same way as speech acts have to satisfy certain "conditions of success", nomenclatural acts only become successful over time if certain "conditions of validity" are fulfilled. These conditions of success/validity are of two kinds: formal (or explicit), and informal (or implicit).

4.3.1   Formal Conditions of Validity

Which I call "formal" are those conditions of validity governed by nomenclatural rules or Codes: respect of these "laws" is mandatory. The validity of a specific or generic name depends on the name itself, and also on the taxon (species or genus) it is naming: so-called "invalid" taxa also render their names invalid. Invalidation of a name does not necessarily entail that of the relevant taxon: if its name is no longer considered valid, a particular taxon receives another name (a new name or an older one, which has to be revalidated); by contrast, if the taxon it refers to is no longer considered valid, the name has no meaning and becomes invalid, even if it is in perfect compliance with nomenclatural rules.

There are four formal conditions of validity:

1. The assumed new name must be "available" in the sense of the *Code*, i.e. it must be neither homonym nor synonym of an older name (see comments below).
2. An assumed new name—either specific or generic—must be printed and published: a publication on paper is still mandatory, but electronic publication (via "internet", CD rom, and other formats) is now accepted as a complement of the printed versions. A manuscript mention, for example on a label in a collection, has no nomenclatural value: manuscript names are called *nomina nuda* ("naked names"), or qualified as *in litteris* ("in handwriting"). Because of the cardinal role of priority (pre-eminence of the oldest name), the publication date must be clearly indicated: day, month, year. If only a month or a year is given, the paper is reputed as being published on the last day of this period. Cases are known of priority by only a few days!
3. Original descriptions must either contain a true diagnosis, or have some diagnostic character: they must include elements enabling an unambiguous definition of the new taxon, and a clear-cut difference from related taxa:
   The principal feature of the description of a new species is not its naming, but rather the working out of the characteristic attributes which will enable us to identify and classify it.[27]
4. A type must be designated: a real and actual specimen for a new species; a species for a new genus. This implies that a dead specimen of a new taxon is available. Such demand has recently been contested, especially in the case of endangered species, which may exist only in very small populations. In these particular cases, a consensus now exists on "indirect proof", like a photograph, a movie, a sound recording (a bird's song), etc., provided that such proof enables a formal recognizing and identification of the new taxon. But in most cases (especially for arthropods, where a species unambiguous description necessitates examination of internal parts, and therefore dissection), availability of a type specimen is still mandatory.

Be that as it may, in the traditional case of a reference specimen, it is necessary that it can be retrieved and recognized. The *Code* prescribes that types must be deposited in public collections, a disposition intended to dissuade amateurs from keeping types in their personal collections. The *Code* also states that they be labeled in a clear and unambiguous way. By convention, each type of a new species is given a red label (bearing its name and other mentions, according to the tradition in the group in question). Vladimir Nabokov, who was an excellent butterfly specialist, left a small poem which refers to this convention:

I found it and I named it, being versed
in taxonomic Latin; thus became
godfather to an insect and its first
describer–and I want no other fame.

Wide open on its pin (though fast asleep),
and safe from creeping relatives and rust,
in the secluded stronghold where we keep
type specimens it will transcend its dust.

---

[27] Mayr (1942), p. 14–15.

> Dark pictures, thrones, the stones that pilgrims kiss,
> poems that take a thousand years to die
> but ape the immortality of this
> red label on a little butterfly.[28]

In his poetic enthusiasm, Nabokov seems to believe that entomological collections will last longer than ancient monuments; but entomological specimens are fragile, and many types have unfortunately been destroyed in the barely 250 years that Linnaean nomenclature has existed!

Conditions of validity (2), (3), and (4) are almost always fulfilled, at least in modern taxonomic papers. But it is not always the case for condition (1), which represents the most frequent cause of failure of nomenclatural acts: the supposed new name must be "available", that is to say, it must be neither a homonym nor a synonym of an older name. Due to the huge number of animal taxa, it is not infrequent that a supposed new name has already been introduced in animal nomenclature. This again is a twofold condition, with two different cases:

- homonymy: two or more taxa (species or genera) were given the same name;
- synonymy: two or more names were given to the same taxon (species or genus).

Although the basic rule is the same (it is mandatory that one name corresponds to one taxon and *vice versa*), these two cases are not completely homologous: homonymy is mostly a problem of name, and synonymy mostly a problem of taxon.

In both cases, the basic principle is that of priority: the valid name attached to a taxon is the oldest one published. Hence the importance of the date of publication (above). By convention, all names introduced by Linnaeus in 1758 (*Systema Naturae*, tenth edition) have priority over all the others. Names posterior to 1758 remain valid for varying lengths of time, but their validity can be checked at any time. Lastly, names which have been invalidated might be revalidated (rehabilitated). But, even if invalidated, a published name will subsist and will be taken into account in the future (with the sole exception of *nomina nuda*); a type is and will be always necessary for a name, even if invalidated: hence Nabokov's poem on near-eternity (at least virtual) of type specimens and their red labels.

Homonymy Homonymy is easily evidenced by its objective character. It is recognized when a supposed new name has been used previously:

- in the same genus for a species name;
- in the whole animal kingdom for a genus name.

Any zoologist is able to call attention to homonymies and to propose new names for the relevant taxa, but courtesy prompts a private mention to colleagues responsible for homonymies in order that they may correct them, if they want. A new homonymy should be marked with the mention "*hom. n.*" (*homonymia nova*) or "new homonymy", just as new species and genera are marked with mentions "*sp. n.*" and "*gen. n.*"

---

[28] 'A Discovery' (December 1941) [published as 'On Discovering a Butterfly' in *The New Yorker* (15 May 1943)] (Nabokov 2000, p. 274).

Due directly to its objective character, homonymy is infrequent. As far as species are concerned, it may occur when a species is transferred from one genus (in which its name was previously not a homonym) to another genus (in which its name becomes homonym of another one). Homonymies are more frequent at the genus level, since it is difficult for specialists to be aware of generic names in groups with which they are unfamiliar. Registers are available, especially on the internet, but they are neither exhaustive nor up to date.

Synonymy As soon as its original description has been published, every taxon is open to discussion and testing by zoologists. Any one of them can claim that a supposed new taxon is in fact the same as another one, which has already been described and named more or less recently. They can then propose a new synonymy: the more recent name becomes synonym of the older one. This is evidenced by the mention "*syn. n.*" (*synonymia nova*), or "new synonymy". In nomenclatural lingo, it is said that a particular taxon (and its name) "falls into synonymy". Such a nomenclatural act is also a speech act, as performative (or illocutionary) as an original description, of which it is somehow symmetrical since it renders it "null and void". Synonymies are more difficult to demonstrate than homonymies, since they are at least partly subjective, and specialists often do not agree on them. In spite of this greater difficulty, they are more frequent than homonymies: indeed they are the true "evil" of nomenclature, which fills and burdens zoological libraries!

4.3.2   Informal Conditions of Validity

When formal conditions of success have been fulfilled, there is only one last test that nomenclatural acts must pass with success: acceptance by fellow specialists. As has been explained, any nomenclatural act is considered successful as long as it is not cancelled by a subsequent act (most often made by another author). In example # 1, the author (Ahyong) considered that three specimens previously considered as belonging to an older species were in fact examples of a new one, which he described. But other specialists in the same zoological group might be of another opinion: they might think that Ahyong is wrong, that his three specimens were correctly identified by earlier authors and do not belong to another, still undescribed species, and therefore that Ahyong's species does not exist. The point is that, in many cases, there are no objective arguments to say that a given specimen either belongs or does not belong to a given species: most arguments are subjective and open to personal interpretation. It is due to this reason especially that all authors (Ahyong in our case) make every effort to convince his or her readers. If readers who have not been convinced are also specialists in a position to question or dispute Ahyong's decision, they are free to do it. In this case, they could write another paper where they would put Ahyong's species in synonymy, even if all formal conditions for success have been respected. As has been said, original descriptions are mere proposals; in the same way, refusals of original descriptions are mere proposals too.

4.3.3   Example of Secondary Nomenclatural Acts

Due to the fact that specialists often do not agree with particular points, there is a huge mass of additional literature, confirming or canceling previously made proposals. This always follows to the same model, with the use of such specialized mentions as *nomen novum, nomen conservandum, combinatio nova*, etc... all of which mean that a suggestion is being made, and that it is proposed for testing by the community of taxonomists. The following text gives examples of some of these acts.

Example # 2 Vorst (2007).

This paper's first aim is to reinstate a species described in 1985, and synonymized in 2003; but various additional problems rendered the whole situation confused. The Introduction casually exposes the case: very small beetles whose taxonomic status—both generic and specific—is still uncertain and need to be clarified. First, the taxonomic rank (genus vs. subgenus) of this group has been questioned, especially since its type-species, namely *Trichopteryx nana*, was a "forgotten name" (*nomen oblitum*), then considered a "senior synonym" (i.e. a remarkable exception to the priority rule) of a more recent species and name: *T. kunzei*. But the latter was designated as the type-species of another genus, namely *Ptiliola*. Another problem is that—according to the opinion of the author under study (Vorst)—two different species were currently listed under the (rectified) name *P. kunzei*, introduced in 1841. The paper discusses this last name especially, which the author (Vorst) thinks to be a mixture of two species. One of these, *Ptiliola flammifera*, was originally described in 1985 in another genus (*Nanoptilium*) [incidentally, the specific name of this species was originally spelled *flammiferum* for reason of grammatical agreement (neuter grammatical gender)], synonymized in 2003 with *P. kunzei*, and is again considered by him as a valid species and name.

Pages 63–67 of the paper give arguments and explanations about the nomenclatural acts being performed. A lectotype is designated p. 65. Eighteenth and early nineteenth century authors often established their new species on a series of specimens they called the "type series"; since the rule is now that only one specimen needs to be selected as a "type" of a name, recent authors select one specimen from a type series and designate it as the unique "lectotype" (from the Greek *lektos*, "elected", "selected") of the species.

The paper ends on p. 67 with a dichotomous key of the three species recognized in the genus *Ptiliola*. This is another example of identification or "determination" keys, supposedly aimed at persuading readers to identify their specimens in the framework established by the author, who in turn commits himself to guarantee them correct identification. But the most important aim of such a key is always to convince the reader that the author's ideas are correct. In the same way, a brief conclusive discussion insists on various arguments already presented, in order ultimately to convince benevolent readers before leaving them. In this discussion, reference is made to the way the relevant specimens have been prepared ("slide mounted" or "dry-mounted"), which again demonstrates that such specimens are not truly "real" objects but artifacts.

4.4   Discussion and Conclusion

After this exposition of the principal kinds of nomenclatural acts, let us again examine the sequence of events which intervene in the most typical of them: the "original description", with the introduction of a new name and taxon.

1. *Creation of a new name.* Before one speaks of or about anything, it is necessary to express what one is talking of or about; this is the aim of names: names are the most direct, least ambiguous way to refer to objects or beings[29]. In original descriptions, the new name precedes the formula "n. sp." Zoologists are not supposed to name a species before they have ascertained its existence with a formal diagnosis and description; but, in practice, it is what they do. There are, in most cases, various preliminary texts (title, summary, introduction, etc.) warning a reader that the author has acquired some indications about the existence of a new species, and that he or she is about to describe and name it (such preliminaries not being parts of the original description proper). New zoological names are not spontaneous creations of a given language, but they are made up on purpose by a particular person. Under the form of a simple binomen *Gs* (i.e. Genus + species), any nomenclatural author seems to imply a complete sentence: "I give the species of which the specimen(s) under study belong(s) the name *Gs*". Therefore, a new name—including the implied sentence—appears to be a speech act of the "declarative" kind, or "declaration" in Searle's taxonomy[30]. But naming is also an adjustment of words to the world: a particular, anonymous object being given, a new name is proposed to designate this object. New names are introduced in writing, more precisely in printed texts. Contrary to spoken utterances (e.g. naming ceremonies, either on religious or secular occasions), written declarations have delayed effects, until the texts which contain them have been published: new names would become "available" only after their publication.

2. *Creation of a new taxon.* By the formula "new species", it is not only a new name, but also a new taxon—i.e. a new concept—which is being created. The formula seems to be a declaration as well; but in fact it is a decision: between two possibilities (the specimen under study either belongs to a new, still undescribed species; or it does not), a zoologist has made a choice: he or she selected the former. Therefore, we may consider—at least tentatively—that nomenclatural acts belong to "assertive declarations", a kind of speech act which does not belong to Searle's taxonomy in 1975, but seems to have been later accepted by him as an intermediate category involving the existence of some institutional authority which might guarantee an actor's decision[31]. The classic example is that of a tennis umpire shouting "Out!", because he or she saw the ball as being out of the prescribed limits. Even if the ball was not out, it will be considered out because of the umpire's decision, and the player charged *de jure* with this fault even if innocent *de facto*. The umpire's decision is imposed on the world; it can

---

[29] See e.g. Kripke (1980), etc.

[30] Searle (1975, 1979).

[31] Searle (1975); Ruiter (1993) (see especially pp. 60–62).

be considered as an adjustment of the world to the umpire's words. In zoological nomenclature, the situation is similar: something new is being created and added to the corpus of the world's objects, or at least the world's concepts, and this creation is imposed by words. But nomenclatural acts are peculiar since there is no external authority governing the decision(s) they entail (codes of nomenclature ruling only the formal conditions): the authors of nomenclatural texts are the sole authority of their acts. When authors consider they are able to take such a decision as "this species/taxon is new", this decision has no institutional or other guarantee (provided the basic rules are respected): it is merely an author's proposal, which can be challenged by another proposal made by any other fellow author.

3. *What exactly is being created?* We may again now ask the question posed above, but with a different meaning. Now, the problem is to acknowledge what exactly is being—or has just been—created. Diagnoses and descriptions claim they fit the natural objects they apply to, and therefore seem to adjust their words to the world; but different authors do not describe the same objects with the same words, which poses the question of the true direction of these speech acts. For example, in a famous case in beetle taxonomy, a new species was described as red; however, on the color plate attached to the description, the beetle appeared green. It turned out that the description was wrong and the plate correct: the author was partially color-blind. Although this case is exceptional, it points out the possible difference between the "true" world (i.e. the world most people agree about) and one's "private" world (i.e. one's particular picture in one's own mind). There are many other reported cases of differences in appreciation of the morphological characters of a species, in fact, the cases that do arise are mostly where a given character is not quantifiable. They may change the decision about a species being new or not, even if they are less remarkable than the green/red beetle. Examples #1 and 2 in the present paper demonstrate such differences in appreciation, later authors discussing characters which—in their opinion—were or were not correctly "seen" (i.e. interpreted) previously by their colleagues. The supposed adjustment of a given description to the world depends in fact on which world is being considered, and the difficulty is that every human being only knows his or her own world. It was hoped that modern methods based on molecular analysis might give objective answers to difficult questions; but currently these methods do not help very much, since in many cases they appear less reliable than the trained eyes of a specialist[32].

4. *Type material designation.* In fact, it is impossible to explain in words what exactly nomenclatural acts create, what a zoological name means, what a precise taxon "is". For example, what does the name *Carinosquilla mclaughlinae* (Example # 1) mean? A layman cannot see any difference from closely—or even distantly—related species! And even a specialist must see with his or her own eyes the object which serves as an example of this name (of this taxon) to recognize exactly what it is. For this reason, type designation was and still is a *conditio*

---

[32] See e.g. Whitworth et al. (2007).

sine qua non for the validity—the success—of a nomenclatural act, since such an act cannot tell a reader what the specimen "is" that it is based on. A reader must see an exemplar, a type ("holotype" or "lectotype" in the Code's words) in order to "realize" both the original description and the meaning of a (new) name. This demonstrates the limits of nomenclatural acts, as well as of any speech or discourse made of words.

4.5 Conclusion

Compared to "classic" speech acts, nomenclatural acts have at least three remarkable peculiarities.

First of all, they belong to a category in-between Searle's assertives and declarations; but even within this category, they are peculiar because—as long as they conform to some basic rules without which they would merely be null and void—there is no authority external to the community of actors (authors rather than speakers) confirming their validity: each actor can contest and even cancel, at any time, any nomenclatural act (this cancellation itself being open to discussion or cancellation).

Second, in a perspective which was first pointed out by Russell, "knowledge by description" is not enough to acknowledge what a zoological name means, what a zoological taxon "is": direct knowledge, or "knowledge by acquaintance", is required[33]. The same idea was expressed rather abruptly by Wittgenstein: "What *can* be shown, *cannot* be said"[34]. It is impossible to express with words any adjustment of a nomenclatural act to a given object, but only to "show" it; hence another dimension of nomenclature: to persuade readers (which would evoke Searle's "assertives") that such an act conforms to the world's reality, either if readers cannot inspect the specimen which is the case (the type of a name)—or even if they can inspect it.

Third and last, this—hopefully reciprocal—adjustment of nomenclatural acts to the world is subject to a certain "vision" of the world: each actor sees in their own way those sections of the world which are the case. The question is thus open on the adjustment of nomenclatural acts, or even of speech acts in general, to a world in which the relevant actors do not see the same things in the same way and therefore do not agree on.

---

[33] Russell (1918).

[34] Wittgenstein (1922): "Was gezeigt werden *kann, kann* nicht gesagt werden" (§ 4.1212).

Appendix 1. An Article About The Taxonomy of Crustaceans as an Example of Text Performing Primary Nomenclatural Acts

A new species of *Carinosquilla* (Crustacea, Stomatopoda, Squillidae) from the Seychelles with a cladistic analysis of the genus

Shane T. AHYONG
Marine Biodiversity and Biosecurity,
National Institute of Water and Atmospheric Research,
Private Bag 14901, Kilbirnie, Wellington (New Zealand)
s.ahyong@crustacea.net

Ahyong S. T. 2006. — A new species of *Carinosquilla* (Crustacea, Stomatopoda, Squillidae) from the Seychelles with a cladistic analysis of the genus. *Zoosystema* 28 (2): 307-314.

ABSTRACT

KEY WORDS
Crustacea,
Stomatopoda,
Squillidae,
Carinosquilla,
Indian Ocean,
new species.

A new species of *Carinosquilla* Manning, 1968, *C. mclaughlinae* n. sp., is described from the Seychelles. It is readily distinguished from its congeners by the spinulose dorsal surface of the sixth abdominal somite, and telson. *Carinosquilla mclaughlinae* n. sp. is the tenth known species of the genus. Phylogenetic relationships within *Carinosquilla* are investigated by cladistic analysis. Results support monophyly of the genus, and the synonymy of *Keijia* Manning, 1995 with *Carinosquilla*.

RÉSUMÉ

MOTS CLÉS
Crustacea,
Stomatopoda,
Squillidae,
Carinosquilla,
océan Indien,
espèce nouvelle.

Une nouvelle espèce de Carinosquilla *(Crustacea, Stomatopoda, Squillidae) des Seychelles et une analyse cladistique du genre.*
Une nouvelle espèce de *Carinosquilla* Manning, 1968, *C. mclaughlinae* n. sp., est décrite des Seychelles. Elle se distingue immédiatement des autres espèces du genre par la surface dorsale spinuleuse du sixième segment abdominal et du telson. *Carinosquilla mclaughlinae* n. sp. est la dixième espèce connue du genre. Les relations phylogénétiques dans le genre *Carinosquilla* sont recherchées par analyse cladistique. Les résultats obtenus appuient la monophylie du genre ainsi que la mise en synonymie de *Keijia* Manning, 1995 avec *Carinosquilla*.

INTRODUCTION

Species of the stomatopod crustacean genus *Carinosquilla* Manning, 1968 are distinctive for the possession of numerous longitudinal carinae covering the entire surface of the body. Until 1983, only three species of *Carinosquilla* were recognised: the type species, *Squilla multicarinata* (White, 1848) (type locality: Philippines), *C. carinata* (Serène, 1954) (type locality: Vietnam), and *C. lirata* (Kemp & Chopra, 1921) (type locality: Singapore). Naiyanetr (1983) added a fourth species, *C. thailandensis*, from the Gulf of Thailand. Moosa (1991) synonymised *C. thailandensis* with *C. carinata* based on apparent morphological overlap between specimens from New Caledonia, the Seychelles and Madagascar identified with *C. carinata* by Manning (1968) and Moosa & Cleva (1984). Manning (1995) subsequently placed *C. lirata* in a new genus *Keijia*, and therefore restricted *Carinosquilla* to *C. carinata* and *C. multicarinata*. Naiyanetr *et al.* (2000) showed that *C. thailandensis* is distinct from *C. carinata*. Ahyong (2001), in a revision of the Australian stomatopod fauna, described three new species of *Carinosquilla* (*C. redacta*, *C. australiensis*, and *C. carita*) and synonymised *Keijia* with *Carinosquilla*. Ahyong & Naiyanetr (2002) subsequently showed that most Indian Ocean records of *C. carinata* are referable to a new species, *C. spinosa*, bringing the number of recognised species of *Carinosquilla* to eight. A ninth species of *Carinosquilla*, *C. balicasag*, was described by Ahyong (2004) from the Philippines. In 2002 and 2004, all specimens of *Carinosquilla* in the Muséum national d'Histoire naturelle, Paris (MNHN) studied by Manning (1968), Moosa & Cleva (1984) and Moosa (1991) from New Caledonia and the western Indian Ocean were reexamined. As suspected by Ahyong (2001), the specimens from New Caledonia reported by Moosa (1991) as *C. carinata* comprised two species, *C. australiensis* and *C. redacta*. The Western Indian Ocean specimens also comprised two species: *C. spinosa*, and an undescribed species resembling *C. thailandensis* and *C. australiensis*. The new *Carinosquilla* is described below, and relationships within the genus are investigated by cladistic analysis.

MATERIALS AND METHODS

Morphological terminology and size descriptors follow Ahyong (2001). Abbreviations used: abdominal somite (AS), thoracic somite (TS). Carapace length (CL) and total length (TL) are measured along the dorsal midline and are given in mm. Corneal Index (CI) is given as 100 times carapace length divided by cornea width.

Type material is deposited in the collections of the MNHN.

Relationships between the nine species of *Carinosquilla* were analysed via cladistic analysis of 28 characters (Appendix 1) and rooted to *Quollastria gonypetes* (Kemp, 1911). *Lophosquilla costata* (de Haan, 1844) was included in the ingroup because of its close relationship to species of *Carinosquilla* (see Ahyong 2005) and to test the monophyly of the genus. Specimens examined are deposited in the collections of Australian Museum and MNHN. The data matrix (Appendix 2) was constructed in MacClade 4.0 (Maddison & Maddison 2000) and analysed in PAUP 4.0 (Swofford 2002) (50 heuristic searches, random input order, all characters unordered). Topological robustness was assessed via jackknifing using 1000 pseudoreplicates with 33% character deletion, implemented in PAUP 4.0.

SYSTEMATICS

Family SQUILLIDAE Latreille, 1802
Genus *Carinosquilla* Manning, 1968

Carinosquilla mclaughlinae n. sp.
(Figs 1; 2)

Carinosquilla carinata – Moosa & Cleva 1984: 427, 428 [part]. — Moosa 1991: 194-196 [part]. *Non* Serène 1954.

TYPE MATERIAL. — **Seychelles.** Holotype: REVES 2, stn 36, 4°40.7'S, 55°03.0'E, 62-55 m, sand, trawl, 10.IX.1980, ♂ TL 71 mm (MNHN-Sto 896).
Paratypes: REVES 2, stn 22, 5°16.3'S, 55°58.2'E, 60 m, sand and shell, dredge, 6.IX.1980, 1 ♀ TL 77 mm (MNHN-Sto 894). — REVES 2, stn 42, 4°31.6'S, 56°09.7'E, 55-60 m, sand and shell, trawl, 13.IX.1980, 1 ♀ TL 83 mm (MNHN-Sto 897).

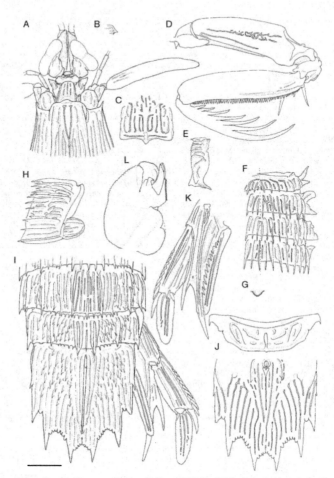

Fig. 1. — *Carinosquilla mclaughlinae* n. sp., ♂ holotype (TL 71 mm) (MNHN-Sto 896), Seychelles: **A**, anterior; **B**, dorsal process of antennular somite, right lateral view; **C**, posteromedian portion of carapace; **D**, right raptorial claw; **E**, TS5, right lateral view; **F**, TS5-8, right dorsal view; **G**, TS sternal keel, right lateral view; **H**, AS1, right lateral view; **I**, posterior abdomen, telson and uropod; **J**, AS6 and telson, ventral view; **K**, right uropod, ventral; **L**, pleopod 1 endopod, left anterior view. Abbreviations: **AS**, abdominal somite; **TS**, thoracic somite. Scale bar: A-K, 4 mm; L, 2 mm.

ETYMOLOGY. — This species is named in honour of Pat McLaughlin for her longstanding contributions to carcinology.

DISTRIBUTION. — Known only from the Seychelles at depths of 55-62 m.

MEASUREMENTS. — Male (n = 1) TL 71 mm, female (n = 2) TL 77-83 mm. Other measurements of holotype: CL 15.61 mm, anterior carapace width 7.48 mm, cornea width 4.10 mm, antennular peduncle length 16.48 mm, antennal scale length 11.10 mm.

DIAGNOSIS. — Eyestalk with irregular dorsal carinae. Ocular scales with apices entire. Raptorial claw merus with vermiform sculpture on outer margin; dactylus with six or seven teeth. Dorsolateral carinae of AS6 and telson divided, forming field of spines.

DESCRIPTION

Eyestalk with short, irregular dorsal carinae; CI 380-409. Ocular scales entire, apices truncate, not bifurcate.

Antennular peduncle 1.02-1.06 CL. Antennular somite dorsal processes with acute, triangular apices; directed anterolaterally. Antennal scale length 0.65-0.72 CL.

Rostral plate trapezoid, about as long as broad; apex truncate to rounded; lateral margins carinate; with long, distinct, median carina flanked by longitudinal supplementary carina (interrupted distally in male specimen).

Carapace anterior width 0.47-0.50 CL; anterior bifurcation of median carina opening anterior to dorsal pit; dorsum covered with numerous closely spaced longitudinal carinae; anterolateral spines reaching anteriorly approximately to level of base of rostral plate.

Raptorial claw dactylus with six or seven teeth; outer surface of propodus with short oblique carina and longitudinal carina subparallel to occlusal margin; carpus with undivided dorsal carina; outer surface of merus with vermiform sculpture.

Mandibular palp 3-segmented. Maxillipeds 1-4 each with epipod.

TS5 lateral process bilobed; anterior lobe a slender spine directed anterolaterally; posterior lobe short, slender, directed laterally. TS6 lateral process bilobed; anterior lobe broad, quadrate, apex truncate; posterior lobe broad, triangular. TS7 lateral process bilobed; anterior lobe triangular, blunt; posterior lobe larger than anterior lobe, broad, triangular, anterior margin convex, apex blunt. TS8 with sharp, triangular, lateral process; sternal keel rounded.

TS5 dorsal carinae longitudinal or oblique, none transverse; supplementary carinae between submedian carinae posteriorly spined. TS6-8 and AS1-5 with most or all supplementary dorsal carinae posteriorly armed above level of lateral carinae. AS1-5 with one to three (usually two) supplementary carinae armed between lateral and marginal carinae. Surface of AS1-5 between intermediate and lateral carinae with some shorter carinae also spined. Articular membrane between AS5 and 6 lined with small spines. AS5 with three or four spinules on posterior margin between submedian and lateral carinae. AS6 submedian carina tricarinate, usually with median spine on posterior margin between submedian spines; with three to six spinules on posterior margin lateral to submedian spines; with short dorsal carinae usually posteriorly spined; sternum with anterior and posterior transverse carina, median carina flanked by four or five sinuous transverse carinae, some uniting laterally; with small ventrolateral spine anterior to uropodal articulation. Abdominal somites with normal complement of carinae spined as follows: submedian 1-6, intermediate 1-6, lateral 1-6, marginal 1-5.

Telson about as long as broad; prelateral lobe longer than margin of lateral tooth, terminating in sharp spine; dorsolateral surface with numerous supplementary longitudinal carinae; proximal supplementary carinae subdivided and posteriorly spined forming field of spines; denticles submedian 4, intermediate 7 or 8, lateral 1; ventral surface with long postanal carina, numerous long supplementary carinae and several short carina and tubercles proximomedially.

Uropodal protopod with smooth outer margin; inner margin with 10-13 slender spines; with short ventral tubercle anterior to endopod articulation; protopod terminal spines with lobe on outer margin of inner spine rounded, narrower than adjacent spine, proximal margin slightly concave. Exopod proximal segment outer margin with 10 or 11 movable spines, distalmost not exceeding mid-length of distal segment; distal margin 2 ventral spines, outer longest. Exopod distal segment slightly longer than proximal segment; unpigmented; dorsally and ventrally carinate. Endopod dorsally and ventrally carinate.

Colour in alcohol

Faded to pale brown. Carinae of carapace dark brown. Posterior margins of thoracic and abdominal somites dark brown. AS2 with dark transverse bar across submedian carinae. AS5 with diffuse pigmentation dorsally between intermediate carinae, and with large black patch below intermediate carinae. Telson with apices of primary teeth and apex of median carina dark brown. Uropodal protopod dark at articulation with exopod; distal half of endopod dark; exopod unpigmented except on articular membrane separating distal and proximal segments.

REMARKS

Carinosquilla mclaughlinae n. sp. is unique in the genus in having the supplementary carinae of AS6 and proximomedial supplementary carinae of the telson broken into short posteriorly spined carinae forming a field of spines. In bearing carinate ocular peduncles and entire, non-bifurcate ocular scales, *Carinosquilla mclaughlinae* n. sp. resembles *C. thailandensis*, *C. australiensis* and *C. balicasag*. *Carinosquilla mclaughlinae* n. sp. further resembles *C. thailandensis* in the vermiform sculpture on the outer surface of the merus of the raptorial claw, but differs by having six or seven instead of five teeth on the dactylus of the claw, and in having the posterior spines of the supplementary abdominal carinae that continue below the level of the intermediate carinae. In having six or seven teeth on the dactylus of the raptorial claw and spined supplementary carinae between the abdominal intermediate and lateral carinae, *C. mclaughlinae* n. sp. resembles *C. australiensis* and *C. balicasag*. *Carinosquilla mclaughlinae* n. sp., however, differs from *C. australiensis* and *C. balicasag* in having vermiform sculpture on the outer surface of the merus of the claw, and in having armed supplementary abdominal carinae not only between the intermediate and lateral carinae, but also between the lateral and marginal carinae. It further differs from both *C. australiensis*, *C. balicasag* and *C. thailandensis* in having many of the supplementary carinae on the lateral surfaces of the abdomen broken into several short posteriorly spined carinae, in having the prelateral lobe of the telson terminating in a sharp spine, in having a carinate outer margin of the propodus of the claw, and in colour pattern. In *C. mclaughlinae* n. sp., a black patch is present laterally on AS5 (absent in *C. australiensis* and *C. thailandensis*) and the segments of the uropodal exopod appear to be unpigmented (entirely or partially dark in other species).

The three specimens of the type series are morphologically uniform but vary in the following features: 1) the male bears seven teeth on the dactylus of the raptorial claw, whereas the two females bear six dactylar teeth; and 2) the carina lateral to the median carina of the rostral plate is interrupted distally in the male, whereas the carina is entire in the two females. These differences are likely to represent individual variation instead of sexual dimorphism. Variation in the number of dactylar teeth on the raptorial claw is present also in *C. redacta*, and similar variation in the development of the rostral plate carinae is occasionally evident in all species of *Carinosquilla*.

KEY TO SPECIES OF *CARINOSQUILLA* MANNING, 1968

1. Eyestalk with irregular dorsal carinae 4
   — Eyestalk without dorsal carinae 2

2. Dorsal carinae on either side of midline of TS5 transverse. Inner margin of uropodal protopod spinose 3
   — Dorsal carinae on either side of midline of TS5 longitudinal or oblique, not transverse. Inner margin of uropodal protopod crenulate .... *C. lirata*

3. Mandibular palp present. Telson prelateral lobe with sharp apex .... *C. multicarinata*
— Mandibular palp absent. Telson prelateral lobe with blunt apex .... *C. carita*

4. Ocular scales with bifurcate apices 5
— Ocular scales with apices entire, not bifurcate 7

5. Posterior margin of AS1-4 between submedian carinae lined with spines ...... *C. spinosa*
— Posterior margin of AS1-4 between submedian carinae unarmed 6

6. AS1-2 with unarmed submedian and intermediate carinae .... *C. redacta*
— AS1-2 with armed submedian and intermediate carinae .... *C. carinata*

7. Merus of raptorial claw with single longitudinal carina on outer margin 8
— Merus of raptorial claw with vermiform sculpture on outer margin 9

8. AS6 without posteriorly armed supplementary carinae between submedian and intermediate carinae. Distal segment of uropodal exopod entirely dark .... *C. australiensis*
— AS6 with one or two posteriorly armed supplementary carinae between submedian and
   intermediate carinae. Uropodal exopod distal segment dark on proximal third ....
   .... *C. balicasag*

9. Dactylus of raptorial claw with five teeth. Dorsal carinae of AS6 and telson entire, not
   forming field of spines ...., *C. thailandensis*
— Dactylus of raptorial claw with six or seven teeth. Dorsolateral carinae of AS6 and telson
   divided, forming field of spines .... *C. mclaughlinae* n. sp.

RELATIONSHIPS IN *CARINOSQUILLA*

Analysis of the dataset resulted in five most parsimonious trees (length 50, consistency index 0.73,
retention index 0.83). The strict consensus is shown
in Figure 2. Jackknife proportions indicate a robust
topology. Monophyly of *Carinosquilla* is strongly
supported with a 100% jackknife proportion.

Carinosquilla falls broadly into four groups. The
first group comprises *C. carita* and *C. multicarinata*,
and is the sister to all other species of the genus. It is
united by the presence of transverse carinae on TS5
and long, straight, carinae flanking the posterior
bifurcation of the carapace that reach the posterior
margin of the carapace.

The second group includes *C. lirata*, differing from
other species of *Carinosquilla* in having a crenulate
instead of spinose inner margin on the uropodal
protopod. Manning (1995) placed *C. lirata* in a
monotypic genus *Keijia*, chiefly on the basis of the
suppression of the mandibular palp in contrast to
the species of *Carinosquilla* known at the time. The
position of *C. lirata*, nested within other species of

the genus, corroborates Ahyong's (2001) synonymy
of *Keijia* with *Carinosquilla*.

The third and fourth groups are united by having
carinate eyestalks, numerous supplementary carinae
on the sternum of AS6 and tricarinate submedian
carinae on AS6.

The third group comprising *C. carinata*, *C. spinosa*
and *C. redacta* is united principally by the possession of bifurcate ocular scales.

The fourth group comprises *C. australiensis*,
C. balicasag, *C. mclaughlinae* n. sp. and *C. thailandensis* and is united by the presence of carinae
on the proximal portion of the antennal scales and
tricarinate submedian carinae on AS5.

The geographical ranges of many species overlap,
but within the aforementioned groups, species have
discrete or near discrete ranges. In group 1, *C. multicarinata* ranges from Japan to the Southern India,
whereas *C. carita* has an adjacent but discrete range
along the coast of northern Australia (Ahyong 2001).
The single species of group 2 ranges from the South
China Sea to Madras, India (Shanbogue 1986). In
group 3, *C. redacta* is known only from northern Aus-

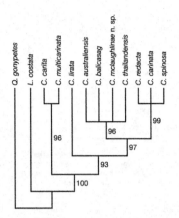

FIG. 2. — Phylogeny of *Carinosquilla* Manning, 1968, strict consensus of five most parsimonious topologies (length 50, consistency index 0.73, retention index 0.83). Jackknife proportions indicated on branches. Characters are listed in Appendix 1; data matrix is presented in Appendix 2. Abbreviations: **L.**, *Lophosquilla*; **Q.**, *Quollastria*.

tralia, *C. carinata* occurs in the South China Sea from Vietnam to the Gulf of Thailand, south to Indonesia, and *C. spinosa* ranges from the western Indian Ocean to the Andaman Sea. In group 4, *C. mclaughlinae* n. sp. is known only from the western Indian Ocean, *C. balicasag* is known only from Balicasag Island (Philippines) and *C. thailandensis* ranges from the Gulf of Thailand to northern Australia, where it overlaps with *C. australiensis* (see Ahyong 2001).

Acknowledgements
I wish to thank Alain Crosnier for his generous hospitality at the MNHN in 2004, during which this study was initiated. Financial support from a Sydney Grammar School Fellowship and a Visiting Fellowship from the MNHN are gratefully acknowledged.

REFERENCES

AHYONG S. T. 2001. — Revision of the Australian Stomatopod Crustacea. *Records of the Australian Museum* Suppl. 26: 1-326.

AHYONG S. T. 2004. — New species and new records of Stomatopod Crustacea from the Philippines. *Zootaxa* 793: 1-28.

AHYONG S. T. 2005. — Phylogenetic analysis of the Squilloidea (Crustacea: Stomatopoda). *Invertebrate Systematics* 19: 189-208.

AHYONG S. T. & NAIYANETR P. 2002. — Stomatopod Crustaceans from Phuket and the Andaman Sea. *Phuket Marine Biological Centre Special Publication* 23 (2): 281-312.

MADDISON D. R. & MADDISON W. P. 2000. — *MacClade. Analysis of Phylogeny and Character Evolution.* Version 4.0. Sinauer Associates, Sunderland, Massachusetts.

MANNING R. B. 1968. — A revision of the family Squillidae (Crustacea, Stomatopoda), with the description of eight new genera. *Bulletin of Marine Science* 18 (1): 105-142.

MANNING R. B. 1995. — Stomatopod Crustacea of Vietnam: the legacy of Raoul Serène. *Crustacean Research* Special No. 4: i-viii + 1-339.

MOOSA M. K. 1991. — The Stomatopoda of New Caledonia and Chesterfield Islands, *in* RICHER DE FORGES B. (ed.), *Le benthos des fonds meubles des lagons de Nouvelle-Calédonie* 1. Orstom, Paris: 149-219.

MOOSA M. K. & CLEVA R. 1984. — Sur une collection de stomatopodes (Crustacea : Hoplocarida) provenant des îles Seychelles. *Bulletin du Muséum national d'Histoire naturelle* Paris, sér. 4, sect. A, 6 (2): 421-429.

NAIYANETR P. 1983. — Two stomatopod crustaceans from the Gulf of Thailand with a key to the genus *Carinosquilla* Manning, 1968. *Senckenbergiana biologica* 63 (5/6): 393-399 (dated 1982, published 1983).

NAIYANETR P., AHYONG S. T. & NG P. K. L. 2000. — Reinstatement of *Carinosquilla thailandensis* Naiyanetr, with a first record of *Alima orientalis* Manning from the Gulf of Thailand, and notes on *Cloridina pelamidae* (Blumstein) (Stomatopoda: Squillidae). *Crustaceana* 73 (10): 1291-1296.

SERÈNE R. 1954. — Observations biologiques sur les stomatopodes. *Mémoires de l'Institut océanographique de Nhatrang* 8: 1-93, pls 1-10.

SHANBOGUE S. L. 1986. — Studies on stomatopod Crustacea from the seas around India, *in* JAMES P. S. B. R. (ed.), *Recent Advances in Marine Biology* (Dr S. Jones 70th Birthday Commemoration Volume). Today & Tomorrow's Printers & Publishers, New Delhi: 515-567.

SWOFFORD D. L. 2002. — *PAUP*. Phylogenetic Analysis Using Parsimony (* and Other Methods).* Version 4.0b10. Sinauer Associates, Sunderland, Massachusetts.

Submitted on 1st October 2005;
accepted on 12 December 2005.

Ahyong S. T.

APPENDIX 1

Morphological characters used for cladistic analysis of *Carinosquilla* Manning, 1968.

1. Ocular peduncles: smooth (0); with short dorsal carinae (1).
2. Ocular scales: entire (0); divided (1).
3. Dorsal processes of antennular somite: non-carinate (0); carinate (1).
4. Distal portion of antennal protopod: non-carinate (0); carinate (1).
5. Carapace carinae density: few, widely spaced (0); numerous, closely spaced (1).
6. Anterior bifurcation length: normal, distant from dorsal pit (0); long, base close to dorsal pit (1).
7. Anterior bifurcation: interrupted basally (0); uninterrupted basally (1).
8. Surface flanking posterior bifurcation of carapace: smooth (0); with straight carina, reaching posterior margin (1); with inward curving carina (2); tuberculate (3).
9. Carapace posteromedian projection: angular, pointed (0); stubby, blunt (1).
10. Raptorial claw dactylar teeth: five (0); six or seven (1).
11. Mandibular palp: present (0); absent (1).
12. Raptorial claw merus lateral: smooth (0); carina (1); sculpture (2).
13. AS1-5 posterior margin between submedian and intermediate carinae: at most with short teeth (0); long spines (1).
14. Posterior margin of AS1-4 below intermediate carinae: unarmed (0); spinous (1).
15. TS5 dorsolateral carinae: longitudinal (0); transverse (1).
16. AS6 sternal carinae: normal (0); few (1); numerous (2).
17. Antennal scale proximal surface: smooth (0); carinate (1).
18. Dorsal abdominal carinae: absent (0); few (1); numerous (2).
19. AS5 submedian carinae: single (0); tricarinate (1).
20. AS6 submedian carina: unicarinate (0); tricarinate (1).
21. AS6 surface between submedian and intermediate carinae: smooth (0); with numerous short carinae (1); long carinae (2).
22. Mid-dorsal telson carinae: short and broken (0); entire (1); spinose (2).
23. Telson lateral denticles: one (0); two (1).
24. Uropodal protopod inner margin: crenulate (0); spinose (1).
25. Uropodal protopod ventral carina of inner terminal spine: unicarinate (0); bifurcate (1).
26. Uropodal endopod dorsal surface: smooth (0); with two carinae (1); with four carinae (2).
27. Uropodal endopod ventral surface: smooth (0); with two carinae (1); with four carinae (2).
28. Multiple short carinae on uropodal exopod distal segment: absent (0); present (1).

APPENDIX 2

Data matrix used for cladistic analysis of *Carinosquilla* Manning, 1968. For characters descriptions see Appendix 1. Abbreviations: **L.**, *Lophosquilla*; **Q.**, *Quollastria*.

	1	1	2	2
	1	0	0	8
C. carinata	11111012010100020201101111111
C. redacta	11111012010100020201101111111
C. spinosa	11111012010100020201101111111
C. lirata	00101113011100010200100011110
C. australiensis	10101112000111021211210111111
C. balicasag	10111112000111021211210111111
C. mclaughlinae n. sp.	10111112110211021211120111110
C. thailandensis	10111112100210021211210111221
C. carita	00001111001110110200210100000
C. multicarinata	00001111000110110200210100000
L. costata	00000003001000001001000000000
Q. gonypetes	00000000000000000000000000000

Appendix 2. An Article About The Taxonomy of Beetles as an Example of Text Performing Secondary Nomenclatural Acts

Zootaxa 1546: 63–68   (2007)
www.mapress.com/zootaxa/
Copyright © 2007 · Magnolia Press

ISSN 1175-5326 (print edition)

ZOOTAXA
ISSN 1175-5334 (online edition)

Ptiliola flammifera (Młynarski) reinstated as a species distinct from *P. kunzei* (Heer) (Coleoptera: Ptiliidae)

OSCAR VORST

National Museum of Natural History, Naturalis, PO Box 9517, 2300 RA Leiden, The Netherlands. E-mail: vorst@naturalis.nl

Abstract

Ptiliola flammifera (Młynarski) is recognised as a species distinct from *P. kunzei* (Heer). Apart from differences in the structure of the aedeagus, both species can be separated by the pronotal surface and the apical fringe of the elytra. The male of *P. flammifera* carries a distinct tuft of hairs on the metaventrite, absent in *P. kunzei*. Currently, *P. flammifera* is only known from Poland and the Netherlands. A key for the identification of all three Palaearctic *Ptiliola* species is presented. A lectotype is designated for *Trichopteryx kunzei* Heer, 1841.

Key words: Ptiliidae, featherwing beetle, *Ptiliola*, Palaearctic region, identification key, lectotype designation

Introduction

The ptiliid genus *Ptiliola* Haldeman, 1848, includes minute beetles measuring about 0.6 mm. For a long time it was named *Nanoptilium* Flach, 1889, and treated as a subgenus of *Ptiliolum* Flach, 1888. Generic rank was attributed by Besuchet (1971), based on elytral structure, spermathecal morphology and the pygidium, which in *Ptiliola* is ornated with a single spine only. *Ptiliola* was erected by Haldeman (1848) for the four species of Gillmeister's "IV. Gruppe" [= fourth group] of *Trichopteryx*, the only ptiliid genus then recognized (Gillmeister 1845). Amongst them is *Trichopteryx nana* Stephens, 1830 (listed as a *nomen oblitum* by Johnson (2004) and generally considered a senior synonym of *Trichopteryx kunzei* Heer, 1841), that was designated as the type species of the genus *Ptiliola* by Motschulsky (1869). Biström and Silfverberg (1979) argued that *Ptiliola* was the valid name of the genus, and that it had to replace the junior synonym *Nanoptilium*, which by original monotypy has *Trichopteryx kunzei* Heer as type species.

In the Palaearctic region two valid species are currently recognised (Johnson 2004): *Ptiliola kunzei* (Heer, 1841) and *P. brevicollis* (Matthews, 1860). Both species have also been reported from North America (Johnson 1990; Sörensson 2003). The first one is considered a true Holarctic element, while *P. brevicollis* is of uncertain origin (Sörensson 2003).

Nanoptilium flammiferum and *N. aequisetum*

Młynarski (1985) described two new species in his treatment of the Polish species of *Ptiliola* (then *Nanoptilium*). He recognized *P. flammifera* (Młynarski, 1985) and *P. aequiseta* (Młynarski, 1985) in addition to *P. kunzei* and *P. brevicollis*. Judging from the original descriptions and especially the pronotal shapes figured, both species should be very close to *P. kunzei*. The species were separated from *P. kunzei* by the type of male metasternal pubescence, subtle differences in pronotal shape and characteristics of the apical fringe of the elytra. Some of these characters are unconventional in ptiliid taxonomy. There are no records published of those two species since.

Recently, Johnson (2003) concluded that both "species" should be considered forms and therefore synonyms of *P. kunzei*. His judgement was based on careful consideration of the original descriptions, which do not mention differences in the male genitalia nor spermathecae, and the observation that "male European specimens of *Ptiliola kunzei* seen by me have pubescence characters of *aequiseta*". Unfortunately, type specimens of both species were not accessible for study.

Recently, I got hold of several specimens of a *Ptiliola* species different from *P. brevicollis* and *P. kunzei*, that matches the description of *P. flammifera*. Close examination revealed additional differential characters in pronotal structure and aedeagus indicating that *P. flammifera* should be reinstated as a proper species.

Depositories

cMS	Collection M. Sörensson, Lund, Sweden
cOV	Collection O. Vorst, Utrecht, The Netherlands
MHNG	Muséum d'Histoire Naturelle, Genève, Switzerland (G. Cuccodoro)
RMNH	National Museum of Natural History, Naturalis, Leiden, The Netherlands (A. van Assen)
WML	World Museum Liverpool, United Kingdom (G. Knight)
ZMAN	Zoological Museum, Amsterdam, The Netherlands (B. Brugge)

Ptiliola flammifera (Młynarski, 1985)
(Figs. 1, 4, 5, 7, 10)

Material studied. THE NETHERLANDS: Prov. of Gelderland: 2 exx, De Imbosch, Nieuwe Aanleg, UTM GT024723, 7.vi.2002 (♀), 28.viii.2002 (♂), carcass of Highland cattle in *Pinus* plantation, O. Vorst (cOV); 2 exx, De Imbosch, Veertien Bunder, UTM GT0472, 30.viii.2002 (♂), 19.xi.2002 (♀), carcass of wild boar in mixed forest, O. Vorst (cOV); ♀(cf *P. flammifera*), Doorwerth, 2.iv.1923, Van der Wiel (ZMAN); ♀, Loenen, Loenermark, UTM GT049742, 7.vi.2002, carcass of Highland cattle in dense *Pseudotsuga* stand, O. Vorst (cOV); 6 exx, Worth-Rhederzand, Tunnekes, UTM KC947701, 18.iv.2003 (♀), 9.v.2003 (2 ♂♂ [Figs. 1, 4, 5, 7, 10], 3 ♀♀), carcass of Highland cattle in open *Pinus* forest, O. Vorst (cOV).

Diagnosis. Very similar in general shape and overall appearance to *P. kunzei*. Size (labrum to apex of elytra): 0.60–0.66 mm (average 0.63 mm, N = 7).

Pronotal pubescence is less dense than in *P. kunzei*; reticulation on the pronotum is somewhat coarser and less pronounced, especially towards the frontal margin (Figs. 1, 2). As a result, the overall appearance of the pronotum, when studied under reflecting light, is more shiny. The same is true for the dorsal surface of the head.

The apical margin of elytra ornated with a regular 2.5-3.0 μm long fringe (Fig. 5). In *P. kunzei* the fringe is finer and more dense, measuring only 1.5-2.0 μm; towards the suture the fringe is fused to form a few characteristic brush-shaped structures (Fig. 6). The structure of the elytral fringe is best studied by transmitted light microscopy at high magnification (300 X) with the object in a matrix (e.g. water or a resin), or by scanning electron microscopy. Spermatheca is very similar to that of *P. kunzei*.

Male aedeagus is smaller than in *P. kunzei* (Figs. 7, 8); the aedeagal sclerites are differently shaped, in ventral view more stout, in lateral view more curved than in *P. kunzei* (Figs. 10, 11). Male metaventrite (in Coleoptera this structure is—erroneously—known as metasternum, cf Beutel & Leschen 2005) is smoothly excavated, apically bordered by a distinct tuft of erect hairs (Fig. 4); the excavation and the tuft are absent in *P. kunzei*.

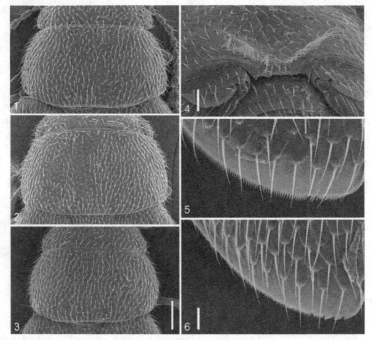

FIGURES 1–6. Scanning electron micrographs of *Ptiliola flammifera* (1, 4, 5), *P. kunzei* (2, 6) and *P. brevicollis* (3). 1–3. Pronotum, scale bar 50 μm; 4. Male metaventrite, scale bar 20 μm; 5, 6. Apical fringe of elytra, scale bar 10 μm.

Although no syntypic material was studied, the identity of this species seems without doubt. The tuft of hairs on the metaventrite in the male (Fig. 4) and the structure of the apical fringe of the elytra (Fig. 5) closely match the figures in the original description (Młynarski 1985).

Bionomics. The amount of material at hand does not allow drawing firm conclusions about the ecological preferences of *P. flammifera*, but it seems to be generally associated with decaying organic material. Although all Dutch records are from carcasses of larger mammals this result is biased by the fact that little other potential habitats were sampled from these localities. In Poland, the species has been reported from decaying hay and horse droppings (Młynarski 1985). Possibly a forest species.

Distribution. So far only known from Poland (Młynarski 1985) and the Netherlands, but probably of wider distribution.

***Ptiliola kunzei* (Heer, 1841)**
(Figs. 2, 6, 8, 11)

Type material. Lectotype (by present designation): ♀: [no original label], "Lectotypus / *Trichopteryx kunzei* Heer, 1841 / design. O.Vorst 2007" (WML, Collection F. Chevrier). **Paralectotypes:** 2 exx: [no original label], "Paralectotypus / *Trichopteryx kunzei* Heer, 1841 / design. O.Vorst 2007" (WML, Collection F. Chevrier).

Additional material studied. THE NETHERLANDS: Prov. of Drenthe: 2 ♀♀, Anderen, Eexterveld, UTM LD4647, 29.v.2005, cattle dung in grazed heath land, O. Vorst (cOV); 2 ♂♂ [Figs. 2, 6, 8, 11], 2 ♀♀, Dwingeloosche Heide, UTM LD2452, 6.ix.2003, cattle dung in grazed heath land, Vorst (cOV); 2 ♀♀, Uffelte, Oosterzand, UTM LD1554, 26.viii.2006, cattle dung in grazed heath land, Vorst (cOV); Prov. of Gelderland: ♀, Apeldoorn, Kerkhoven (RMNH); Prov. of Utrecht: ♀, Doorn, .viii.1888, Neervoort van de Poll (RMNH); Prov. of Zuid-Holland: ♂, Den Haag, v [May], Everts (RMNH); Prov. of Noord-Brabant: ♂, Westelbeers, Landschotsche Heide, 11.vi.2006, horse dung in forest, Vorst (cOV); Prov. of Limburg: ♂, Neercanne, 17.x.1950, Excursion St. Pietersberg (RMNH); 1 ♂ 2 ♀♀, St. Pietersberg, 23.iii.1950, Excursion St. Pietersberg (RMNH); 2 ♂♂, Weert, .vi.1919, MacGillavry (ZMAN). **GERMANY:** Niedersachsen: 3 exx, Aldrup, 25.vi.1959 (♂), 28.iv.1962 (♀), 6.x.1962 (♂), Kerstens (RMNH); Saarland: ♀, Nennig, 8.vi.1996, heap of decaying hay, Vorst (cOV). **SWITZERLAND:** Genève: ♂, 3 exx, La Plaine, 19.ix.1983, rotting plant material, Besuchet (MHNG); Vaud: ♂, ♀, 6 exx, Bussigny, 6.vi.1953, horse droppings, Besuchet (MHNG). **LATVIA:** Valga Distr.: ♀, Mežmuiža, Rauza River, UTM MD3959,14.vi.2005, river bank in *Alnus* forest, Vorst (cOV). **FINLAND**: Oulu Prov.: ♂, 2 ♀♀, Hiidenportti National Park, UTM PL0286, 2.viii.2006, decaying boletes in mixed forest, Vorst (cOV); Prov. of Western Finland: ♂, 2 ♀♀, Lahdenkylä, UTM MJ1070, 4.viii.2006, horse droppings in forest, Vorst (cOV).

Remarks. Heer (1841) attributed *Trichopteryx kunzei* in his "Fauna Coleopterorum Helvetica" to Chevrier, who, however, never published this manuscript name. The type locality was cited as Genf (= Geneva). No syntypic material of this species could be traced in the collections of the Eidgenössische Technische Hochschule Zürich (M. Schmid, personal communication), where the Heer collection is preserved, nor in MHNG (G. Cuccodoro, personal communication). In the Chevrier collection, kept at WML, there are three unlabelled specimens standing as *Ptilium kunzei*. From Chevrier's catalogue that accompanies his collection it becomes clear that these should be treated as syntypes. The entry under "Kunzei Mihi" reads: "G. HFH.", where "G." is used as shorthand for Geneva and "HFH." most likely stands for Heer's Fauna Helvetica. The three specimens are conspecific and fit the current interpretation of *Trichopteryx kunzei*. A female specimen, whose identity could be confirmed by the apical fringe of the elytra, is herewith designated as lectotype.

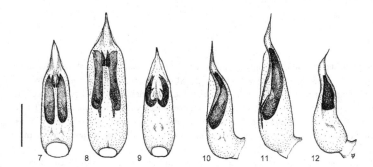

FIGURES 7–12. Aedeagi of *Ptiliola flammifera* (7, 10), *P. kunzei* (8, 11) and *P. brevicollis* (9, 12) in ventral (7–9) and lateral (10–12) view. Scale bar 50 μm.

Ptiliola brevicollis **(Matthews, 1860)**
(Figs. 3, 9, 12)

Material studied. THE NETHERLANDS: Prov. of Groningen: ♂. [Figs. 3, 9, 12], 4 ♀♀, Overschild, Schildmeer, UTM LE5506, 18.ix.2004, heap of decaying grass, Vorst (cOV); Prov. of Overijssel: ♀, Fortmond, Duursche Waarden, UTM LD0306, 24.iv-14.v.1998, window trap at river meadow, Vorst (cOV); ♀, Weerribben, Woldakkers, UTM GU0153, 31.viii.2001, heap of old hay and reeds, edge of *Alnus* forest, Vorst (cOV); Prov. of Limburg: ♀, Urmond, UTM FS9552, 22.viii.2000, heap of horse manure, forest edge, Vorst (cOV). **GERMANY:** Saarland: 2 ♀♀, Nennig, 8.vi.1996, heap of decaying hay, Vorst (cOV). **LATVIA:** Krāslava Distr.: 1 ex, Šķeltiņi, 1.viii.1995, A. Barševskis (cOV). **POLAND:** Śląsk Voivodship: ♀, "Silesia, Teschen" [= Cieszyn], Th. von Wanka (RMNH). **SWEDEN:** Uppland: ♂, Storvreta, 30.x.1993, A. Lindelöw (cMS).

Key to the Palaearctic species of *Ptiliola*

1. Smaller, length (labrum to elytral apex) usually less than 0.59 mm; pronotum less transverse, sides more rounded, posterior angles little defined (Fig. 3). Male: aedeagal sclerites not reaching half the length of the aedeagus, strongly sclerotized (Figs. 9, 12); metaventrite simple .... *P. brevicollis* (Matthews)
-   Larger, length usually more than 0.59 mm; pronotum more transverse, sides less rounded, somewhat narrowed anteriad (Figs. 1, 2). Male: aedeagal sclerites larger, more than half the length of the aedeagus, sclerotization weaker (Figs. 7, 8, 10, 11); metaventrite simple or ornated with a tuft of hairs (Fig. 4) 2
2. Apical fringe of elytra regular and somewhat wider (Fig. 5) (this character is best appreciated by transmitted light microscopy). Pronotal reticulation somewhat coarser and less pronounced, especially towards the frontal margin (Fig. 1); as a result the overall appearance of the pronotum, when studied under reflecting light, more shiny. Male: aedeagus smaller; aedeagal sclerites in ventral view more stout (Fig. 8); metaventrite smoothly excavated, apically bordered by a distinct tuft of erect hairs (Fig. 4) ....
    ....*P. flammifera* (Młynarski)
-   Apical fringe of elytra more narrow, towards the suture fused to form a few brush-shaped extensions (Fig. 6). Pronotal reticulation somewhat finer and more pronounced (Fig. 2); as a result the overall appearance more dull. Male: aedeagus larger; aedeagal sclerites in ventral view more slender, somewhat contracted in the middle (Fig. 7); metaventrite simple..*P. kunzei* (Heer)

Discussion

The present study shows that the recently synonymized *Ptiliola flammifera* is a valid species distinct from *P. kunzei*. Młynarski (1985) used somewhat unconventional characters when describing *P. flammifera* and *P. aequiseta*, including the apical fringe of the elytra. At the same time other, widely accepted ones, like the excavation of the male metaventrite and aedeagal and spermathecal characters were ignored. The main reason for this apparently is that he based his studies on slide mounted rather than dry-mounted specimens, a method which is unusual for European workers. Only in material prepared this way the apical fringe of the elytra is easily observed.

Ptiliola aequiseta, the second species described by Młynarski (1985), seems to be correctly synonymized with *P. kunzei* (Johnson 2003). It is clear from the original description that the supposed differences from *P. kunzei* are much more subtle than in *P. flammifera*. The pronotal shape and the apical fringe of the elytra are described as identical in both species. The slight differences mentioned are in pubescence of both male and female metaventrite only (Młynarski 1985). The genitalia were not illustrated.

Acknowledgements

I would like to thank A. van Assen (RMNH), B. Brugge (ZMAN), G. Cuccodoro (MHNG), G. Knight (WML) and M. Sörensson (Lund) for the loan of material; J. Goud and C. van den Berg (RMNH) for preparing the scanning electron micrographs, and A. ten Hoedt, J. Potkamp and H. van Dijk of Vereniging Natuurmonumenten for giving permission to study the beetle fauna at "Veluwezoom" and providing assistance afield.

References

Besuchet, C. (1971) 21. Familie: Ptiliidae [excl. *Acrotrichis*]. *In*: Freude, H., Harde, K.W. & Lohse, G.A. (Eds.), *Die Käfer Mitteleuropas. Band 3. Adephaga 2, Palpicornia, Histeroidea, Staphylinoidea 1.* Goecke & Evers, Krefeld, pp. 311–334.

Beutel, R.G. & Leschen, R.A.B. (Eds.) (2005) *Handbook of Zoology. Volume IV, Arthropoda: Insecta. Part 38, Coleoptera, Beetles. Volume 1: Morphology ans systematics (Archostemata, Adephaga, Myxophaga, Polyphaga partim).* Walter de Gruyter, Berlin, xi + 567 pp.

Biström, O. & Silfverberg, H. (1979) The type species of the European genera of Ptiliidae (Coleoptera). *Annales Entomologici Fennici*, 45, 12–15.

Flach, [C.] (1888) 34. Fam. Trichopterygidae. *In*: Seidlitz, G., *Fauna Baltica. Die Kaefer (Coleoptera) der deutschen Ostseeprovinzen Russlands. Zweite neu bearbeitete Auflage.* Hartungsche Verlagsdruckerei, Königsberg, pp. [Arten] 289–295.

Flach, C. (1889) *Bestimmungs-Tabellen der europäischen Coleopteren. XVIII. Heft. Enthaltend die Familie der Trichopterygidae.* E. Reitter, Wien, 54 pp., plates x–xiv.

Gillmeister, C.J.F. (1845) Trichopterygia, Beschreibung und Abbildung der haarflügeligen Käfer. *In*: Sturm, J., *Deutschlands Fauna in Abbildungen nach der Natur mit Beschreibungen. V. Abtheilung. Die Insecten. Siebzehntes Bändchen. Käfer.* Selbstverlag, Nürnberg, pp. i–xviii + 1–98, plates cccxx–cccxxviii.

Haldeman, S.S. (1848) Descriptions of North American Coleoptera, chiefly in the cabinet of J.L. Le Conte, M.D., with references to described species. *Journal of the Academy of Natural Sciences of Philadelphia*, (2)1, 95–110.

Heer, O. (1841) *Fauna Coleopterorum Helvetica. Pars I.* [Fasciculus III]. Orelii, Fueslini et Sociorum, Turici, pp. 361–652.

Johnson, C. (1990) New observations on Canadian Ptiliidae (Coleoptera). *The Coleopterists Bulletin*, 44(3), 267–268.

Johnson, C. (2003) Further notes on Palaearctic and other Ptiliidae (Coleoptera). *Entomologist's Gazette*, 54, 55–70.

Johnson, C. (2004) Ptiliidae. *In*: Löbl, I. & Smetana, A. (Eds.), *Catalogue of Palaearctic Coleoptera. Vol. 2.* Apollo Books, Stenstrup, pp. 122–131.

Matthews, A. (1860) Notes on British Trichopterygidae, with descriptions of some new species. *The Zoologist: A Popular Miscellany of Natural History*, 18, 7063–7068.

Młynarski, J.-K. (1985) Les espèces polonaises du genre *Nanoptilium* (Coleoptera, Ptiliidae). *Polskie Pismo Entomologiczne*, 55, 255–264.

Motschulsky, V. de (1869) Énumération des nouvelles espèces de coléoptères rapportés de ses voyages. 6-ième article. *Bulletin de la Société Impériale des Naturalistes de Moscou*, (1)41[1868](2), 170–201, plate viii.

Sörensson, M. (2003) New records of featherwing beetles (Coleoptera: Ptiliidae) in North America. *The Coleopterists Bulletin*, 57(4), 369–381.

Stephens, J.F. (1830) *Illustrations of British entomology; or, a synopsis of endigenous insects: containing their generic and specific distinctions. Mandibulata. Vol. III.* Baldwin and Cradock, London, [i] + 369 pp., plates xvi–xix.

References

Ahyong, S. T. 2006. A new species of *Carinosquilla* (Crustacea, Stomatopoda, Squillidae) from the Seychelles with a cladistic analysis of the genus. *Zoosystema* 28 (2): 307–314.

Austin, John L. 1962. *How to do things with words. The William James lectures delivered at Harvard University in 1955*. Oxford: Clarendon Press.

Blunt, Wilfrid. 1971. *The compleat naturalist: A life of Linnaeus*. London: Collins.

Daston, Lorraine. 2004. Type specimens and scientific memory. *Critical Inquiry* 31 (1): 153–182.

Farber, Paul Lawrence. 1976. The type-concept in zoology during the first half of the nineteenth century. *Journal of the History of Biology* 9 (1): 93–119.

Hennig, Willi. 1966. *Phylogenetic Systematics*. Urbana: University of Illinois Press.

Huber, J. H. 2007. Non-availability of a name electronically published: The case of *Adamas* Huber, 1979 (Pisces, Cyprinodontiformes, Notobranchiidae), invalidly replaced on the Internet. *Zoosystema* 29 (1): 209–214.

International Code of Zoological Nomenclature. 1985. *Third Edition adopted by the XXth General Assembly of the International Union of Biological Sciences*. London: The International Trust for Zoological Nomenclature, c/o British Museum (Natural History).

International Code of Zoological Nomenclature. 1999. *Fourth Edition adopted by the International Union of Biological Sciences*. London: The International Trust for Zoological Nomenclature, c/o The Natural History Museum.

Kripke, Saul. 1980. *Naming and necessity*. Cambridge: Harvard University Press.

Lherminier, Philippe, and Michel Solignac. 2005. *De l'espèce*. Paris: Syllepse.

Li, Chenyang. 1993. Natural kinds: direct reference, realism, and the impossibility of necessary *a posteriori* truth. *The Review of Metaphysics* 47:261–276.

Linnaeus, Carolus. 1751. *Philosophia botanica in qua explicantur fundamenta botanica cum definitionibus partium, exemplis terminorum, observationibus rariorum, adjectis figuris aeneis*. Stockholmiae: apud Godofr. Kiesewetter.

Linnaeus, Carolus. 1753. *Species plantarum, exhibentes plantas rite cognitas, ad genera relatas, cum differentiis specificiis, nominibus trivialibus, synonymis selectis, locis natalibus, secundum systema sexuale digestas*. Holmiae: Laurentii Salvii.

Linnaeus, Carolus. 1758. *Systema Naturae: per regna tria Naturae, secundum classes, ordines, genera, species, cum characteribus, differentiis, synonymis, locis. Edition decima, reformata. Regnum animale*. Holmiae : Laurentii Salvii.

Mayr, Ernst. 1942. *Systematics and the origin of species from the viewpoint of a zoologist*. Cambridge: Harvard University Press.

Mayr, Ernst. 1969. *Principles of systematic zoology*. New York: McGraw Hill.

Minelli, Alessandro. 2005. Publications in taxonomy as scientific papers and legal documents. *Proceedings of the California Academy of Sciences* 56:225–231.

Nabokov, Vladimir. 2000. *Nabokov's Butterflies: Unpublished and uncollected writings*. Brian Boyd and Robert Michael Pyle, eds. Boston: Beacon.

Putnam, Hilary. 1973. Meaning and Reference. *Journal of Philosophy* 70:699–711.

Réaumur, R. A. 1734. *Mémoires pour servir à l'Histoire des Insectes*. tome 1. Paris: Imprimerie royale.

Recanati, François. 2008. *Philosophie du langage (et de l'esprit)*. Paris: Gallimard (collection Folio Essais).

Ruiter, Dick W. P. 1993. *Institutional legal facts: Legal powers and their effects*. Dordrecht: Kluwer Academic Publishers.

Russell, Bertrand. 1918. Knowledge by acquaintance and knowledge by description. In *Mysticism and Logic, and other essays*, 209–232. London: Longmans, Green, & Co.

Schwartz, Stephen A., ed. 1977. *Naming, necessity, and natural kinds*. Ithaca: Cornell University Press.

Searle John R. 1969. *Speech acts: An essay in the philosophy of language*. Cambridge: Cambridge University Press.

Searle John R. 1975. A taxonomy of illocutionary acts. In *Language, mind and knowledge, Minnesota Studies in the Philosophy of Science*: 344–369.
Searle John R. 1979. *Expression and meaning: Studies in the theory of speech acts*. Cambridge: Cambridge University Press.
Simpson, George G. 1961. *Principles of animal taxonomy*. New York: Columbia University Press.
Vorst, O. 2007. *Ptiliola flammifera* (Młynarski) reinstated as a species distinct from *P. kunzei* (Heer) (Coleoptera: Ptiliidae). *Zootaxa* 1546:63–68.
Whitworth, T. L., R. D. Dawson, H. Magalon, and E. Baudry. 2007. DNA barcoding cannot reliably identify species of the blowfly genus *Protocalliphora* (Diptera: Calliphoridae). *Proceedings of the Royal Society* B 274:1731–1739.
Wittgenstein, Ludwig. 1922. *Tractatus logico-philosophicus* (German text online edition: http://tractatus-online.appspot.com/Tractatus/jonathan/D.html; online English translation: voidspace.org.uk/psychology/wittgenstein/tractatus.shtml).

Chapter 5
Ordering Operations in Square Root Extractions, Analyzing Some Early Medieval Sanskrit Mathematical Texts with the Help of Speech Act Theory

Agathe Keller

Abstract Procedures for extracting square roots written in Sanskrit in two treatises and their commentaries from the fifth to the twelfth centuries are explored with the help of Textology and Speech Act Theory. An analysis of the number and order of the steps presented in these texts is used to show that their aims were not limited to only describing how to carry out the algorithm. The intentions of authors of these Sanskrit mathematical texts are questioned by taking into account the expressivity of relationships established between the world and the text.[1]

List of Abbreviations

Ab	Āryabhaṭa I 's *Āryabhaṭīya* (fifth century)
APG	The Anonymous and undated commentary on the *Pāṭīgaṇita* of Śrīdhara
BAB	Bhāskara I's commentary on the *Āryabhaṭīya*: *Āryabhaṭīyabhāṣya* (seventh century)
PG	The *Pāṭīgaṇita* of Śrīdhara (tenth century)
SYAB	Sūryadeva Yajvan's commentary on the *Āryabhaṭīya*: *Bhaṭaprakāśikā* (twelfth century)

---

[1] This study was undertaken within the *History of Science, History of Text* Seminar in Rehseis in 2007. It was completed with the help of the algo-ANR. I would like to thank J. Virbel, K. Chemla, C. Proust, F. Bretelle-Establet, J. Ritter, C. Singh, A. Bréard, K. Vermeir, M. Keller, C. Montelle, K. Plofker, C. Singh and R. Kennedy: their thoughtful comments and encouragement have been woven into this article and have brought it into existence.

A. Keller (✉)
SPHERE (ex-REHSEIS; CNRS & University Paris Diderot),
Université Paris 7, UMR 7219, Paris Cedex 13, France
e-mail: kellera@univ-paris-diderot.fr

© Springer International Publishing Switzerland 2015
K. Chemla, J. Virbel (eds.), *Texts, Textual Acts and the History of Science*,
Archimedes 42, DOI 10.1007/978-3-319-16444-1_5

5.1 Introduction

The Sanskrit scholarly tradition of composing compact procedural *sūtras* with hair-splitting prose commentaries offers a fertile field for reflecting, as Speech Act Theory (SAT) does, on how prescriptive discourses relate to the real world.[2] Studying the construction and composition of cryptic statements of procedures and the way they are unraveled in commentaries sheds light on just how diverse the relationships are between the texts that refer to algorithms and the actual physical execution of the algorithms. In other words the way the procedure is stated and the way the procedure is executed are two different realities whose relationships are studied here in the specific case of a connected set of two treatises and three commentaries.

Most available Sanskrit sources on mathematics provide procedures for extracting square and cube roots (*vargamūla, ghanamūla*). Such rules were part of the set of elementary operations (*parikarma, vidha*) that formed the basis of arithmetic and algebra. The square root procedure remained unchanged, except for small details in the inner workings, from the end of the fifth century to at least the beginning of the twelfth century and probably later.[3]

Texts that hand down these rules are of two, tightly linked, kinds: treatises and commentaries. This study then will bring to light different ways in which a treatise and its commentary handled the tension of how a procedure is stated versus how the procedure is executed. This analysis is part of a larger endeavor, with the aim of studying descriptive practices in Sanskrit mathematical texts, while focusing on how commentaries relate to their treatises.[4]

In the following, the spotlight will be on how the different steps in square root extraction are presented in both the treatises and their commentaries. Attention will be paid to how different actions are stated and explained, in order to unravel the intentions with which these texts were composed.

5.1.1 Corpus

Five Sanskrit mathematical compositions serve as the basis for this study, as illustrated in Fig. 5.1. The first, an astronomical *siddhānta* (that is, a theoretical text)[5] from the fifth century, the *Āryabhaṭīya* (Ab),[6] and two of its prose and rather prolix

---

[2] For a more general study on how Austin's work could help contextualize Sanskrit scholarly knowledge see Ganeri (2008).

[3] Starting with the procedure given by Āryabhaṭa (499), remaining virtually unchanged in Bhāskarācārya's (fl. 1114) *Līlāvatī and Bījagaṇita*, and thus still in use in later commentaries of these texts.

[4] Keller (2010).

[5] Pingree (1981, p. 13).

[6] Shukla (1976).

Fig. 5.1 A connected set of
five Sanskrit texts dealing
with Mathematics: 2 treatises,
3 commentaries

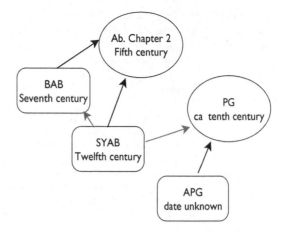

commentaries: Bhāskara's *Āryabhaṭīyabhāṣya*[7] (BAB), from the seventh century, and Sūryadeva Yajvan's twelfth century *Bhaṭaprakāśikā* (SYAB).[8] Then, Śrīdhara's "practical" *Pāṭīgaṇita* (PG) (ca. mid eight century-tenth century) and its anonymous and undated commentary (APG).[9] These texts belong to the early medieval period of Indian mathematics: after the ancient tradition of ritual geometry stated in the *Sulbasūtras* and before Bhāskarācārya's (twelfth century) influential and synthetic works, such as the *Līlāvatī* and the *Bījagaṇita*.

The corpus consists of a set of connected texts, although they were composed at different times and in different places. As seen in Fig. 5.1, commentaries are linked, naturally, to the text they comment on, here symbolized by black arrows. Furthermore, a commentator of the Ab, the author of SYAB, has read the PG, and quotes it. He also often paraphrases BAB. These relationships are symbolized by grey arrows. Considered together, these texts belong and testify to the cosmopolitan Sanskrit mathematics culture of the fifth to twelfth centuries.[10] However the two treatises examined here are different in nature: as stated previously the Ab is a theoretical astronomical text, with only one chapter devoted to mathematics (*gaṇita*), while the PG is solely a mathematical text, devoted to wordly earthly (eg. everyday) practices

---

[7] Shukla (1976, pp. 52–53). A translation of Bhāskara's commentary on Āryabhaṭa's verse on root extraction can be found in Keller (2006 Volume 1, pp. 20–21), and an explanation of the process in Keller (2006, Volume 2, pp. 15–18).

[8] Sarma (1976) A translation of his commentary on Āryabhaṭa's verse for square root extractions is given in Appendix C.

[9] Shukla (1959) A translation of the anonymous and undated commentary on the Pāṭīgaṇita's rule for extracting square roots is given in Appendix D. An explanation of this rule is given in Shukla's translation.

[10] Pollock (2006, Part I).

Table 5.1 Two rules for extracting square roots

Treatise	Sanskrit transliteration	English translation[a]
Ab.2.4.ab[b]	*bhāgaṃ hared avargān nityaṃdviguṇena vargamūlena|	One should divide, repeatedly, the non-square [place] by twice the square-root|
Ab.2.4.cd	*vargād varge śuddhe labdhaṃ sthānāntare mūlam|	When the square has been subtracted from the square [place], the result is a root in a different place|
PG.25.abcd[c]	*viṣamāt padas tyaktvā vargaṃ sthānacyutena mūlena| dviguṇena bhajec cheṣaṃ labdhaṃ viniveśayet paṅktau|	Having removed the square from the odd term, one should divide the remainder by twice the root that has dropped down to a place [and] insert the quotient on a line|
PG.26.abcd	*tadvargaṃ saṃśodhya dviguṇaṃ kurvīt purvaval labdham| utsārya tato vibhajec śeṣaṃ dviguṇ īkṛtaṃ dalayet|	Having subtracted the square of that, having moved the previous result that has been doubled, then, one should divide the remainder. [Finally] one should halve what has been doubled.|

[a] [] indicate my own completions
[b] (Shukla 1976, p. 52)
[c] (Shukla 1959, Sanskrit text, 18; English Translation, 9)

(*lokavyavahāra*)[11]. Practices employed for stating procedures changed from one type of text to another. The aim of this study then is to forge tools to better describe and understand such differences.

The rules given by Āryabhaṭa and Śrīdhara are shown with their Sanskrit transliteration in Table 5.1. In the following, various analyses of these rules implicitly suppose that the reader has this table to hand, and can compare and analyze the graphics, the lists etc. with the texts presented here.

The procedure used for extracting square roots will not be discussed in what follows. Appendix A lists steps for extracting a square root, Appendix B illustrates this with the extraction of the square root of 186 624, an example carried out in the APG. A visualization of the process is given in Fig. 5.2.[12]

---

[11] Thus Śrīdhara starts his treatise with the following statement, Shukla (1959, Sanskrit: i, English: 1):

PG. 1cd (aham) lokavyavahārārthaṃ gaṇitam saṃkṣepato vakṣye\|

I will briefly state mathematics aiming at wordly practices

[12] This diagram should not be seen as an attempt to formalize the algorithm: it is only a heuristic illustration.

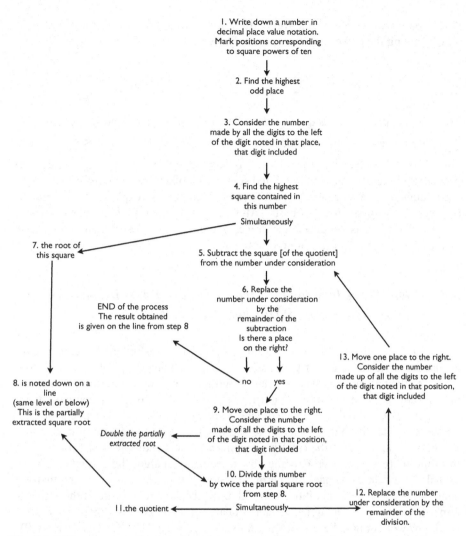

Fig. 5.2 Different steps in the extraction of a square root. Here we follow Ab. See Appendix 1.

5.1.2   The Mathematical Ideas Underlying Square Root Extraction Procedures

The process for extracting square roots relies on the decompositional nature of decimal place-value notation: the number, say 186 624, whose square root is to be extracted is considered to be the numerical square of another number. That is $186624 = b^2$. Extracting its square root means recovering the different elements of the developed square. In other words, if we take the numerical example from the

APG, 186 624, the process uncovers different b_i values (that is both the values of b and i, i giving the powers of ten concerned) such that

$$186624 = \left(\sum_{i=0}^{p} b_i 10^i \right)^2 = \sum_{i=0}^{p} b_i^2 10^{2i} + \sum_{\substack{0 \leq i, j \leq p}}^{i+j \leq 2p} 2b_i b_j$$

To do so, the process takes decimal development of 186 624 as the sum of squares, $\sum_{i=0}^{p} b_i^2 10^{2i}$, and of double products of the type $2b_i b_j 10^{i+j}$, for $0 \leq i, j \leq p$ and $i + j \leq 2p$. Consequently, the process of extracting square roots, an iterative process is characterized as the repeated subtraction of squares and division by doubled numbers. The repeated division by a doubled number explains the difference between the process provided by the Ab and that given in the PG: the PG arrives at a doubled root (useful during the process), while the Ab describes a process that enables one to obtain the square root immediately.

5.1.3   The Procedure for Extracting Square Roots in Sanskrit Texts: The Difficult Question of Description

Trying to determine the intentions and meanings of Sanskrit mathematical texts is made difficult by the fact that, as historians, we know little of the context in which mathematical texts were produced and used. Furthermore, I do not possess native knowledge of Sanskrit. To put it with Austin's words, the accompaniments and circumstances of the utterance of *sūtras* are largely lost to us as readers today.[13] Or, to state the difficulties inherent to the historian's trade according to Searle's categories,[14] and as described by Virbel in this volume, condition 1 of Searle's "how to promise" (e.g. in this case, being able to execute an algorithm) involves native knowledge of the language. Furthermore ignorance of the context means that we cannot satisfy Searle's conditions 4, 6 and 9. Indeed, we are not sure of the author's aim (6), nor that of his imagined reader or hearer (4) and can thus only be poor judges of how well, or not, the author's intentions are conveyed by the texts we read (9). As pointed out by Virbel then, certain conditions on the possibility for communication (1 and 9) and for making commitments (4 and 6) are not fulfilled. Nonetheless, in the following, treading carefully, the intentions of the Sanskrit authors of these statements of mathematical algorithms will be discussed. To do so, the light shed on the authors by their commentators will be used.

The procedure for extracting square roots has consistently attracted attention from historians of Indian mathematics. It testifies to an early use of decimal place-value notation. Furthermore, the process found here is very similar to the one taught until the middle of the twentieth century in secondary schools in Europe, the United States and probably elsewhere in the world. However, how the procedures were

---

[13] Austin (1962, p. 76).
[14] Searle (1969, pp. 57–61).

originally carried out, practically, step by step, remains obscured by variations developed over time and the concision of the rules. Various reconstructions have been offered by secondary sources, from Singh,[15] to, more recently Plofker.[16] How such processes were executed in practice is, however, rarely discussed or justified by a direct quotation of sources.

Indeed, there are several layers of difficulties in such reconstructions.

Even if we set aside the muddles inherent to the historian's trade, the reconstruction of a procedural text is made arduous because of what one may term, following K. Chemla in this volume, the granularity of steps. This problem is certainly familiar to anyone who has had to describe an algorithm: what is stated as one step can often hide several others. For instance when "one should subtract the square from the square", given that the numbers and place where the subtraction should be carried out are detailed, this operation is considered as a single step, although a subtraction or a squaring might involve many steps. Thus "elementary" operations in a more complex algorithm are stated without being described.

Of course, part of this granularity may have to do with "tacit knowledge". Thus some steps may have been considered so obvious that they did not need to be stated. For instance, none of the authors considered here specify that the remainder of the division should replace the initial dividend, or that the remainder of the subtraction replaces the minuend. Similarly, they do not state explicitly that after each arithmetical operation (division or subtraction) one needs to move one place to the right. Since all the texts are silent on these steps that however are required, they may thus be considered tacit, as illustrated in Fig. 5.3. Furthermore, commentators on theoretical *siddhāntas* may have considered, tacitly, that the"practical" steps of the process were not to be specified. This may explain why so few layouts are indeed provided in the texts handed down to us. The following study will focus on the steps that are actually stated, leaving the tacit in the shade.

However, concentrating on the steps "actually stated" in the corpus only helps to bring out the difficulties in defining and specifying what "detailing an algorithmic step" actually means. Indeed, obstacles in recovering "the algorithm" may be inherent to the complexity of the relationships between what is stated about an algorithm, and the algorithm's execution. A symptom of this difficulty has pervaded the writing of this article: each new approach to (the texts on) square root extraction induced a new representation of the algorithm. Each new representation never exactly coincided with the others. Of course, I could try endlessly to coordinate such different representations: checking that they keep the same number of steps, respect the same identified actions and hierarchies between different steps. But I finally decided, on the contrary, to leave each description with its singular expressivity: none are wrong or faulty in respect to the text it illustrates, or the algorithm it refers to. But none coincide exactly either: each representation gives only part of the information. *No two representations coincide with each other*. Indeed these

---

[15] Singh (1927).
[16] Plofker (2009, pp. 123–125).

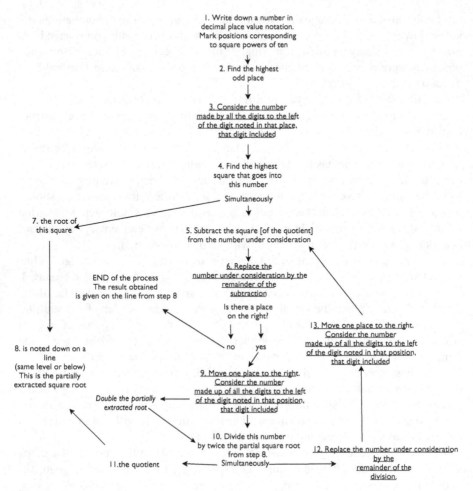

Fig. 5.3 The tacit steps in a square root extraction underlined

multiple representations demonstrate and illustrate how complex the relationships are between the executed algorithm and the way it can be referred to with words or figures. Each heuristic representation we forge to explain one or other aspect of the algorithm adds yet another layer to this complexity. In other words, there is no single, absolute way of describing the algorithms for extracting square roots and the different ways they are stated, whether it is to express the different ways it can be executed, or the different statements that can be made about it.

The analysis in this article will be restricted to three elements that are usually associated with algorithms. If a procedure is thought of as (1) an ordered (2) list of (3) actions to be carried out: the kinds of actions, the way the steps are listed and ordered will be discussed here. More specifically, in the following, first the kinds of statements Āryabhaṭa's and Śrīdhara's rules provide will be discussed, noting the paradox of *sūtras* which both prescribe and are cryptic. What the different texts tell us of the algorithm will be studied, looking at how they detail and order actions and treat the procedure's iteration. In the end, the relationships these texts weave with the real world will help provide hypotheses on their different intentions.

In order to understand how and why an author "states" the steps in an algorithm, the focus needs to be on the kind of text that transmits the procedure. What kinds of statements on procedures are produced in Sanskrit mathematical texts: Descriptions? Incentives to actually carry out the procedure?

5.2   The Prescriptive Paradox of Compact Procedures

Procedures are transmitted through rules (*sūtras*) and their commentaries. A *sūtra*, as has been noted in some detail by Louis Renou, is a complex linguistic object used in a great diversity of communication acts.[17] In the following, the focus will be on how this complexity is given an additional twist as mathematical algorithmic *sūtras* are analyzed.[18] As *sūtras* are often described as being cryptic, let us look closely at what this means in the case of *sūtras* providing a procedure for extracting square roots.

5.2.1   Being Cryptic

As seen in Table 5.1, read in isolation, the rules given by Āryabhaṭa and Śrīdhara are difficult to understand:[19] the Ab and the PG do not specify what is produced if the rules are followed, nor do they specify what an odd term/square place is. None of the verses[20] indicate how to start the procedure, nor how to end it. Only some of the steps allowing the procedure to be carried out are given. This ellipse is illustrated in Figs. 5.4 and 5.5.

In both Figures, the steps as given are contrasted with the actual steps required to carry out the process as analyzed in Fig. 5.2. The underlined "tacit" steps of Fig. 5.3 are left out. The specificities of Śrīdhara's extraction of a double square root are not represented in Fig. 5.4 on Āryabhaṭa's rule. Āryabhaṭa's *sūtras* gives only part of the process, its core: reduced to four steps, the algorithm is given in an

---

[17] Renou (1963).

[18] Incidentally, this study shows that these mathematical rules do not correspond either to Group A or Group B as defined by Renou in Renou (1963), part C. Features of group A, such as the use of the optative, are combined here with the prescriptive norms of group B.

[19] The cryptic character of this rule has been analyzed in Keller (2006, p. xvii; Keller 2010, pp. 235–236) and is noted in Plofker (2009, pp. 123–125). Some of its characteristics are described in Singh (1927).

[20] The Ab and the PG provide rules for extraction in a verse form that counts the number of syllabic units, the *āryā*. This is a very common verse form for prescriptive texts.

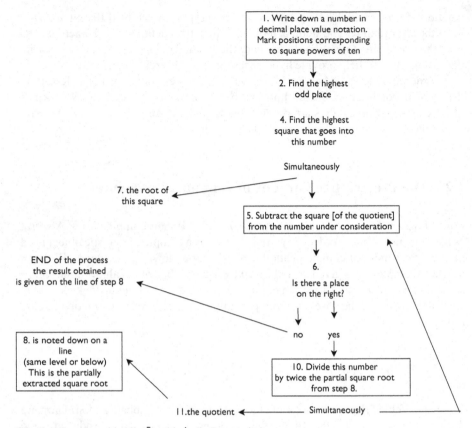

Fig. 5.4 Steps provided by Āryabhaṭa (in Rectangles)

unspecified order and seems restricted to a succession of divisions and subtractions around which other steps gravitate. Śrīdhara's rule, although more detailed, also gives only part of the process: reduced to seven steps, unspecified in order (how does one go from step A to step B?), the emphasis is less on the heart of the iteration and more on the detail of what is done to the number "inserted on a line".

Thus, a first level of reading immediately reveals the ellipses of the rules when contrasted with the execution of the algorithm.

Another way to state the same fact consists in listing the detailed steps. With arbitrariness and limitations in mind, the steps in the procedure given in Āryabhaṭa's verse can be listed as follows:

i.   Divide the non-square place by twice the square root
ii. Iterate ("repeatedly")
iii. Subtract the square from the square place
iv. The result, a quotient, noted in a "different place" is the (square) root

One could add an implicit step, the one which notes down the number to be extracted in a grid which identifies even powers of ten. This step can also be considered as

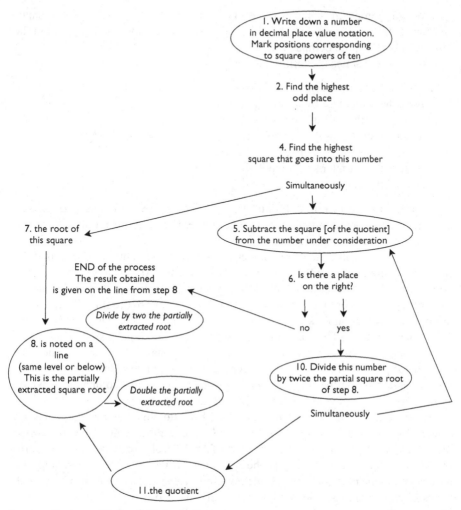

Fig. 5.5 Steps provided by Śrīdhara (in Ovals)

included in the subtraction step. Similarly step i and step iii could also be interpreted as, in fact, including two steps each.[21]

Whatever the nuances we might want to add, this enumeration highlights how compact Āryabhaṭa's verse is. Indeed the square root extraction as reconstructed in Appendix A in order to carry it out includes 13/14 steps, while Āryabhaṭa states between 3 and 8 steps.

---

[21] The difficulty of actually singling out the steps in Āryabhaṭa's verse, addressed in the next section, can be seen when this enumeration is compared with Fig. 5.4. With less contrast, the same can be seen for Śrīdhara's rule as well.

Although, compared to the Ab, the PG is less concise- indeed Śrīdhara states the process in two verses while Āryabhaṭa uses only one—the process given in the *Pāṭīgaṇita* is also quite compact.

Śrīdhara's rule states the following steps:

i. Remove the square from an odd place
ii. Divide the remainder by twice the root
iii. The digits of the partial root are placed on a line below
iv. The square of the quotient is subtracted (from what is not specified)
v. Double the quotient and place it on a line
vi. Divide the remainder as in step 2, that is: Iterate
vii. The final result is divided by two

As can also be seen in Fig. 5.8, doubling and dividing by two adds two steps to the process described in Āryabhaṭa compact verse. Furthermore Śrīdhara indicates more explicitly how the process ends.

This first analysis of the different steps provided by the authors shows that the *sūtras* considered here—whether overtly short as in the Ab, or more explicit as in the PG- are not sufficient to actually carry out the algorithm. If these rules aim to describe the process or prescribe actions, then some steps are missing. If these rules do not have such an aim, we can only remark that their initial intention is, at this stage, unknown to us. Thus, in both cases some information is lacking. These rules are so compact as be difficult to understand as they stand: they are cryptic. The difficulty of properly isolating the different steps stated in each rule shows that the tools necessary for further, rigorous description of the kind of compactness which characterizes different mathematical *sūtras* elude us.

Elliptic formulations are often understood by Indologists as recalling the oral sphere. The enigma of cryptic *sūtras* could have a mnemonic value unraveled through oral explanation. For instance, part of Ab.2.4's obscurity is rooted in word-play on the word 'square' (*varga*). The Ab gives this name to both the square of a number and to the places in place-value notation having an even power of ten. Such places have the value of a square power of ten. It is also from such 'square places' that we find the 'square numbers' the process tries to bring out. Thus such wordplay recalls the main mathematical idea behind the procedure while simultaneously giving rhythm to the verse and making it confusing. Ab.2.4 can thus be understood as a mnemonic "chimera".[22] Other reasons commonly advanced for using short forms include secrecy, the desire to emphasize the difficulty of the given technical knowledge to add prestige for a profession living on patronage.

While these rules are compact to the point of being cryptic, they nonetheless prescribe an action to be taken. This prescription is voiced by an optative.

---

[22] To apply in this context the concept that Severi (2007) uses to denote pictorial mnemonic artifacts mostly used by North American Indians: the important idea is stamped into the artifact by relating two things that normally should not be connected. This association works like a knot in a handkerchief: something should be remembered here. In this case: where a digit is noted down and a square quantity are given a common name. They are associated in a confusing way in the verse, creating such a chimera.

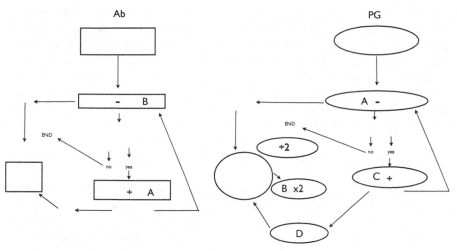

Fig. 5.6 Comparing the Ab and the PG. Step order is indicated by capital letters.

5.2.2 Using Optatives

Sanskrit uses nominal forms extensively. Therefore, the use of conjugated forms is in itself an expressive statement. Conjugated verbs in mathematical *sūtras* indicate a prescription. Indeed, most *sūtras* of *jyotiṣ a* texts (astral science including mathematics) use the optative.[23] Theoretically it is an equivalent to our conditional: it expresses doubt. However, it should be understood here as expressing requirement.[24]

In Ab. 2.4, "one should divide" is the translation of an expression which uses an optative: *bhāgaṃ hared*, 'one should withdraw the share'- the usual expression of a division. The verb to withdraw (*hṛ*) is in the optative voice. PG 25-26 is a succession of optatives: one should divide (*bhajed*), place (*viniveśayet*), make (*kurvāt*) the double, divide (*vibhajet*), then halve (*dalayet*).

The commentaries follow closely the use of optatives given in the treatises. Thus Bhāskara comments on Āryabhaṭa's optative by providing a synonym (*gṛh-*), con-

---

[23] Note that in grammar (*vyakaraṇa*), according to an oral communication by Jan Houben, the optative belongs essentially to the commentary.

[24] Renou (1984, § 292):

> L'optatif exprime les nuances variées d'un optatif propre- souhait, hortatif, délibératif, éventualité, prescriptif, hypothétiques (...). La coexistence de ces divers emplois n'est relevable que dans la poésie littéraire; dans les textes techniques prédomine la valeur prescriptive.

That is, in English (my translation):

> The optative expresses diverse nuances of a true optative: wish, hortative, deliberative, possibility, prescriptive, hypothetical (...) voices. The coexistence of these various uses are only found in literary poetry; in technical texts a prescriptive value prevails.

jugated as an optative:[25] "One should remove the part, that is, one should divide". Sūryadeva (whose commentary is translated in Appendix C) does not comment upon Āryabhaṭa's terms for division but repeats the verb in the optative form, while commenting on what a non-square is:[26] "One should divide by the (last) non-square place". The APG (a translation of the commentary is given in Appendix D) preserves Śrīdhara's optatives sometimes supplying a synonym for others. Thus it uses *bhāgam apaharet for bhājet* (one should divide). The APG provides optatives for a number of actions: the first subtraction of a square (*tyajet*), the placement of the root of this first square (*sthāpayed*), the subtraction of the square of the quotient (*śodhayet*), the fact that results should be considered as a unique quantity (*jñāyet*), etc.

Commentaries also use other moods to voice prescriptions: Imperatives when inviting one to solve a problem, obligational verbal adjectives when describing the steps to be taken. In the APG the optative is only used while commenting directly on Śrīdhara's verse. When solving the problem, actions are given with absolutives (which give precise temporal orders), such as *śodhayitvā* ("having subtracted") used twice, and by verbal obligation adjectives (such as *kartavya*, "one should carry out").

Therefore, such algorithmic Sanskrit texts are prescriptive. Their prescription is first voiced in the treatises by an optative. These optatives are also taken up and declined in other prescriptive forms in commentaries. A cryptic sutra prescribing a procedure to be carried out is a paradox: Indeed, why elaborate short cryptic prescriptions, if the aim is to have them followed? In other words, if the aim is to have a procedure applied, the directive character of an algorithmic rule is contradicted here by its cryptic form. What then were the intentions of the authors of such rules? As this question cannot be replied to directly, how the commentators understood the authors' intentions will be observed. But to do so requires further unraveling of the complexity of statements in mathematical *sūtras*.

5.2.3   Stating a Procedure

The commentators are quite explicit on how they understand the kinds of statements the treatises provide. All the commentators consider the rule primarily as a linguistic assertion: a text whose language is the primary subject of the commentary (which kind of verb(s) it uses, what it means and how it is constructed syntactically). In this respect, all three commentators refer to the text they explicate as a *sūtra*. They also sometimes refer to it as a verse, *kārikā*.[27]

Furthermore, the commentaries use vocabulary that relates the verse to mathematical procedures. Thus, *ānayana*, "computation", derived from the verb *ā-Nī*,

---

[25] tam bhāgam haret gṛhṇīyāt.

[26] *avargasthānād bhāgam haret.*

[27] This cross-reference may refer to the merging of both forms as referred to by Renou (1963) who considers that real *sūtras* are non-versified. The etymology of *kārikā*, derived from the verb *kṛ-.*, "to make", can maybe be understood in this context as "(verse) for action".

to lead towards, is used by our three commentators to refer to the mathematical content of the rule. Bhāskara writes as an introductory sentence:[28] "In order to compute (*ānayana*) square roots, he says:" Later in the commentary he uses the word *gaṇitakarman* "mathematical process". Similarly Sūryadeva uses almost the same words, but different declensions to introduce the verse in this way:[29] "He states a square root computation with an *āryā*". The anonymous and undated commentary on the *Pāṭīgaṇita* starts by specifying:[30] "A two *āryā* algorithmic rule (*karaṇasūtra*) concerning square roots". He later refers to the process using the expression *ānayana*.[31]

Thus the commentators understand the rules as primarily being about mathematical procedures. Because the rules are also prescriptive, they contain a "commitment", that of stating an algorithm that provides a correct answer to a given problem. Note that the commentary's first move is to provide the procedures' intended result. Expecting commentaries, the authors of the *sūtras* may not have felt it necessary to specify the result of the procedure in the rules they composed.

The commentators thus refer to the rules on both levels: as a statement (on whose language one may comment) and as a procedure (on whose steps one may comment). How do the commentators deal with the *sūtras* on these two levels? In addition, if the rules for extracting square roots are thus understood as prescribing a process that should be executed, does this mean that they provide a list of steps to carry out?

5.3 Detailing Steps for Extracting a Square Root

As noted earlier, the speech act "stating an algorithmic step" is complex. Two aspects of this act, the distinction between a certain number of steps and their subsequent ordering, are studied now.

5.3.1 Expressing Actions and Enumerating Steps

Earlier in this article, in an attempt to show that rules provided both by Āryabhaṭa and Śrīdhara were compact, they were crudely restricted to a list of steps. Indeed, our contemporary representation of what a good prescription should be involves

---

[28] *vargamūlānayanāyāha.*

[29] *vargamūlānayanam āryayāha.*

[30] *vargamūle karaṇasūtram āryadvayam.*

[31] Standard vocabulary is used throughout Sanskrit mathematical texts to refer to computations, methods and algorithms. We do not know if there was any difference in meaning between these different words, if their meaning changed over time, according to authors. We have adopted the following translations here: "computation" for words derived from *ānī*; "method" for *karman*; algorithm or procedure for *karaṇa*.

listing actions. But how then do the rules given here fare in this respect? Are they lists of actions? And if not, does this imply that they do not describe an algorithm?

Recall Śrīdhara's statement of the procedure, as given in Table 5.1. A certain number of steps are expressed by a succession of optatives. The essential ordered backbone of operations to be carried out is conveyed in this way: a division, the insertion of a quotient on a line, a doubling, another division and a halving.[32] If we understand this succession as being a list,[33] thus we can take each action as being on the same level of co-enumerability.

The impression that Śrīdhara provides a list of actions is emphasized by the APG's way of taking each optative and following it through twice. Thus in the general commentary:[34]

> And one should divide (*bhāgam apaharet*) from above by twice this, just there. The result should be inserted on a line (*viniveśayet*), one should subtract (ś) the square of that from above that, and this should be doubled (*dviguṇīkuryāt*). If when this is doubled an additional place is created (*jāyet*),[35] then it should be used as before (*yojayet*) when it is a result. (...) One should repeat (*utsārayet*) this, thus one should divide (*vibhajet*), one should insert (*viniveśayet*) the result on a line, etc. as before in as much as the serpentine <progression> is possible, when finished one should halve (*dalayet*) the whole result.

The optative is used when commenting directly on Śrīdhara's verse. Moreover, the APG takes elements that Śrīdhara did not formulate with conjugated verbs, and transforms them into conjugated, optative forms. Thus the subtraction, expressed by Śrīdhara with an absolutive (*saṃśodhya*), becomes an optative in the commentary (*śodhayet*). The repetition, an absolutive in the verse (*utsārya*), is an optative in the APG (*utsārayet*). In the resolution of the problem, the APG uses a conjugated verb, to make a quantity slither onto a line (*sarpati*), where the Pāṭīgaṇita uses a non-conjugated form to describe a quantity that has been dropped down (*cyuta*). Both Śrīdhara and his commentator seem to consider the rule provided as a list of steps, identified by the use of verbal forms, conjugated or not. Consequently, unraveling here how the authors "detail steps" seems fairly simple and straight-forward: conjugated verbs give us the clue.

---

[32] The fact that the PG's process provides a doubled root that needs to be halved is highlighted (by mistake?) in SYAB. Indeed, this commentator on the *Āryabhaṭīya* notes in Sarma (1976):

labdhe mūlarāśau dviguṇī kṛ taṃ dalayet
When the root quantity has been obtained, having multiplied it by two, it should be halved.

However, Āryabhaṭa's rule does not provide a double root and therefore does not request a halving at the end.

[33] Note that the Indian subcontinent's diversity of manuscripts presents a great variety of material settings; its scholarly texts, a large number of lists. However, there seems to have been no specific typographical layout for lists in mathematical manuscripts in the Indian subcontinent.

[34] *dviguṇ ena ca tena tatraiva sthitena upariṣṭāt bhāgam apaharet labdhaṃ paṇ ktau viniveśayet tatas tad vargam upariṣṭāc chodhayet tac ca dviguṇīkuryāt tasmin dviguṇe kṛte yadi sthānam adhikaṃ jāyet tat prāglabdhe yojayet (...) tam utsārayet tato vibhajet labdhaṃ paṅktau viniveśayed ity ādi pūrvavat yāvat utsarpaṇasambhavaḥ / samāptau sarvaṃ labdhaṃ dalayed/.*

[35] In the corpus looked at here, this is the only use of an optative in a non-prescriptive form.

Table 5.2 Expressing actions in Śrīdhara's verse and in its anonymous commentary (APG)

Śrīdhara	APG
Subtract	**Subtract**
Noting the number	*Noting the number*
Drop down	**Place (under)**
Divide	**Divide**
Insert on a line	**Insert on a line**
Subtract	**Subtract**
Move	**Lead, Slither**
Double	**Double**
Divide	**Repeat**
Halve	**Halve**

Āryabhaṭa and his commentators Bhāskara and Sūryadeva provide a stark contrast to this attitude. Indeed, Āryabhaṭa's verse itself cannot be reduced to a list of steps of actions to be carried out. It uses only one conjugated verb, referring to a division. Furthermore, the final assertion in the verse is a description of the fact that a result gains a new status by changing place: this declaration has a reflexive character. Such a statement is partly what makes this rule neither a prescription nor an enumeration. Bhāskara's and Sūryadeva's readings of the actions in the rule emphasize division. Thus, the only conjugated verb referring to an action in the algorithm used by Bhāskara concerns this action.

In both cases, in the PG and in the Ab, the presence of conjugated verbs on one side, and of verbal non-conjugated forms on the other, constructs a hierarchy among the various steps of the stated procedure. In the PG the actions, to subtract/ remove (*tyaktvā, saṃśodhya*), to drop down (*cyuta*), and to evoke the past doubling (*dviguṇīkṛta*) are stated with verbal but non-conjugated forms (absolutives and verbal adjectives). They also seem to provide a list of actions of lesser importance, thus creating a second level of co-enumerability. In other words, the two kinds of verbal form form a hierarchy in the actions to be carried out, as seen in Table 5.2. Verbs in italics represent non-conjugated forms. Bold verbs represent conjugated forms.

As the APG treats all actions on the level of execution, it does not reproduce Śrīdhara's hierarchy of actions.[36]

In Āryabhaṭa's case, the voicing of steps to be carried out cannot be restricted to verbal forms. Aside from the division another action, a subtraction (*śuddha*), is stated with a non-conjugated verbal form, a verbal adjective. Other parts of the algorithm that could be expressed by actions, such as squaring and multiplying by two, are not described in that way: the square of the number (*varga*) is considered

---

[36] Except for two ambiguous elements: the semi-tacit use of decimal place-value notation, and when APG considers the case of a two digit result.

Table 5.3 Expressing actions

Āryabhaṭa	Bhāskara	Sūryadeva
		Noting the number
		Setting aside
Divide	**Divide**	**Divide**
	Noting the number	
Subtraction	*Subtraction*	*Subtraction*
		Double
		Halve

directly, as if it had already been computed. Multiplication by two is described with an adjective meaning "having two for multiplier" (dviguṇa). Among all the actions to be carried out to extract a square root, two main actions emerge from Āryabhaṭa's verse, those given in verbal form: the action of dividing (first in the verse, and by the fact that it is conjugated) and secondly, the action of subtracting, as illustrated in Table 5.3. Both commentators of the *Āryabhaṭīya* further respect Āryabhaṭa's use of a verbal adjective to refer to subtraction.

Bhāskara does not introduce new intermediary steps with conjugated verbs of action. For instance, he does not comment upon the subtraction, nor on the multiplication, but underlines how (by contrast with square places), the numerical square in Āryabhaṭa's verse refers to an action:[37] "When subtracting the square, a computed square is the meaning". But the aim here is to accentuate Āryabhaṭa's wordplay, while raising its ambiguities: a square operation (vargagaṇita) is not to be confused with a square place (*vargasthāna*).

Bhāskara, Sūryadeva and the APG take care to emphasize the use of decimal place-value notation, especially when describing the grid that is used to carry it out.[38] The two later commentaries, SYAB and APG, express the use of the formal features of decimal place-value notation as an action. Thus, in the APG one should "make (*kṛ*)" (marks for the abbreviations of) even and odd places before noting down the number. In SSūryadeva, numbers are set down, placed (*vinyas-*), and then noted down (*cihn-*). This is even more the case in Bhāskara who only refers indirectly to the notation: the settings of the two solved examples in the commentary involve writing numbers; decimal place-value notation also appears when a distinction between "odd" and "even" places is required. However, Bhāskara does not specify this as an action. In the description he makes of the process as the answer to a question, decimal place-value notation just seems to be the natural background:[39]

---

[37] *śuddhe varge vargagaṇita iti arthaḥ.*

[38] We studied this aspect of the process, and what it means for the concept of decimal place-value notation in Keller (2010), we will thus not dwell on this aspect here.

[39] *kasmāt sthānāt prabhṛtīty*
āha − avargāt (...) atra gaṇite viṣamaṃ sthānaṃ vargaḥ (...) avarga iti samam. sthānam, yato hi viṣamaṃ samaṃ ca sthānam/.

(One should divide) beginning with which place?
He says: 'From the non-square ⟨place⟩' (...). In this computation, the square is an odd place.
Therefore a non-square (...) is an even place, because, indeed, a place is either odd or even.

However, this is not the central step of the process.

Because of the diverse ways of expressing algorithmic steps, rules do not appear directly in the form of a list - a format quite usual in Sanskrit technical texts (*śāstra*). However, loose enumerations of conjugated verbs, such as those given by Śrīdhara, can quite easily be interpreted as a list of steps to be carried out.[40] When a rule only has one, unique conjugated verb, as in the case of Āryabhaṭa, then this interpretation, although possible (as seen in our first section), distorts the statement of the rule itself.

Nonetheless, conjugated verbs by their contrast with the other verbs do tell us something of the hierarchization of the different steps of an algorithm. In the rule for extracting square roots studied here, the optative can be seen as first ordering the enumeration of steps contained in the algorithm. In this ordering, the optative provides the action around which the others are structured. This feeling may emerge from commentaries, which always carefully preserve the different ranges of voices: they do not transform the optatives or conjugate the nominal forms- except where the APG focuses on describing on an equal level each effective action of the process.

Therefore, while rules do not necessarily provide lists, they do transmit a hierarchy for the steps. The question now is, what order does this hierarchy reveal?

5.3.2   Ordering Steps

One of the difficulties of reconstructing algorithms concerns the temporal order in which the different steps of a procedure are to be carried out. Mathematical constraints might sometimes impose a temporal order, but not always. Thus in the procedure for extracting square roots, once the defined (largest) square has been found by trial and error, two actions then have to be carried out: the square has to be subtracted from the number under consideration and the root of the square noted down on a separate line. The order in which the actions are performed does not change the final result. This question of order can be seen as a consequence of having several implicit steps contained within one given step: when several actions are lumped together in a same step, the order in which they could be carried out, if there are no mathematical or exterior constraints, remains ambiguous. Does the hierarchy of steps observed in the previous section correspond to a temporal order for carrying out the algorithm?

A specific verbal form is used in rules to order a set of actions in time. Absolutives are indeclinable. They are built on a verbal root and mean 'having carried

---

[40] Although, even in this case, there is still a great disparity between the representation of the action given in Fig. 5.5 and in Table 5.2.

out the action of the verb concerned'. Absolutives thus indicate an action to be carried out before the main action indicated by a conjugated verb. Śrīdhara uses absolutives. In PG.2.25 one must subtract (*tyaktvā, sam. śodhya*), before dividing or doubling; one should move before dividing. This does not mean that the order of all different steps is elucidated in Śrīdhara's formulation. As illustrated in Fig. 5.5, the placing of the quotient which has dropped down on a line is situated ambiguously in time during the process, as well as the doubling of its digits situated "after the subtraction".

The APG describes, in great detail, the part of Śrīdhara's process which seemed ambiguous: To do so, as noted previously, it does not use absolutives, but first the order in which conjugated verbs are enumerated, to which spatial modifiers (*upari* "above", *adhas* "below") are added. In its solved example the APG uses a wide variety of verbal forms: absolutives (*apāsya*, "having subtracted"), verbal adjectives (*vyavasthita, sthita*, "placed"), obligational adjectives (*neya*, "one should lead"). Thus not one device but many different types are used in this commentary to express with precision the temporal order of execution.

In certain specific parts of his commentary Sūryadeva uses absolutives such as *cihnayitvā* meaning "having noted", *dviguṇī kṛtya* (having multiplied by two), *vibhajya* (having divided), apasya (having removed). In these sections, he spells out precisely the order in which different steps are to be carried out. Note that the first description concerns the use of decimal place-value notation, and the grid of even-odd places that is applied to it. The second part describes what happens when one computes the square root of fractions. In both cases then, the order does not concern Āryabhaṭa's rule directly. The first case supplies actions that enable one to initialize Āryabhaṭa's rule. In the second case, Āryabhaṭa's rule provides the essential steps for another algorithm. The order expressed articulates steps additional to Āryabhaṭa's procedure. In the general commentary[41] Sūryadeva more or less follows the appearance of a first, second and third digit of the square root being extracted. This is how the order to perform the process is specified. In this part of his commentary a large diversity of verbal forms are used: verbal adjectives, absolutives and optatives. Although, this ordering of actions is certainly not the main part of his text, Sūryadeva can focus on one element of the process to detail its temporal order. Overall, Sūryadeva does not seem preoccupied by the order displayed in Āryabhaṭa's rule.

Thus the ambiguous temporal order for different steps in the procedure is not always clarified by the commentators. They do not always use standard devices, such as the absolutive form of the verb. However there is no temporal ambiguity. To extract the square root you need to start with a subtraction. Āryabhaṭa's verse starts with a division, the subtraction is stated at the end of the verse. That is, *in the* Āryabhaṭīya steps to carry out the algorithm are given in reverse order. This is underlined in Fig. 5.6, by the letters A and B, which respectively denote the first and second steps to be carried out.

---

[41] And if my interpretation of the use of "both" and "three" in this text is correct, as noted in the footnotes of Appendix C.

Note that as Sanskrit is a declensional language, a strict order for the words does not need be given. Even though a colloquial word order does exist, *sūtras* often scramble them. The two actions stated in this rule are given in two successive verses: the action of division is emphasized by the fact that it is the first word, while Sanskrit usually positions the conjugated verb at the end of the sentence.

Neither Āryabhaṭa nor Bhāskara use absolutes. They thus show that the hierarchization implied by the use of conjugated and non-conjugated verbal forms, which gives emphasis on one action over another, may not concern the temporal order.

In BAB, the steps are spelled out in a succession of questions and answers:[42]

> One should take away [in other words] one should divide this [square].
> Beginning with which place?
> He says: 'From the non-square [place]' (...) In this computation the square is the odd place.
> (...)
> By what should one divide?
> He says: 'Repeatedly, by twice the square root'. (...)
> How, then, is this square root obtained?
> He says: 'When the square has been subtracted from the square [place], the result is a root in a different place'.

Bhāskara reads Āryabhaṭa's verse as being entirely structured around the division. The dialog argues that the order given by Āryabhaṭa is logical from this operative emphasis. Indeed, to carry out a division, a place to carry it out, a divisor and dividend are needed. By unraveling where the division is performed and what the divisor and dividend are, the steps are thus re-ordered and specified. The use of this staged dialog simultaneously emphasizes that the verse's steps are disordered while at the same time making an argument for its coherence.

Different orders then can be layered in the statement of a single rule: a temporal order, a logical operative order, or even the order for different cases in which a rule could be applied. Furthermore, the statements of procedures do not necessarily list all the actions that are to be performed, and those listed are not necessarily in temporal order. But what then do they do?

---

[42] *taṃ bhāgam, haret gṛhṇīyāt/*
  kasmāt sthānāt prabhṛtīty
  āha – avargāt (...) atra gaṇite viṣamaṃ sthānaṃ vargaḥ
  (...) kena bhāgaṃ haret ity
  āha – nityaṃ dviguṇena vargamūlena
  (...)
  kathaṃ punas tat vargamūlaṃ labhyate ity
  āha – vargāt varge śuddhe labdhaṃ sthānāntare mūlam

5.4   Back to the Prescriptive Paradox of Procedural *Sūtra*s

Bearing Austin's "descriptive fallacy"[43] in mind, it is (sometimes) difficult not to consider procedural *sūtras* as descriptions of procedures. It is also tempting to understand the prescription they voice literally. However, Āryabhaṭa's rule is a paradoxical act of communication: a cryptic scrambling of the algorithm's steps. This is a clue, that Āryabhaṭa's verse contains an indirectly stated intention (an illocutionary force), which may be evading us.

In the following, the initial questions will be raised again: What kinds of statements does a sutra provide when it refers to a procedure? How *sūtras* and commentaries deal here with the commitment contained in the incentive to perform the procedure will be examined. Afterwards, each author's relation to language- the meta-textual part of the rules- and each author's relation to the world where the algorithm is performed will be analyzed.

5.4.1   Commitments and Iteration

These *sūtras* invite one to carry out a process (one or several operations): they thus contain a more or less implicit commitment, that of obtaining a result. The word used for "result" is a substantivated verbal adjective, labdha, literally meaning "what has been obtained". It is sometimes translated, as in PG.2.25, as "quotient", being the result of a division.

As noted before, the rules examined here do not state explicitly what the procedure produces[44]. This is the heart of the paradox of prescriptive *sūtras*: suggesting an action to be carried out, but being evanescent in the commitment the action will fulfill. The result literally shifts repeatedly. In Ab.2.4, *labdhaṃ sthānāntare mulam*, the quotient/the result (a digit of the partial square-root) is the root in a different place. In PG.2.25, *purvaval labdham* the previous result/quotient is doubled and moved. In both cases, we are implicitly in the midst of a process in which one result will produce another. The condition of success for the procedure, we understand, has less to do with "obtaining" a particular result, than with repeating the process until it is completed.[45]

Śridhara and Āryabhaṭa do not express the iteration in the same way. Āryabhaṭa states the procedure by beginning with the middle of the process. Moving the

---

[43] (Austin 1962, p. 100).

[44] Thus Sūryadeva needs to explain:

taṃ saṅkhyāviśeṣaṃ mūlatvena gr hnīyāt / tadatra vargamūlaphalam ity ucyate /
"This special number is referred to (in the rule) as a root. Consequently, here, the result which is a square root (vargamulaphala) has been mentioned (ucyate)."

[45] The difficulty of pinpointing exactly how the iteration is given in the verses explains why it appears and disappears in the previous illustrations we have given of the algorithm.

quantity from a line where it is a quotient, to a line where it is a digit of the square root, is what enables the procedure to be executed repeatedly. Śrīdhara repeats the process twice using different words: In the first verse it seems that he indicates how the process starts, while in the second verse, a second or final execution of the procedure is described. The rule ends with an evocation of the termination of the process. The authors of both these *sūtras* use a literary device to explain how the procedure should be repeated: they do not so much describe the performance as offer a textual imitation of it.

Bhāskara and the APG do not reproduce this imitation of the process. Thus Bhāskara states plainly:[46] "This very rule is repeated again and again, until the mathematical process is completed (*parisamāptaṃ*)." Bhāskara uses the expression *āvartate* (from *āvṛt-* to revolve, to turn). As seen previously, the APG takes up Śrīdhara's expression using the verb *utsṛj-* (to turn), and adds to it an ordered list of actions to be carried out. By contrast, by repeating the process for several digits, and then describing how it ends, Sūryadeva actually seems to use the *Pāṭīgaṇita*'s device.

Stating a rule as if it had already started, repeating it twice: textual devices are used by treatises to offer an imitation of what should be taking place on a working surface, where the algorithm is carried out. Thus for the iteration, the treatises observed here are intent on making their text and the world in which the algorithm is completed coincide not by describing what is going on *in words*, but by describing the repeated algorithm *with words*. Let us look more closely at how the treatises and their commentaries play with the world of the text and the world in which the algorithms are performed.

5.4.2 World and Text

Although Sanskrit mathematical texts may not list or give a temporal order for actions to be carried out, *sūtras* and commentaries do nonetheless state something about the algorithm. The kind of adjustments between the world and the text included in the corpus will now be considered.

At times Śrīdhara gives a momentary description of what the computation should be at a given moment. Verbal adjectives seem, in this case, to indicate where the text and the algorithm should coincide. Thus Śrīdhara uses the expression *sthānacyuta*, "that has dropped to a place", *dviguṇīkṛtam*, "that has been doubled". An action is not spelled out but a description is made of the state of the working surface on which the process is performed. These descriptions enjoin the person performing the procedure to adjust the world to the statements in the rule.

Specifically because this is a part of the process which requires know-how that belongs to the world of algorithm execution, the APG is careful to describe, digit

---

[46] *etat eva sūtraṃ punaḥ punararāvartaye yāvatparisamāptaṃ gaṇitakarmeti.*

by digit, how numbers should be moved around on the working surface. Certain expressions are invitations to verify that at a given moment, the result obtained coincides with the text. Like for instance, in the purely descriptive:[47] "When twenty-four is subtracted by three from below, above two remains." In other cases, the text describes the temporary state of the working surface, followed by a disposition:[48]

> Below, 86 is produced. This quantity slithers (*sarpati*) on a line. Below two, [there is six], below seven, eight. Setting down:
> $$\begin{array}{cccc} 1 & 7 & 2 & 4 \\ & 8 & 6 & \end{array}$$

Thus the *Pāṭīgaṇita* and its anonymous undated commentary are intent on making the statement of the procedure and its performance- on a working surface using tabular dispositions- coincide.

Understanding conjugated verbs as an expressive device reveals how Bhāskara's commentary on Āryabhaṭa's rule is mainly on the level of the language Āryabhaṭa uses. Bhāskara uses the expression "he says" (*āha*)[49] four times. His answers to the questions in the dialog always refer to Āryabhaṭa's statements. Bhāskara then is not describing how the process is to be carried out independently from Āryabhaṭa. He is not adjusting Āryabhaṭa's statements to how the process should be executed. Indeed, he is just modifying Āryabhaṭa's statements in an attempt to show the internal coherence of their arrangement. He explains that this arrangement makes sense if the division is taken as the core from which all other steps in the process derive.

In three instances Bhāskara explores the limits of the mathematical reality expressed by Āryabhaṭa. First, when he specifies evenness as the opposite of oddness, then when the process ends because no other space to carry it out can be found, and finally as he gives an example concerned with fractions in which he then introduces his own rule. In other words, the world of Bhāskara is not like that of the APG, not a world of algorithm execution. His is one of mathematical objects.

Indeed, the iteration in the process is voiced by Āryabhaṭa as a change of status: as a quantity changes place, it becomes another quantity. Such a change needs to be properly identified. This is done by a name change. This way of expressing the iteration is repeated by Bhāskara as he explains that a result changes place, becomes a root, and re-enters the process. The end of the process appears when this change of status becomes impossible. Bhāskara states it as follows:[50]

---

[47] *tribhiḥ patanāt caturviṃśatauśuddhāyāṃ upari dvau śeṣaḥ.*

[48] *adhaḥ ṣāḍaśītijayīte / eśa rāśiḥ sarpati, paṅktyāṃ dvayor adhaḥ (ṣāṭkaṃ) bhavati, aṣṭakaṃ saptādhāh / nyāsaḥ-* $\begin{array}{cccc} 1 & 7 & 2 & 4 \\ & 8 & 6 & \end{array}$.

[49] He also uses once each labh (to obtain), bhu (to be, have, produce), vidyate (to exist, discern). The three only other conjugated verbs of this part of his commentary are: (1) the optative used for division, and (2) verbs used while solving examples at the end of the commentary.

[50] *yadatra labdhaṃtat sthānāntare mūlasaṃjñaṃ bhavati/ (...) tasmin sthānāntare tasya labdhasya mūlasaṃjñā/ yatra worldplayaḥ sthānāntaram eva na vidyate, tatra tasya tatraiva mūlasaṃjñā.*

The quotient here becomes, in a different place, what has the name root (*mūlasaṃjñā*). (...) In this different place, this quotient has the name root. However, precisely when a different place is not found, there, in that very place, that [result] has the name root.

The change of place which simultaneously is a change of status is acted out by a formal action: a name change. This action is very literally an attempt to adjust the statement of the process, to the mathematical world the quantity belongs to. In this case, Bhāskara emphasizes how what can seem a confusing change is actually what explains how the process works: each repeated division provides the digits of the square root. The centrality of the division is thus once again stated, even as the quotient disappears to leave space for the root. Finally, this name change is also what signals the end of the process.

All three commentaries link the movement of the quotient to a separate line to a status change. In the anonymous commentary on the *Pāṭīgaṇita* a digression discusses the status of the quantity that has been moved and noted down on a separate line. The double square root is called "the result" (*labdha*). After having inserted the result/quotient of the division on the line, having subtracted its square and having doubled it, the APG examines the case where the doubling provides a number bigger than ten:[51] "If when this is doubled an additional place is created, then it should be used as before when it is a result (*praglabdhe*). Both have the quality of being a unique quantity (*rāśitā*). This quantity has the name "result" (*labdhasaṃjñā*)." And when one arrives at the end of the process, the "result" appears again, and has to be halved.

Sūryadeva starts by considering the first digit of the square root obtained by trial and error. He calls this quantity a "special number" (*saṃkhyāviśeṣa*) and notes:[52] "this special number is referred to (in the rule) as a root". Then commenting on the last quarter of Āryabhaṭa's verse, he adds[53] "this ⟨quotient⟩, in the next square place, becomes (*bhavati*) the root." Although, the change of status is the same, here Sūryadeva does not change the name. He does so earlier, when he establishes the equivalence between "square, non-square places" and "odd, even" ones:[54] "In places where numbers are set-down, the odd places have the name (*saṃjñā*) 'square'. Even places have the name 'non-square'."

Therefore, naming appears as a central commentarial activity: when a quantity is renamed by a commentator it reveals how the statements concerning an algorithm are adjusted to coincide with the world they refer to. This world can be on the level of performance (APG), of mathematical objects (Bhāskara), or a combination of both (Sūryadeva).

Therefore, the way the authors relate mathematical statements to the working surface on which a procedure is being carried out provides us with a clue to what is

---

[51] *tasmin dvigune kṛte yadi sthānam adhikaṃ jāyet tat prāglabdhe yojayet, tayor ubhayor ekarāśitājñeyā / tasya rāśer labdhasaṃjñā.*

[52] *taṃ saṃkhyāviśeṣaṃ mūlatvena gṛhnīyāt*

[53] *tat purve vargasthāne mūlaṃ bhavati.*

[54] *saṃkhyāvinyāsasthāneṣu viṣamasthānāni vargasaṃjñāni / samasthānāny avargasaṃjñāni.*

important to them: the APG develops Śrīdhara's brief descriptions in order to give an algorithm to carry out that is as unambiguous as possible. Bhāskara highlights the fact that when the quotient is moved to a separate line, its name changes: what happens on the working surface is always coherent in the world of Āryabhaṭa's statements. Finally, Sūryadeva, not surprisingly for a commentator on Āryabhaṭa quoting Śrīdhara, seems to try to position himself between both approaches. Thus, he is intent on adjusting decimal place-value grid to Āryabhaṭa's statements, and specifies how, in practice, the process should start by trial and error.

In other words, all the analyses carried out thus far shed light on the different intentions of the various authors of the corpus.

5.4.3   Intention

Specific tools for describing ways of making a text compact, expressing iteration, listing some actions and not others, and relating language and practice can help us infer authorial intention with greater assurance. For instance, by paying attention to which "essential" elements of the algorithm a rule states and how different hierarchies of actions are imbedded in a *sūtra* provides us with each author's interpretation of the important points in his algorithm. In the following, the different kinds of statements on algorithms unraveled here will be re-examined focusing on the intentions of their authors.

We have thus seen the use of conjugated verbs (especially the optative) in the *sūtras* as an expressive device. Śrīdhara singles out a certain number of actions (division, inserting the number on a line) over others (subtracting, doubling). The APG, on the other hand, does not follow such a hierarchy. This is consistent with the aim of the commentary to treat each action in the execution of the algorithm on an equivalent level. Thus the hierarchy of steps in Śrīdhara's verse, not being included in the APG, sheds light on the aims of both. The APG describes how the algorithm is carried out on a working surface: all actions are equivalent from this point of view. Śrīdhara states (and maybe orders) the required actions. The APG with the dynamic image of a slithering snake enters into the detail of the process unraveling the ascending and descending operations, unwinding the intricate temporal order in which each step of the process should be carried out. In other words, the APG also sees the verse as evoking a dynamic process.

Śrīdhara and his anonymous commentator present a stark contrast to the intellectual couple formed by Āryabhaṭa and Bhāskara with their sparse number of conjugated verbs. Bhāskara's emphasis on the operative logic of Āryabhaṭa's verse shows that his aim is to comment the coherence of Āryabhaṭa's verse, not on how it should be carried out. Similarly, Sūryadeva's relative indifference to Āryabhaṭa's scrambled order directs us towards another aim. Indeed, as in the processes described by Karine Chemla in this volume, Sūryadeva integrates Āryabhaṭa's in specific cases where the algorithm can be applied. The commentator describes how the rule is situated within other algorithms: root extractions which arrive at double roots and root extractions of fractions. Āryabhaṭa's rule then is a general rule, whose temporality and logic is not in question. The hierarchy of operations to be carried out uses Āryabhaṭa's rule as a central nod against which further operations are assessed. Sūryadeva's endeavor as a commentator is to make sure that the process covers all possible cases.

Finally, looking at how different authors treat the world of linguistic statements and the world in which a procedure is carried out on a working surface confirms these conjectured intentions. In places Śrīdhara seems to describe the working surface at specific tricky points that are detailed by his anonymous commentator. On the contrary, Bhāskara does not comment at this level, but rather on providing a name at the right time, for the right quantity: re-naming a quantity that has been moved assures us that the process is coherent with the stated rule. Sūryadeva does a bit of both, renaming the decimal place-value notation grid used in the process, and describing how it should be used.

For both BAB and SYAB, the *Āryabhaṭīya* gives the main mathematical ideas behind this procedure. Three hypotheses can be drawn on the intentions behind Āryabhaṭa's way of describing the procedure: His first aim could be to establish the procedure (both by explaining it, and providing a way of being able to recall it easily), giving its gist. A second aim could be to transmit a reflection on what the procedure is about (how does one undo a squaring in decimal place-value notation) and what this tells us about numbers. Most probably the aim was to add together all the above, e.g., prescribing a procedure, giving its gist, and hinting that this is less about doing than reflecting: an effort to be as general as possible. Since the square root process comes after the definition of a square, since Bhāskara contrasts operations of increase which include squaring with operations of decrease which includes seeking square roots,[55] and since his general commentary on the sutra is followed by the resolution of an example which calculates the roots of previously computed squares, we might conclude that, for Bhāskara the square root procedure was less a procedure to follow, and more a reflection on how one dismantles a squaring operation using decimal place-value notation. In a mathematical tradition where the correction of an algorithm was sometimes verified by inverting it, and finding the initial input, square root extraction may have been seen as inverting the squaring procedure. Āryabhaṭa's rule then would seem to exhort one to carry out the process whose steps are described, but his real aim (as seen through Bhāskara's eyes at least) would be to transmit a reflection on the procedure's mathematical grounding. He might actually be suggesting that the process itself is not only useful for extracting square roots but also as a reflection on what undoing a square operation using decimal place-value notation involves.

---

[55] Keller (2007).

5.5 Conclusion

Part of the *sūtras* perlocutory or contextual effect is irremediably lost to us, as is the case for all historical texts, but even more so on the Indian subcontinent, where so little is known about the context in which mathematics was practiced. The cryptic algorithmic statements of mathematical *sūtras* are, to put it in Austin's words, neither unambiguous nor explicit. At first reading they can seem strangely vague and full of uncertain references. However this detailed study of rules for square root extraction gives us hope that we can uncover elements of how past milieus created, read and understood mathematical *sūtras*.

Maybe the "descriptive fallacy" of statements on algorithms is to consider that all such statements aim to describe the way algorithms should be carried out; and more specifically that all invitations to carry out an algorithm include a more or less explicit description of how to do so. Indeed, this study has shown firstly that an invitation to carry out an algorithm does not necessarily describe literally how to do so. Secondly, what can be classified as the description of an algorithm can be very diverse. Thirdly, that an invitation to execute an algorithm can also be a coded invitation to reflect on it.

Verses stating algorithms are not neutral descriptions of how to carry out an algorithm. They indicate what the authors wanted to transmit and to emphasize concerning these rules. Due to their expressivity, the procedural statements may also then be read using techniques usually ascribed to reading literature. The commentators' readings of these rules show clearly that ambiguous expressions are doors opening onto several specific meanings; the obscure phrases are those that in the end highlight the meaning of the rule.

Conjugated verbs tell us here something of the emphasis, or not, which each text puts on action: Āryabhaṭa's theoretical rule uses one conjugated verb, while Śrīdhara's practical rule gives several. Bhāskara is intent on commenting on statements, and thus frequently conjugates the verb "to speak, state", while Sūryadeva who reflects on different forms of square-root extractions conjugates the verb *bhū*, "to produce, become, be". Finally the anonymous commentator of Śrīdhara's *sūtras*, intent on reworking and specifying different steps, supplies many optatives. For different aims, different practices of algorithmic statements can be used. Practices of algorithmic statement appear to vary according to the type of text (theoretical, practical).

Thus, the authors did not necessarily list actions. The hierarchy of steps they do provide does not always represent a temporal order. The authors could specify actions, describe a working surface with a dynamic tabular layout, formulate relationships between *sūtra* statements and the world of mathematical objects, and elucidate different mathematical objects to which the procedure could be applied.

This study has used different tools to describe and understand how processes were made compact in *sūtras*. Whether represented as a flow chart or as a list, Āryabhaṭa's verse does not appear to be an arbitrary fragment of the algorithm. The study of how language related to the world of algorithm creation helped in understanding how the iteration of the process was expressed by the authors of the *sūtras*. The iteration of the process, given by a repetition by Śrīdhara, is shown by Āryabhaṭa by stating the procedure as if it had already started. In both cases, the literary device of imitation is used to describe a complex reality. Such processes, like the play on the word varga used by Āryabhaṭa, can be seen as striking stylistic idiosyncrasies- specific to each rule and to each author- which may have had the role of the "knots in one's hankerchief", if such rules were meant to be learned by heart.

The authors' main aim then would not have been to describe an algorithm, but rather to comment on it, that is, to emphasize a point in the procedure: its mathematical grounding for Āryabhaṭa, its coherence for Bhāskara, the fact that it was worked out on several connected horizontal lines for Śrīdhara as understood by his anonymous commentator, and finally as a fundamental operation which can be carried out on both integers and fractions for Sūryadeva. The compression of the *sūtras* then would have less to do with mnemonics and secrecy than with the expressive granularity of algorithm statement.

Finally, for Āryabhaṭa and Bhāskara, stating such rules seems to have had the aim of indicating how the algorithm was constructed, and the mathematical properties it was based on. In other words, they may have intended to highlight that such a procedure gave insights into the properties of numbers written in decimal place-value notation, and into what made them perfect squares or not. Maybe the procedure itself was thought of as a reflexive algorithm.

The definition of the *sūtra* as recalled by Renou (1963) is a paradox: a self-sufficient compact verse but also one of a series:

> The word sutra or 'string' refers sometimes to a rule stated as a more or less short proposition (...), sometimes as a set of propositions forming a collection. The sutra genre is defined by its relationships rather than its content, a sutra (understood as a 'rule ' or an 'aphorism') is first and foremost an element dependent on its context, even if it is autonomous grammatically; it is determined by the system and is correlated to the group that is around it.[56]

Indeed, this study has shown that *sūtras* and commentaries are deeply intertwined. The (authors of the) *sūtras* expected commentaries to provide the mathematical context, the procedure's result and the detail required for the execution of the algorithm. Looking at the statements and the way they are formulated and interpreted has, no doubt, underlined the technical reading a *sūtra* requires. If there is expressivity in a sutra we need its commentary to reveal it. There is a specific rhythm to reading a *sūtra* and its commentary: knowing the text of the *sūtra* by heart, understanding it, which means unfolding its meanings and understanding the text's expressivity. Possibly, neither the *sūtra*, nor the commentary were intended to be read just once in a linear way, but masticated over in the way Nietzsche defines aphoristic reflections... the way iterative algorithms are executed.

---

[56] Le terme de *sūtra* ou "fil" désigne tantôt une règle énoncée sous la forme d'une proposition (...) plus ou moins brève, tantôt un ensemble de propositions concourant à constituer un même recueil. (...) Le genre du *sūtra* se définit par sa relation plutôt que par son contenu: un *sūtra* (au sens de "règle" ou "aphorisme") est d'abord un élément dépendant du contexte, même s'il est grammaticalement autonome; il est déterminé par le système et (...) corrélatif au groupe qui l'environne.

A   Appendix 1: Different Steps in the Algorithm for Extracting Square Roots as Spelled Out in the Corpus

Taking into account all the steps detailed by the authors considered here (with an arbitrary filter—the mesh of the net may at times seem too small and at others too large- that is underlined in paragraph 5.2.3), thirteen steps for extracting a square root can be listed. Step 3, 6 and 12 state common tacit steps. The algorithm may be more efficiently illustrated in Fig. 5.2, although they are not equivalent renderings of the process.[57]

1. The number whose square root is to be extracted is noted down in decimal place-value notation. Places are categorized with a grid that enables one to identify square powers of ten. Either positions for square and non-square powers of ten are listed or the series enumerating positions starting with the place with the lowest power of ten is considered. This list categorizes places as even or odd places.
2. The highest odd/square place is identified.
3. Consider (tacitly) the number made by all the digits to the left of the digit noted down in that place, that digit included.
4. Find the largest square contained in the number noted down to the left in the last/highest odd place.
   From here, onwards, one could also start by considering step 8, before turning to step 5 to 7.
5. Subtract the square from the number under consideration.
6. Replace (tacitly) the minuend by the remainder of the subtraction.
7. The root of the subtracted square is the first digit of the square root being extracted.
8. The root of this square (Ab family)/The double of the root of this square (PG family) is noted on the same line, to the left of the whole number/ on a line below the line of the number whose root is being extracted. In the PG family then, the doubling of the digit is a separate step in the process. The doubling does not necessarily need to take place immediately, one can note down the digit, and then double it just before it enters the division described in Step 10. This is what the APG recommends.
9. Consider the number whose highest digits are the previously noted remainder and which includes the next digit to its right.
10. Divide this number by (twice) the partial square root from Step 8.[58] In the following,
11. Replace the dividend with the remainder of the division. Then one should consider the next place on the right, which is a square/uneven place.

---

[57] The "reconstruction" of these variants of the different steps of the process is not discussed here. Hopefully this issue will be tackled in a forthcoming article.

[58] Although this is never mentioned in the ancient texts, the quotient obtained needs to be sufficiently small. This sometimes requires a subtraction by 1 or 2 (and a change in the remainder of the division) to find the adequate digit.

12. The quotient is the next digit in the partial square-root. It (Ab), or its double (PG), is thus noted down next to the previously found digit, as in Step 8. Its square is what will be subtracted from the number with the next digit as the process is iterated here from Step 5.
13. When there is no place on the right, the algorithm is finished. Examples only consider a process that extract a perfect square, consequently, either the square-root, or its double is obtained, according to the procedure followed. If we are in the latter case, the number obtained is halved.

B Appendix 2: Extracting the Square Root of 186 624

This is a numerical example addressed in APG. Footnotes and asterisks indicate non-attested forms.

1. The number whose square root is extracted is noted in decimal place-value notation. These decimal places are categorized using a grid: Square (*varga*), non-square (*avarga*) powers of ten (Ab), or even (*sama*, abr. sa) and odd (*visama*, abr. *vi*) place ranks - counted starting with the lowest power of ten- (BAB, PG, SYAB, APG).

avarga	varga	avarga	varga	avarga	varga
sa	*vi*	*sa*	*vi*	*sa*	*vi*
10^5	10^4	10^3	10^2	10^1	10^0
1	**8**	**6**	**6**	**2**	**4**

$$186624 = \mathbf{1}.10^5 + \mathbf{8}.10^4 + \mathbf{6}.10^3 + \mathbf{6}.10^2 + \mathbf{2}.10^1 + \mathbf{4}.10^0$$

2. Subtract the square from the highest odd place
   The highest "odd" (*viṣama*) place or "square" (*varga*) place is 10^4. The process starts by finding, by trial and error, the highest square number contained in the number noted to the left of this place. In this example, one looks for the highest square that will go into 18. And thus $18 - 4^2$ is the operation carried out.
3. Replace the minuend with the reminder, 4, (BAB) or place the reminder, 4, below (APG).
4. The root, 4, of the subtracted square (16) is the first digit of the square root being extracted. The root of this square (Ab family-4)/ The double of the root of this square (PG family- 8) is noted down on the same line (BAB)/ or a separate line (PG).

Bhāskara might have written:[59]

$$*10^5 \quad 10^4 \quad 10^3 \quad 10^2 \quad 10^1 \quad 10^0$$
$$\mathbf{4/2} \quad 6 \quad 6 \quad 2 \quad 4$$

While the APG writes:

2   6   6   2   4
  8

Because $1.10^5 + 8.10^4 = [4.10^2]^2 + 2.10^4$,
$186624 = [\mathbf{4.10^2}]^2 + 2.10^4 + 6.10^3 + 6.10^2 + 2.10^1 + 4.10^0$.

5. Moving one place to the right, one should divide by twice the root.
   In this example, 26 is divided by 8: $26 = 8 \times 3 + 2$. The quotient is 3, 2 is the remainder.
   This is then set down. Bhāskara's style

$$* \quad \mathbf{4} \quad \mathbf{3}/2 \quad 6 \quad 2 \quad 4$$

APG style

$$*2 \quad 6 \quad 2 \quad 4$$
$$\mathbf{8} \quad 3$$

In other words, because $2.10^4 + 6.10^3 = 8 \times 3.10^3 + 2.10^3, 186624 = [\mathbf{4.10^2}]^2$
$+[2 \times (4.10^2)(3.10^1)] + 2.10^3 + 6.10^2 + 2.10^1 + 4.10^0$.

6. Moving one place to the right, iterate. That is "subtract the square" again. This time the square of the quotient is subtracted. In this example 3^2 is subtracted from 26: $26 - 9 = 17$. The remainder is 17. This is noted down again:
   Bhāskara style:

$$* \quad \mathbf{4} \quad \mathbf{3}/1 \quad 7 \quad 2 \quad 4$$

APG style: 1   7   2   4
                                                  8 6

In other words, writing $26 - 9 = 17$ according to the corresponding powers of ten.
$186624 = [\mathbf{4.10^2}]^2 + [2 \times (4.10^2)(3.10^1)] + [\mathbf{3.10^1}]^2 - [3.10^1]^2 + 2.10^3 + 6.10^2$
$+2.10^1 + 4.10^0 = [\mathbf{4.10^2}]^2 + [2 \times (4.10^2)(3.10^1)] + [\mathbf{3.10^1}]^2 + 1.10^3 + 7.10^2$
$+2.10^1 + 4.10^0$.

7. Moving one place to the right, divide by twice the root. In this example one should divide 172 by $2 \times 43 = 86$: $172 = 2 \times 86$. The quotient is 2 and there is no remainder. This is noted

---

[59] * mark non-attested forms.

Bhāskara style:

$$* \quad \mathbf{4} \quad \mathbf{3} \quad \mathbf{2} \qquad 4$$

APG style:

$$* \qquad 4$$
$$\mathbf{8} \quad \mathbf{6} \quad 2$$

$$186624 = [\mathbf{4.10^2}]^2 + [2 \times (4.10^2)(3.10^1)] + [\mathbf{3.10^1}]^2 + 2 \times (2.10^0)(4.10^2 + 3.10^1)$$
$$+ 4.10^0 = (\mathbf{4.10^2 + 3.10^1 + 2.10^0})^2$$

8. The square root is 432. To end the procedure, moving one step to the right, one can "subtract the square of the quotient" (2^2):

Bhāskara style:

$$\mathbf{4} \quad \mathbf{3} \quad \mathbf{2}$$

APG style:

$$\mathbf{8} \quad \mathbf{6} \quad \mathbf{4}$$

$$186624 = (432)^2$$

C   Appendix 3: SYAB.2.4

[60] He states (*āha*) a square root computation (*vargamūlānayana*) with an *āryā*[61]:

One should divide, repeatedly, the non-square [place] by twice the square root |
When the square has been subtracted from the square [place], the result is a root in a different place ||

In places where numbers are set-down (*vinyāsa*), the odd places have the technical name (*samjñā*) "square". Even places have the technical name "non-square". In this verse, when a square quantity is chosen (*uddiṣṭa*), having initially started by marking (*cihnayitvā*) the square and non-square places, when one is able to subtract (*śodhayitum śakyate*) the square of a special number- among those [squares of the digits] beginning with one and ending with nine- from the last square place, having subtracted (*apāsya*) that square; this special number is referred to [in the rule] as a root (*mūlatvena gṛhṇīyāt*). Consequently, here, the result which is a square root (*vargamūlaphala*) has been mentionned (*ucyate*). One should divide (*bhāgam haret*) the next adjacent non-square place by twice that [root]. In this verse, when the square of this quotient has been subtracted (*śuddhe*) from the next adjacent square place, that quotient from the non-square place, in a different place, in the

---

[60] For a translation into English of BAB.2.4, see Keller (2006).
[61] (Sarma 1976, pp. 36–37).

next square place, that [quotient] becomes (*bhavati*) the root.[62] Also, when one has multiplied it (the quotient) by two (*dviguṇīkṛtya*), dividing (*bhāgahāraṇa*) in due order both [digits] from its adjacent non-square, as before, the computation of the third root [is accomplished]. Once again with three [digit numbers, the process is carried out].[63] In this way, one should perform (*kuryād*) [the process] until no square and non-square [place] remain (*bhavanti*). When the root quantity has been obtained (*labdhe*), having multiplied it by two (*dviguṇī kṛtam*), it should be halved (*dalayet*). Concerning fractions also, having divided (*vibhajya*) the square root of the numerator by the square root of the denominator the quotient[64] becomes (*bhavanti*) the root. One states (*āha*) in this way:

> **When the square root of the numerator has been extracted, and the root born from the denominator [also,] the root [is obtained]**(PG 34) |

In order to obtain the roots of the square which were previously explained (in the commentary of verse 3 which is on squares), setting down: 15 625. The result is the square root 125. Setting down the second: $\frac{4}{9}$. The root of the numerator 2, the root of the denominator 3, having divided (*vibhajya*) the numerator by that, the result is the square root of the fraction: $\frac{2}{3}$. Thus the fourth rule [has been explained].

D Appendix 4: APG

An algorithmic rule (*karaṇasūtra*) of two *āryas* for square roots:[65]

> **PG.25. Having removed the square from the odd term, one should divide the remainder by twice the root that has dropped down to a place [and] insert the quotient on a line||**
>
> **PG.26. Having subtracted the square of that, having moved the previous result that has been doubled, then, one should divide the remainder. [Finally] one should halve what has been doubled.||**

What is the root of a given quantity whose nature is a square? This is the aim of that procedure. One should subtract (*tyajet*) a possible (*saṃ bhavina*) square, from the odd (*viṣama*) ⟨place⟩ of the square quantity, from what is called odd (*oja*), that is from the first, third, fifth, or seventh etc., ⟨place⟩; the places for one, one hundred, ten thousand, or one million, etc.; from the last term (*pada*), that is from

---

[62] This long sentence has an equivocal expression: is *sthānātare* (in a different place) glossed into *pūrve vargasthāne* (in the next square place), or should one understand that two actions are prescribed, first setting aside the quotient as a digit of the root, and then that its square enters an operation in the next square place?

[63] This is a mysterious cryptic expression, it is thus my interpretation that the three here, as the "both" (*ubhaya*) used in the sentence before, refers to the number of digits of the square root being extracted.

[64] Reading labdham instead of the misprinted ladhdham.

[65] Shukla (1959, pp. 18–19).

⟨the last⟩ among other places. This should provide (*syāt*) the root of that square which one should place (*sthāpayed*) beneath the place of decrease, ⟨under⟩ the place ⟨where⟩ the possible square is subtracted (*śodhita*) from that, ⟨the place for⟩ one, a hundred, ten or thousands, etc., the last among the other places. And one should divide (*bhāgam apaharet*) from above (*upariṣṭāt*) by twice that, just there. The result should be inserted (*viniveśayet*) on a line, one should subtract (*ś*) the square of that from above that, and this should be doubled (*dviguṇīkūryāt*). If when this is doubled (*dviguṇe kṛte*) an additional place is created (*jāyet*), then it should be used (*yojayet*) as before when it is a result. Both have the quality of being a unique quantity (*raśita*). This quantity has the name "result". One should repeat (*utsārayet*) this, thus one should divide (*vibhajet*), one should insert (*viniveśayet*) the result on a line, etc. as before in as much as the serpentine ⟨progression⟩ is possible (*sarpaṇasambhava*)[66]. When finished (*samāpta*) one should halve (*dalayet*) the whole result, thus obtaining the square root.

Thus for 186624, for which quantity is this a square?

In due order starting from the first place which consists of four, making (*karaṇa*) the names: "odd *viṣama*), even (*sama*), odd (*viṣama*), even (*sama*)".

Setting down:

sa	vi	sa	vi	sa	vi
1	8	6	6	2	4

In this case, the odd terms which are the places for the ones, hundreds, and ten thousands, consist of four, six and eight. Therefore the last odd term is the ten thousand place which consists of eight. Then, the first quantity is 18. Having subtracted (*apāsya*) 16 since it is a possible square for these quantities, the remainder is two. That last quantity is placed separately (*vyavatiṣṭhate*) above. Thus, where it is placed (*sthite sati*) the root of 16, 4, times 2, 8, is to be led (*neyaḥ*) below the place where the square was subtracted (*vargaśuddhi*), which consists of six for the place of decrease. And then division (*bhāgāpahāraḥ*) of twenty six led above (*uparitanyā*). Setting down:

2 6 6 2 4

8

When 24, which is eight multiplied by 3, is subtracted from below (that is from 26), above two remains. Below, the quotient which is three should be inserted (*niveśya*) on a line, they (e.g., these three units) should be placed (*sthāpya*) under ⟨the place⟩ consisting of six. Its square is nine. Having subtracted (*śodhayitvā*) this from above, these (1724) ⟨are placed above⟩, three is multiplied by two, six is to be made (*kartavya*). Below, eighty six is produced. This quantity slithers (*sarpati*) on a line.

Below two, there is [six], below seven, eight. Setting down:

1 7 2 4

8 6

---

[66] Reading here as in the manuscript rather than the *utsarpaṇa* suggested by Shukla (1959), p. 18, footnote 9.

Division above of a 100 increased by 72 by that 86. Decreasing from above the dividend without remainder by two, the result is two, having inserted (*niveśya*) those two on a line, having placed (*sthāpyau*) both below four, its square is four; having subtracted (*śodhayitvā*) from above, those two multiplied by two should be made (*kartavyā*) four, therefore 800 increased by 64 is produced (*jayite*).

Since above the quantity subtracted has no remainder, there is no sliding like a snake etc. process, it remains just the halving of the quantity obtained which should be carried out. Thus, when that is done (*kṛta*), the result is 432. Its square is 186624.

References

Austin, J. L. 1962. *How to do things with words*. Clarendon Press, 1971, Oxford: Oxford University Press. paperback; second edition published 1975 by the Clarendon press; second edition published 1976; new edition 1980 with new index.

Ganeri, Jonardon. 2006. Contextualism in the study of Indian intellectual cultures. *Journal of Indian Philosophy* 36:551–562.

Keller, Agathe. 2006. *Expounding the mathematical seed, Bhāskara and the mathematical chapter of the Āryabhaṭa*. 2 volumes. Birkhaüser: Basel.

Keller, Agathe. 2007. Qu'est ce que les mathématiques? les réponses taxinomiques de Bhāskara *Sciences et Frontières*, ed. Marcel Hert, Philippe Paul-Cavalier, 29–61. Bruxelles: Echanges, Kimé.

Keller, Agathe. 2010. On Sanskrit commentaries dealing with mathematics (fifth-twelfth century). In *Looking at it from Asia: the Processes that Shaped the Sources of History of Science*, ed. F. Bretelle-Establet, 211–244. Boston Studies in the Philosophy of Science, volume 265. Dordrecht: Springer. URL http://halshs.archives-ouvertes.fr/halshs-00189339. Accessed June 2015.

Pingree, David. 1981. Jyotih&astra: *Astral and mathematical literature*. Wiesbaden: Harrassowitz.

Plofker, Kim. 2009. *Mathematics in India*. Princeton: Princeton University Press.

Pollock, Sheldon. 2006. *The language of the gods in the world of men: Sanskrit, culture and power in premodern India*. University of California Press.

Renou, Louis. 1963. Sur le genre du *sūtra* dans la littérature sanskrite. *Journal Asiatique* 251: 165–261.

Renou, Louis. 1984. *Grammaire Sanskrite*. Paris: J. Maisonneuve.

Sarma, K. V. 1976. *Āryabhaṭīya of Āryabhaṭa with the commentary of Sūryadeva Yajvan*. New-Delhi: INSA.

Searle, John R. 1969. *Speech acts, an essay in the philosophy of language*. Cambridge: Cambridge University Press.

Severi, Carlo. 2007. *Le Principe de la chimère, une anthropologie de la mémoire*. Paris: Aesthetica. Aesthetica, Editions de la rue d'Ulm, musée du quai Branly, Paris.

Shukla, K. S. 1959. *Paṭıganita of Śrıdharacarya*. Lucknow: Lucknow University.

Shukla, K. V., and K. S. Sharma. 1976. *Āryabhaṭīya of Āryabhaṭa, critically edited with translation*. New-Delhi: Indian National Science Academy.

Singh, Avadhesh Narayan. 1927. On the Indian method of root extraction. *Bulletin of the Calcutta Mathematical Society* XVIII (3): 123–140.

Part II
Enumerations as Textual Acts

Chapter 6
Textual Enumeration

Jacques Virbel

Abstract This chapter pursues a two-level objective.

On one hand, it presents a study of the textual object of an enumeration type, from four aspects: syntactical structures, semantic operations and functions, logical aspect and pragmatic performatives.

The first point is addressed, from the point of view of internal structures, drawing on specific concepts such as the initializer, the classifier, the item, etc.; and from the contextual point of view by shedding light on relations specific to the enumeration on the level of both sentence and text structures.

The second aspect considers the different types of operation an enumeration allows or requires, for example in terms of definition, description, categorization, etc. With regard to the latter operation, enumeration seems to be a form of presentation of what cognitive psychology considers to be functional categorization. The logical standpoint considers the conditions of identity and existence of an enumeration as a textual object, and other aspects related to reference and categorization.

Finally, from the pragmatic point of view, we aim to characterize the type of speech act that the textual act of enumeration performs: importance is given to the 'co-enumerability' of constitutive items. This notion can have an explicative relation with the properties of syntactical and functional non-parallelism of some enumerations, as well as with the functional aspect of enumerative categorization.

On the other hand, we would like this study to offer the reader, through the example of one type of textual object, a reusable model for the analysis of specifically textual objects, as well as an illustration of various points of view that could be relevant to the study of these objects and for the objectives of textological research.

J. Virbel (✉)
IRIT, CNRS, 1, rue du Commerce, 31540 Saint-Felix Lauragais, France
e-mail: virbel@irit.fr

© Springer International Publishing Switzerland 2015   221
K. Chemla, J. Virbel (eds.), *Texts, Textual Acts and the History of Science*,
Archimedes 42, DOI 10.1007/978-3-319-16444-1_6

6.1 Presentation

Textual enumeration[1] is sometimes defined as a kind of syntactical coordination enhanced by devices such as temporal (first….then…finally), spatial (on one hand… on the other hand) or numeral adverbs (firstly… secondly…). This is, for instance, the thesis of (Damamme-Gilbert 1989). In this conception of enumeration, we mainly consider it in the framework of the sentence, and furthermore we allow numerous aspects to escape our attention: the fact that an enumeration can have several sentences; that its elements (which we call 'items' below) can also have syntactical links between them other than coordination (which relies on the identity of grammatical function); that an enumeration does not only include a mass of enumerated items but also, sometimes, a kind of preface (which we call an 'initializer' below) which indicates which kind of items constitute the enumeration ('John likes the following kinds of fruit: apples, pears and oranges.'), or that it could even include a kind of final conclusion separate to the items. In addition, there are contemporary theories of discourse which aim to characterize phenomena, some of which can be far reaching, and especially whose characterizations go beyond the simple syntactic functions within the framework of the sentence, and endeavour to reach a more 'holistic' level (for example, centering theory, framing theory and, Rhetorical Structure Theory—RST). One other conception, the Text Architecture Model (TAM), which we will follow and develop here, considers enumeration to be a type of textual object. The latter idea could be considered by some as more or less trivial (what can there possibly be in a text other than textual objects or entities?), or as a vague terminological facelift. We must, nonetheless, stress the fact that we believe it covers an authentic novelty: a text is not made up of (only) sentences, nor of (only) parts such as chapters or sections, and nor does it put into play the same type of syntactical organization that can be found in a sentence. We must then accept that texts are made up of specifically textual objects, just as sentences are not made of (only) words, but of specifically sentence objects, of syntagms (of a given type). Methods of inventorying types of textual objects have been proposed before (Virbel 1985, 1989). It seems that tens, even hundreds, of textual objects are identifiable in this way, even if it is just because they are susceptible of being given a label (essentially: a type of title) indicating their nature or textual function (and not their specific content or textual position), such as: thesis, proposition, argument, conclusion, preface, introduction, draft, refutation, termination, adaptation, translation, summary, transliteration, commentary, annotation, demonstration, extract, remark, answer, etc. While such objects are always read for their propositional content, their textual function must also be considered to determine the overall structures of the texts where they appear, and allow a good understanding of the given text. We can see that the concept of 'textual object' is not an ad hoc way of approaching a given

---

[1] 'Enumeration', as is often the case with the nominalization of a verb, means both the act (the act of enumerating) and the final product of the process (a kind of list). We believe that the context, in this article, rules out any ambiguity, and we did not consider it necessary to separate the two uses by two different terms (for instance, 'enumeration' for the act and 'list' for the object).

textual phenomenon (like enumeration), but that it immediately raises the question of the constitution and organization of the text, and thus, in perspective, the question of its identity. Additionally, as far as TAM is concerned, textual objects, since they are textual, necessarily have a formal spatial inscription: we will see that this position is particularly pertinent for enumerations. More generally, we tried in the framework of TAM to provide a thorough explanation of the conditions of existence, conceptual and formal as well as cognitive, of such objects.

6.2 Some Elements of the TAM

In order to succinctly summarize a few elements that we will need in this chapter[2], let us say that TAM aims to incorporate the visual dimension of texts in their global logico-linguistic description. For this, it relies on two groups of both explicative and operatory hypotheses with respect to the meaning, the communicational reach and the function of this dimension:

- visual properties—essentially: morphological (especially typographical) and spatial (positional) properties, incorporated in the term 'typo-dispositional' used hereafter—that are significant are related to developed discursive formulations and explicit pursuant to a principle of functional equivalence. For example, the title of this chapter, with its particular typo-dispositional realization corresponds to a family of utterances such as p: 'The object of this chapter pertains to textual enumeration'. Thus, we recognize that there is, within the group of typographic and spatial properties of a text, a subgroup which plays an important part: we call it the *'formatting device'* (or: 'material format') and we identify through the expression *text architecture* which one from the group of structures of text can be realized (= visually perceptible) by its formatting device[3].
- syntactically, these developed discursive formulations are reducible, the impact of these reductions being manifested by typo-dispositional traces. Semantically and logically, the latter traces are expressions of a meta-linguistic kind (meta-textual in this case): their referents do not exist in the world of which texts talk but in the texts themselves[4]. Finally, from an illocutionary point of view, p carries out a particular performative of the declarative type as Searle describes it. This identification allows a strong and natural pragmatic interpretation of the reductions

---

[2] For a more complete presentation, see (Virbel 1989; Pascual 1996; Pascual and Virbel 1996; Luc et al. 1999; Luc and Virbel 2001; Virbel et al. 2005).

[3] We must add that these properties are entirely interlinked with others, for instance properties of communicational efficiency or the physical properties of the supports. This point is not discussed here.

[4] One can describe and interpret the relations between language, sublanguage and metalanguage within a text in a more complete way than in this chapter, by drawing inspiration from the analyses of these concepts that Z. Harris (Harris 1991) carried out for the level of language in general (language as a whole).

of meta-language, already characterized in syntactical terms, to be given. An illocutionary act can, indeed, be realized implicitly by a simple assertion of its propositional content, in contrast to a realization by an explicit performative, if, within the conditions of success, the essential condition is assumed to be fulfilled for the interlocutors: 'I will come' (in contrast to 'I promise you I will come') can be interpreted as a commitment and not as a simple assertive (of a predictive type) if it is obvious for the interlocutors that the essential condition of a commitment (i.e.: to place oneself under the obligation to come) is fulfilled[5].

And yet, the essential condition of success of a declarative is that the speaker has the authority (institutional or moral) needed to cause changes in the world just by uttering them (explicitly: 'I hereby declare this meeting open'; or implicitly: 'The meeting is open'). Except for specific cases (for example, a form of which the structure and section titles are completely fixed) a writer has this kind of authority over his text.

In the following we will present a study of enumeration according to four dimensions (one section for each): syntactical, semantic, logical and pragmatic. We are convinced of the artificiality of these divisions and are also convinced that major practical and theoretical issues on the very nature of language come into play, for example, at pragmatic-semantic or logical-semantic interfaces. In other words, our choice is one of convenience for an analytical presentation, and not an implicit affirmation of a point of view on major questions, still largely open in current research.

6.3 Syntactical Aspects

The most widely studied aspect of enumeration pertains to its syntactical structures. Some works are more empirical studies (Péry-Woodley 1998; Bouffier 2006), others use language theories, particularly text theories, to approach them: we can cite the theory of centering (Grosz et al. 1995), the theory of framing (Charolles 1997; Jackiewicz 2002), the Rhetorical Structures Theory (Mann and Thompson 1992) and the Textual Architecture Model—TAM- (Pascual 1991, 1996; Virbel 1989; Luc 2000; Luc et al. 1999, 2000). Within the framework of this model, we pay particular attention to enumeration, because of its particular position in written language. Indeed, enumeration is a remarkable case of exploitation of the possibilities linked to writing. On one hand, the written form allows the development of enumerations to be as long and embedded as needed and on the other hand, these enumerations are a particularly clear case of correspondence between developed discursive forms based on adverbial expressions (firstly, secondly, then, finally...) and syntactically reduced forms with typo-dispositional traces of these reductions (use of numeratives, diacriticals, etc.). Thus, enumeration participates in both entirely visual structures which have no sources in oral speech and in entirely discur-

---

[5] Thus defined, implicit illocutionary acts are distinguishable from indirect acts which are realized by means of an act of a different illocutionary type; cf.: 'Count on me to come and help you tomorrow' (in which a commitment is indirectly realized by means of a direct directive act).

sive structures. The directly significant reach of the former appears in this example: the enumeration is ambiguous: item 3 can either belong to the principal enumeration (first level) or to the secondary enumeration within item 2.

```
1.M————  ———

————  ———; 

2.M————  ———

————  ———; 

1.M————  ———

————  ———

————  ———

————  ———; 

2.M————  ———

————  ———

—; 

3.M————  ———

————  ———

————  ———

————  ———

————  .
```

The typo-dispositional properties of the two following realizations express the choice of either one or the other possible interpretations above:

```
1.M————  ———

————  ———

————  ———; 

2.M————  ———

————  ———; 

1.M————  ———

————  ———

————  ———

————  ———; 

2.M————  ———

————  ———

—; 

3.M————  ———

————  ———

————  ———

————  ———

————  .
```

226 J. Virbel

1.M————————

————————— ;
2M————————
————————— ;
1.M————————

————————— ;
2.M————————

— ;
3.M————————

——— .

A sad, yet famous, example also illustrates these aspects (André 1989): a dramatic accident at the 'Gare de Lyon' train station in Paris on 27th June, 1988 caused the death of 56 people, after a collision between a stopped train and runaway train with no brakes. An investigation put the blame on, among others, the driver's engine maintenance manual. Indeed, the section on brake incidents the manual was presented as follows:

b) Plusieurs véhicules sont bloqués, le mécanicien :
S'assure que ce blocage n'est pas la conséquence de la fermeture d'un robinet d'arrêt de la conduite générale situé avant la partie de train bloquée :
1er CAS : Aucun robinet d'arrêt CG n'est fermé :
    Il actionne la commande de la valve de purge le temps suffisant pour
    provoquer le desserrage sur chaque véhicule bloqué.
2e CAS : Un (ou plusieurs) robinet d'arrêt est trouvé fermé :
    Il ouvre le robinet.
    Dans les 2 cas, le mécanicien :
    — ouvre le robinet d'arrêt CG situé en arrière du dernier véhicule relié à la CG.
    — vérifie le serrage des freins du dernier véhicule freiné.
    — referme le robinet CG
    — vérifie en se dirigeant vers la tête du train :
        • le desserrage des freins du tous les véhicules,
        • que le blocage n'a pas provoqué d'avarie aux roues.
Il applique les mesures concernant le signalement et la reprise de marche (article 385).

The text presents two different cases (1er CAS and 2e CAS) of brake incidents and the actions the engineer has to perform. The third line of the second case (2e CAS) reads: 'In both cases, the engineer:' (what follows is the four actions the engineer has to perform). The problem is the following: visually, these two cases are presented as internal to the second case (2^e CAS) which also enumerates two cases in its first line: the engineer finds one or more closed 'taps'. This is not an anomaly; a part of a list can be a list as well. Yet, semantically, it is about what the engineer must do in both cases; in the first case (1er CAS) and the second one (2e CAS). The expression 'In both cases…' should have been horizontally aligned with '1er CAS' and '2e CAS', and have the same typographical attributes: DANS LES 2 CAS, le mécanicien:… (IN BOTH CASES, the engineer:…).

A recurrent question is that of the definition of such a type of textual object, or maybe, if not a definition, at least a threshold on a given continuum where we recognize quite clearly some extremes: if on one hand we can maybe 'accept' that 'John buys and sells furniture' does not contain an enumeration, for this utterance uses none of the procedures or markers that seem to belong to enumerations (cf. 'John buys on one hand and on the other hand sells furniture'), on the other hand, we can also hesitate to call an enumeration the series of chapters or sections of a book, even if they are designated by numbers, just as can happen in the case of an enumeration.

We propose below[6] to address and illustrate this question by means of three series of examples: one example described according to a group of variations created for this illustration; another from the two books by J. Searle; and finally a last group from the works of Aristotle. The reasons for the last two choices are simple: we have paraphrased Searle's texts abundantly in our study of speech acts and it seemed a good idea to offer direct quotations from the author to read; secondly, we were looking for an author who would be universally known, translated and accessible in order to use examples which would be entirely convincing for any reader: we chose Aristotle for this universal cultural role[7].

Let us consider the following series of utterances:

[0] John likes apples.
----
[1] John likes apples and pears.
[2] John likes apples, pears and oranges.
[3] John likes apples but only Golden Delicious, and if they are very ripe, pears, especially Boscs, and oranges, but not Navels.
[4] John likes apples, pears, peaches, grapes, pineapples, strawberries and oranges.
----
[5] John likes on one hand apples, on the other hand pears, and finally oranges.
[6] John likes: on one hand apples, on the other hand pears, and finally oranges.
[7] John likes: apples, pears and oranges.

---

[6] This will be the same approach for the other three sections of the chapter.

[7] We do not claim that our work compares to that of philosophers or philologists: we are neither.

[8] John likes
  • apples
  • pears and
  • oranges
[9] John likes
  • on one hand apples
  • on the other hand pears
  • and finally oranges

----

[10] John likes apples, oranges, mandarins and grapefruits.
[11] John likes on one hand apples, and on the other hand oranges, mandarins and grapefruits.
[12] John likes:
  - on one hand apples
  - on the other hand oranges, mandarins and grapefruits.

----

[13] John likes fruit such as apples, pears and oranges.
[14] John likes fruit such as: apples, pears and oranges.

----

[15] John likes numerous kinds of fruit such as apples, pears and oranges.
[16] John likes numerous kinds of fruit: apples, pears and oranges.

----

[17] John likes three kinds of fruit: apples, pears and oranges.
[18] John likes the following kinds of fruit: apples, pears and oranges.
[19] John likes the following three kinds of fruit: apples, pears and oranges.

[20a] (the list + inventory + group + enumeration) of fruit that John likes are the following: apples, pears and oranges.
[20b] Here is the (list + inventory + group + enumeration) of fruit that John likes: apples, pears and oranges.

From [0] to [20] we pass from a sentence where we can see no enumeration, despite the plurality of kinds of fruit, to a sentence that undoubtedly does contain an enumeration.

[0] cannot invoke the notion of enumeration, if we consider it solely from the syntactic point of view. On the logical-semantic plan, we could say that this sentence, unlike others such as: 'John likes some apples'; 'John likes all apples' (i.e.: John likes all the kinds of apples) implicitly includes all or some of the apples that John likes. Yet, this plurality of referents does not give the status of enumeration to the sentence.

From [1] to [2] we pass from two to three kinds of fruit; in [3] each kind of fruit is the object of a particular description; in [4] the inventory of fruit that John likes is a lot bigger. We can thus see that [1] to [4] express an explicit plurality but without any marker of enumeration; in some treatises on rhetoric, a case such as [4] is recognized to be an 'accumulation': we intuitively see that the longer the accumulation, the stronger the impression can be that we are dealing with an enumeration, even if nothing explicit indicates one (except this accumulation).

From [5] to [9] we get the feeling to be moving closer and closer to an enumerative form. Indeed, even if this is not explicitly announced, we can recognize the use of symbols which are markers used in explicit enumerations:

- a marker of the beginning of an enumeration /:/
- markers of the beginning and/or ending of items ('on one hand', hyphens, indentations)

Only some cases are presented here (cf. for example: 'John likes: -apples; -pears; -oranges.')

In [10, 11] and [12], we shed light on a common phenomenon in enumeration: the possibility of implicitly exploiting existing categories. If [10] does not particularly attract attention on the 'semantics' of the types of fruit mentioned, [11] and even more [12] mark the existence of two groups among the four types of fruits, the second appearing as belonging to the citruses.

We can, thus, see that from [0] to [12] we pass from an absence of enumeration to situations where, contrarily, it is quite natural to invoke such a type of textual object. Yet, starting from [13] there is the new fact of the presence of a term, which we identify as a type of classifier, playing the role, in this case, of a hypernym (kind of fruit). It can be combined (or not) with diverse markers of quantity (some, a few, numerous, a lot, three, etc.) and/or explication of its relation to the objects in question (kinds, species, varieties...) and/or supplementary specifications aiming to a description (for example: summer fruit, fruit from the southern hemisphere, stone fruit, etc.). The formal phenomena concerning the initializer (or: introducer) and those already analyzed concerning the enumeration itself are largely independent. Thus, we only included a limited number of possibilities in these examples. Thus:

[16] John likes numerous kinds of fruit: apples, pears and oranges.

Has variants such as:

[16a] John likes numerous kinds of fruit: on one hand apples, on the other hand pears and finally oranges.

[16b] John likes numerous kinds of fruit:

- apples,
- pears
- and oranges

[16c] John likes numerous kinds of fruit:

- on one hand apples,
- on the other hand pears
- and finally oranges.

We see that differences between these sentences are systematic: from [16] to [16a] there are new lexical markers of the beginning of items (on one hand, on the other hand, and finally); from [16] to [16b] the markers are typo-dispositional; in [16c] markers are lexical and typo-dispositional.

Finally, [20] is an example of an utterance where there is a term which indicates the nature of the textual object. Some types of objects correspond to some specialized classifiers: notebook, inventory (for what is in a house for example), register, catalogue, etc.

A possible conclusion to this overall examination is the following: different procedures, both lexical and typo-dispositional, can be used to indicate a series (plurality or accumulation): presence or non-presence

i. of a classifier, and possibly of a generic of classifiers ('kinds', 'species', 'varieties', etc.);
ii. of a mention of quantity (numerous, a certain number) or of a number (two, three);
iii. of a marker of the beginning of enumeration (/:/, following);
iv. of a marker of the beginning and/or ending of an item (indentation + hyphen, 'on one hand'), and of the last one in particular (for example: 'and finally');
v. of a minimum number (to be specified) of listed elements.

It seems to be clearly established that starting from a fixed number (v), the simultaneous use of all of the procedures (i) to (iv) undoubtedly indicates an enumeration, and surely with some redundancy. Nevertheless, other than this case, and in relation to a criterion (i.e.: necessary and sufficient conditions) of identification of the enumeration, numerous attitudes are possible. One of the interests of the approach previously illustrated is that it allows them to be specified. For example, we could think that the presence of a classifier (i) and of a marker of the beginning of enumeration (iii) is a sufficient condition (as in [14]). Or that an accumulation beginning with a certain number of elements (to specify) will be considered an enumeration, even without the presence of classifiers (as in [4]). What can be considered or not an explicit or implicit enumeration, or even just an accumulation or a simple plural series, varies according to the type of text, the writing style, periods and cultures, languages, writing, etc., even possibly according to the objectives of reading[8].

---

[8] For instance, in the domain of literature, and particularly in lyric poetry, the recurrence of syntactical structures can be a sufficient marker. Thus, a poem like *Childhood III*, by Arthur Rimbaud, is universally considered to be created on an enumeration although it does not have any other marker:

Childhood — III

In the woods there is a bird, its song stops you and makes you blush.
There is a clock that does not strike.
There is a pothole with a nest of white beasts.
There is a cathedral that descends and a lake that rises.
There is a little carriage abandoned in the coppice or which descends the path, beribboned.
There is a troupe of little actors in costume, glimpsed on the road through the edge of the wood.
There is, at last, when you are hungry or thirsty, someone who chases you off.
(translated by Christopher Mulrooney)

There are numerous poems by Guillaume Apollinaire constructed on the same principle, which indicates that it was considered an efficient procedure.

It is important to note now that the example developed above is of the simplest form: the enumeration is included in the framework of a simple sentence (not composed of other sentences), the elements listed correspond to simple substantives, and the order of appearance of enumeration markers and the structure of the enumeration are the same. And yet, on these three aspects, numerous variations are possible. For example, on the third point, we can complete our example as follows:

[21] Apples, pears and oranges are the (three) kinds of fruit John likes.

[22] John likes apples, pears and oranges. These kinds of fruit are those that Mary likes too.

where we see that the classifier is introduced *after* the series of fruit.

A new series of examples, taken from real texts, will allow us to illustrate some more complex situations.

Any reader of Searle[9] who reads:

It seems to me there are (at least) twelve significant dimensions of variation in which illocutionary acts differ one from another and I shall—all too briskly—list them:
1. *Differences in the point (or purpose) of the (type of) act.* The point or purpose of an order can be specified by saying that [...].
2. *Differences in the direction of fit between words and the world.* [...]
3. *Differences in expressed psychological states.* [...]. The psychological state expressed in the performance of the illocutionary act is the sincerity condition of the act [...]

[...]
12. *Differences in the style of performance of the illocutionary act* [...]

[S1] pp. 2–8

---

gets the feeling of dealing with an enumeration for the following reasons:

- the number of items (12) is announced
- and we also know, due to the presence of a classifier, to what each item corresponds here: 'significant dimensions of variation in which illocutionary acts differ one from another'. We can also say more precisely that the introductory sentence includes a subpart: 'there are (at least) twelve significant dimensions of variation in which illocutionary acts differ one from another' which is syntactically autonomous and is the initializer of the enumeration, by indicating the classifier: 'dimensions of variation' and by specifying this classifier: 'significant' and 'in which illocutionary acts differ one from another',

---

[9] Compare Searle, *Expression and Meaning* (Searle 1979, pp. 2–8).

- even though it is 'differ' under the nominalization 'Differences' which appears in the title of each item, in other words an element of the specification of the classifier and not the classifier itself.

Finally, let us note that depending on the definition of the sentence which we use (from the point of view of the punctuation in particular), the entire twelve point enumeration (which represents almost six pages) could be part of the introductory sentence, while all these points are themselves multi-sentence, and even multi-paragraph.

When in the third item Searle summarizes:

These three dimensions—illocutionary point, direction of fit, and sincerity condition—seem to me the most important, and I will build most of my taxonomy around them, but there are several others that need remarking.

[S2] pp. 5

the reader is still sure he/she is dealing with an enumeration, even though different from the previous enumeration in terms of appearance: here, each of the three items is given as a 'dimension', and does not give rise to any textual development: it is reduced to its name. It is, in fact, more a recapitulation of a part of an enumeration (its three first items).

The following example allows us to address other cases:

Chapter I
A Taxonomy of Illocutionary Acts
I Introduction
The primary purpose of this paper is to develop a reasoned classification of illocutionary acts [...]Since any such attempt to develop a taxonomy must take into account Austin's classification of illocutionary acts into his five basic categories of verdictive, expositive, exercitive, bahabitive, and commissive, a second purpose of this paper is to assess Austin's classification to show in what respects it is adequate and in what respects inadequate. Furthermore, [...] a third purpose of this paper is to show how these different basic illocutionary types are realized in the syntax of a natural language such as English.

In what follows, I shall presuppose a familiarity with the general pattern of analysis of illocutionary acts offered in such works as *How to do things with words* (Austin 1962), *Speech Acts* (Searle 1969), and "Austin on Locutionary and Illocutionary Acts" (Searle 1968).

[S3] p 1.

1. On a first level, we have an inventory of three 'purposes' which are not announced in any way (as opposed to a chapter which would begin by 'This chapter pursues

three objectives'). This absence can lead us to ask if it really is an enumeration, at least in the same sense as before, even if the presence of the numeral 'primary' allows one to suppose series of 'purposes'. A. Tadros noted this predictive aspect, or the aspect of commitment of a 'contractual" type between the writer and the reader, that exists in enumeration; she places these with other categories sharing this predictive characteristic: advance, labeling, reporting, recapitulation, hypotheticality and question (Tadros 1994, pp. 70, cited by S. Porhiel in (Porhiel 2007)).

Our example with fruit would here take the following form:

[24] The first type of fruit that John likes is apples; the second variety, pears; and the third, oranges.

2. The presentation of the second purpose is preceded by a justificative argument which invokes 'Austin's classification of illocutionary acts into his five basic categories of verdictive, expositive, exercitive, bahabitive, and commissive': we see that it would be artificial to speak here of an enumeration which Searle would have written, since he merely recalls Austin's categories. We seem to be closer to a quotation of an enumeration already existing elsewhere.
3. Let us note in passing that the content of the second purpose includes two aspects: 'to show in what respects it is adequate and in what respect inadequate', but that these, as they are presented, cannot be considered to constitute an enumeration.
4. Finally, we can be even more reticent to recognize an enumeration resembling one of the previous examples in the last sentence of the passage: indeed, while it does include an inventory of three texts, these are presented as 'giving' 'the general pattern of analysis of illocutionary acts', which determines their role in the presentation of the SAT very well, but without establishing a definition of the items as constituted by their mention in this sentence.

Finally, this last example from Searle reveals other aspects of these limit cases:

> The distinction as I have tried to sketch it is still rather vague, and I shall try to clarify it by commenting on the two formulae I have used to characterise constitutive rules: "The creation of constitutive rules, as it were, creates the possibilities of new forms of behaviour", and "constitutive rules often have the form: X counts as Y in context ".
>
> *"New forms of behaviour"*: There is a trivial sense in which the creation of any rule creates the possibility ... [...].
> [...]
> *"X counts as Y in context C"*: This is not intended as a formal criterion for distinguishing constitutive and regulative rules ... [...]
> [...]
>
> [S4] pp 35–36[10]

---

[10] Searle, *Speech Acts* (Searle 1969).

The first sentence in this extract does not allow the reader to clearly predict that it could be an introductory sentence of a later enumeration: indeed, even if it does mention two formulae that have been used previously by the author, and also the intention of commenting on them, the sentence lacks the presentational force of an enumeration. The reader is lead to reconsider the sentence as an initializer sentence of an enumeration only retrospectively, by becoming aware of the two self-quotations and associated commentaries.

The last group of examples presented belong to Aristotle[11].

Whereas:

[A3] "There are six sorts of movement: generation, destruction, increase, diminution, alteration, and change of place." (*Categories*, part 14)

is indubitably an enumeration, and while:

[A2] "Quantity is either discrete or continuous. Moreover, some quantities are such that each part of the whole has a relative position to the other parts: others have within them no such relation of part to part." (*Categories*, part 6),

does not seem to be able to be considered as such, the status of:

[A1] "Forms of speech are either simple or composite. Examples of the latter are such expressions as 'the man runs', 'the man wins'; of the former 'man', 'ox', 'runs', 'wins'." (*Categories*, part 2) is not really clear and depends on the status given to 'examples' and also to 'either...or', 'the latter...the former', as well as 'such...as'.

Other uncertainties exist in the following extract:

[A4] "Since definition is said to be the statement of a thing's nature, obviously one kind of definition will be a statement of the meaning of the name, or of an equivalent nominal formula. A definition in this sense tells you, e.g. the meaning of the phrase 'triangular character'. [...] That then is one way of defining definition. Another kind of definition is a formula exhibiting the cause of a thing's existence. Thus the former signifies without proving, but the latter will clearly be a quasi-demonstration of essential nature, differing from demonstration in the arrangement of its terms. [...] Again, thunder can be defined as noise in the clouds, which is the conclusion of the demonstration embodying essential nature. On the other hand the definition of immediates is an indemonstrable positing of essential nature. We conclude then that definition is (a) an indemonstrable statement of essential nature, or (b) a syllogism of essential nature differing from demonstration in grammatical form, or (c) the conclusion of a demonstration giving essential nature."[12] (*Posterior Analytics*, II, 10)

As we can see, this extract is not presented as an enumeration—it does not have a phrase or a sentence as initializer. Nonetheless, it lists and illustrates four different definitions of 'definition'. Above all, it concludes with a sentence which is fairly similar to an organizing initializer, yet which plays the role of a summary, although there is no clear one-to-one correspondence between the elements of the

---

[11] We present the same examples with different formatting in the Annex. For references, see, Aristotle. *Categories*; Aristotle. *Metaphysics*; Aristotle. *Posterior Analytics*.

[12] In this example, and numerous others below, item markers such as (a) or (i) have been added by translators and cannot be attributed to Aristotle.

two extracts (not least because the second only includes three elements, Aristotle not repeating the definition of the sense of the terms given previously).

Except for the case where one of the items of an enumeration is also an enumeration, the most frequent relationship between two enumerations is that where an enumeration is introduced by an organizer which is also an enumeration announcing its structure (generally on the basis of an isomorphism of structure)[13].

This example is very representative:

> [A10] "In establishing a definition by division one should keep three objects in view: (1) the admission only of elements in the definable form, (2) the arrangement of these in the right order, (3) the omission of no such elements. The first is feasible because one can establish [...] The right order will be achieved if the right term is assumed as primary, [...] Our procedure makes it clear that no elements in the definable form have been omitted [...]" (*Posterior Analytics*, II,13).

Aristotle offers a more complex example on three levels:

> [A11] "We must next explain the various senses in which the term 'opposite' is used. Things are said to be opposed in four senses: (i) as correlatives to one another, (ii) as contraries to one another, (iii) as privatives to positives, (iv) as affirmatives to negatives;
> Let me sketch my meaning in outline. An instance of the use of the word 'opposite' with reference to correlatives is afforded by the expressions 'double' and 'half'; with reference to contraries by 'bad' and 'good'. Opposites in the sense of 'privatives' and 'positives' are 'blindness' and 'sight'; in the sense of affirmatives and negatives, the propositions 'he sits', 'he does not sit'.
>
> > i.   Pairs of opposites which fall under the category of relation are explained by a reference of the one to the other [...]
> > ii. Pairs of opposites which are contraries are not in any way interdependent, but are contrary the one to the other. [...]
> > iii. 'privatives' and 'Positives' have reference to the same subject. [...]
> > iv. Statements opposed as affirmation and negation belong manifestly to a class which is distinct [...]." (*Categories*, part 10)

As we can see, the beginning of this extract is made up of an enumeration with a complete initializing sentence (Things are said to be opposed in four senses). This sentence is followed by a paragraph which is not presented as an enumeration, but in which the author repeats the four items of the first enumeration, explaining their meaning. We could say that we are dealing with an implicit enumeration here because it does not have an explicit enumeration marker, but its structure refers to the previous one. Finally, a new enumeration of four items begins which repeats the items of the previous one. This enumeration does not have its own initializer, but this is because the first enumeration acts as initializer to the whole.

---

[13] We must cite here the work of S. Porhiel (Porhiel 2007) who analysed in detail a related type of enumerative structures: two-step enumerative structures.

6.4   Semantic Aspects

To the best of our knowledge, few previous studies have directly addressed the semantic aspect of enumeration. C. Bush (Bush 2003), alone and with Ch. Jacquemin
(Jacquemin and Bush 2000), were the first, as far as we have been able to ascertain,
to define the kinds of functional semantic roles of the various components of an
initializer: "In order to fully understand the structure of a trigger, one must first understand its role. The four functions of a trigger are firstly to indicate the presence of
an enumeration, secondly to describe the nature of the structure of the enumeration,
thirdly to describe the **hypernym**, and fourthly to describe the **differentia**, in other
words to characterize the items of the enumeration." (Bush 2003, pp. 49; bold in
the text). Thus, we can combine these four functions, in the order (1) to (4), in an
initializer such as:

'Here is (1) the list (2) of universities (3) receiving public funding in Indonesia
(4).'

An additional advantage of this approach is to allow functional qualification of
the cases where some elements of the initializer are absent through ellipsis, as for
example in:

'List (2) of universities (3) receiving public funding in Indonesia (4)'
'Here are (1) the universities (3) receiving public funding in Indonesia (4)'.

A very interesting fact to consider, still in relation to the functional approach
used by C. Bush and Ch. Jacquemin, is the following: theoreticians of "Conceptual
Role Semantics" noted (in an article published in the *Handbook of Philosophy of
Language* (Greenberg and Harman 2006)) the use of lists as a typical example of
the role of the use of symbols on their meaning. Conceptual Role Semantics (CRS)
supposes that meaning is determined by the different ways in which symbols can be
used in communication and in thinking. This use includes for instance representations and forms of inference such as perceptual representation, modeling, labeling,
categorization, etc. In (Greenberg and Harman 2006, pp. 302), the examples of uses
of symbols listed are:

- maps (on "paper" or "on computer screen, and also internal 'mental maps'");
- gauges ("fuel gauge in order to make sure there is enough gas in the gas tank,"
  or "internal gauges that indicate via hunger and thirst when people need food or
  drink)";
- "models and diagrams to help in planning;"
- representation of numbers and measure in mathematical reasoning;
- "envisioning possibilities" ("representation of various possible scenarios)";
- "implication" and "reasoning;"
- "mental models" ("representation to think about implications," given some information);
- labels (used "in order to recognize things later"). A label may be a mark on something, but "one might even use an actual feature of an object as a label;"
- categorization;

- and lists (in particular 'to do' lists. "People make shopping lists," "they keep diaries and schedules". People also solve problems "by listing initial possibilities and ruling as many out as they can. Some puzzles can be solved in their heads, others require writing things down on paper. In trying to decide what to do, people make lists of considerations, trying to correlate those supporting one decision with others supporting another decision, so that the considerations can be crossed off, leaving easier problems. CRS might suggest that what makes these things lists is at least in part that they are used in such ways. In support of this, note that it is not the case that every sequence of representation is a list. For instance, a sentence or a mathematical proof is not a list"[14].

The interest of C. Bush's approach is undeniable, and it is impossible to ignore it in her field of application: automatic analysis of Web documents to identify named entities. At the same time this approach has limits because of its very nature, that is, that it limits itself to the functions formally marked in the initializer. And yet, it appears that the study of enumeration demands the consideration of the cognitive operations made necessary or possible by the activity of enumeration, which in general cannot be reduced to (or be inferred from) the initializer alone. We are thus lead to formulate hypotheses of a cognitive nature: to enumerate includes, surely, one way or another, in one proportion or another, activities that we usually denote by terms such as: to describe, to define, to refer, to class, to group, to categorize, to number, to list, to identify (and one should make sure of the differences, should any exist, between to make out an inventory, to index, to list, to enumerate, to define nomenclature, to catalogue).

Research carried out by Ballmer and Brennenstuhl (Ballmer and Brennenstuhl 1981) can give another perspective on the same question. The authors propose a classification of 4880 English speech activity verbs, separated into 8 'models' (groups), 24 types and 600 categories.[15]

Among the eight models, there are the three following: 'Discourse Models', 'Text Models' and 'Theme Models'. In one of the types of 'Text Models', we can find a 'Construct' category which includes the following verbs:

Amalgamate, arrange, collate, combine, compose, concatenate, connect, construe, coordinate, design, form, form into files, form into rank, invent, join, link,

---

[14] It is remarkable that this list of uses of symbols constructed by the authors and which includes the lists does not seem to respond clearly to any of the examples of the use of lists that they indicate in this last (it is true that it is not a 'to do' list). We should probably accept that the use of lists is even larger than is said.

[15] For the sake of comparison, Austin's lists include 188 examples—but he was saying that he believed the number of speech act verbs in English to be between 1000 and 9999... Wierzbicka (1987) counts about 300 speech act verbs. We can note here the notion of speech act for these two authors owes nothing to that of Austin-Searle, and that the categorisations they carry out are largely based on semantic intuition (they want to shed light on a kind of folksemantics of common sense). For instance, the authors integrate in their list all the verbs linked to emotion in the expression of an utterance, such as frown, howl, shout, moan, scream, tremble, etc. which in SAT do not realise illocutionary acts.

make up a list, organize, parallelize, place together, plan, project, put together, rank, summarize, systematize, synthesize.

Another category, belonging to a different type, 'Categorize' includes almost fifty verbs, among which: enumerate, list, inventory, identify, locate, order, rubrify, schedule, register, tabulate, typify, attach to.

We can further find 'enumerate' and 'make series' in another category: 'Connect', accompanied by other verbs such as: bind together, chain together, combine, conglomerate, connect, interweave, juxtapose, link together, tie together, count up.

The names of categories used by the authors, and maybe even more so the verbs they include in the categories, are indicative of the dimensions of the meaning of 'to enumerate', which consists of a form of construction of categories, and thus carries out categorization and a form of connection between the terms (items) concerned.

Another possibility for penetrating the semantic of 'to enumerate' consists in noting that the items are listed *as being* related to the classifier: in 'John likes the following kinds of fruit: apples, pears and oranges', we can say that 'apples', 'pears' and 'oranges' are cited *as being* the fruit that John likes and not for instance because they have a botanic property. This observation presents the interest of referring verbs such as 'to list' or 'to enumerate' to a group of verbs which satisfy the following syntactical schema:

N0 V N1 (like + as + for + Ø) N2[16].

'The doctor interprets this fever and these spots as symptoms of measles.'

'The population (elected + chose + designated) John as Mayor.'

'Jack hired John as a driver.'

'Bill listed apples and pears as kinds of fruit that John likes'

What follows is a more complete, though far from exhaustive, list of such verbs:

Analyze

Categorize

Characterize

Choose

Class

Conceive

Decode

Define

Describe

Designate

Diagnose

Elect

Employ

Engage

Enumerate

Hear

---

[16] In this notation 'Ni' denotes nominal groups, 'V' a verb, '+' notes an 'or', '*' a form which does not belong to English and 'Ø' the empty word. The choice of the preposition could be not free, but on the contrary dependent on the verb. Compare 'We elected Jack (as + to be + *for + Ø) mayor in contrast to' 'John interprets these symbols (as + *being + *for + * Ø) Greek.'

Identify
Interpret
List
Name
Perceive
Promote
Propose
Qualify
Read
Receive
Regard
Represent
See
Understand
Use

Examination of this list reveals some relatively homogenous subgroups from the point of view of the sense:

i. analyze, understand, interpret, regard, represent, conceive, hear, see, read, receive
ii. describe, define, decode, designate
iii. characterize, categorize, class, identify, qualify, diagnose
iv. choose, engage, elect, name, propose, promote, employ, use
v. enumerate, list

Subgroup (i) groups verbs which indicate an activity of interpretation; (ii) of description; (iii) of categorization; (iv) of promotion to a role or function; (v) of enumeration. One can think that the signification of the activity of enumeration includes a certain composition of these semantic characteristics.

These approaches to the lexicon lead to a question: what are we enumerating in an enumeration? The question could seem arbitrarily provocative but certain shifts in meaning that appear in the literature prompt it to be asked. For example, the answer to the question 'What is being enumerated in the sentence 'John likes apples, pears and oranges?', could spontaneously be: 'Apples, pears and oranges' or 'kind of fruit'. These would not be acceptable: there are no apples, pears oranges or fruit in the text; we can only find them in the world outside the text. On the other hand, in the text one can find words, or units constructed from words, such as items. It would, thus, be logical to maintain a clear distinction in order to avoid confusion of the categories, and maybe to specify the terminology: We make an inventory, class, tally (for instance[17]) objects of the world (such as fruit); we list, group, number (for instance) items in an enumeration. Sentences including factorization by coordination indicate this necessity more clearly. In p: 'John buys on one hand and on the other hand sells furniture as well as books', should we need to answer this simple

---

[17] Our objective here is not to regulate the use of words, nor establish technical terminology, but to indicate a problem and an example of a possible solution.

question 'How many enumerations are there in the sentence p?', by referring to the form of the sentence, one could consider that there are two (that of the verbs and that of the objects), while by distribution there are four <activities—goods> couples which in fact are real entities existing in the extra-textual world. This existence is expressed by a syntactical form of factorization. We can see in this simple example the necessity of distinguishing between the entities existing in the extra-linguistic world of which an utterance speaks and the linguistic units included in and making up an enumeration.

However, example [A12] indicates a case with a double hierarchy:

[A12] "The pre-existent knowledge required is of two kinds. In some cases admission of the fact must be assumed, in others comprehension of the meaning of the term used, and sometimes both assumptions are essential." (*Posterior Analytics*, I, 1)

There are two kinds of pre-existent knowledge, but there are three cases, each allotted to a specific item, marked by 'some cases', 'in other' and 'sometimes'.

These details allow us to re-examine the presentation we made above of the example S2 in a self-critical way. We reproduce here for the sake of convenience the appropriate part of extract S2 from Searle, and our commentary:

S2': "It seems to me there are (at least) twelve significant dimensions of variation in which illocutionary acts differ one from another and I shall—all too briskly—list them:"

- the number of items (12) is announced
- and we also know, thanks to the presence of a classifier, to what each item here corresponds: 'significant dimensions of variation in which illocutionary acts differ one from another'.

And yet, strictly speaking, Searle does not announce the items, as we said, but affirms the existence of twelve dimensions of illocutionary act, in other words it is an inventory of objects or entities from the extra-textual world (outside the text itself). Another thing that we discover by reading the rest of the text is that Searle pairs an item with each dimension, but that is not what this sentence is about. And furthermore, the fact that in the text, there are as many items as there are objects identified in the world is certainly a common case, but is by no means a necessity. See for example:

This problem has three solutions which we shall call solutions A, B and C, and which we shall examine below.
1. Solutions A and B
[…]
2. Solution C
[…]

We can also use the [11a] variant of our example [11].

[11a] John likes four kinds of fruit:

- on one hand apples,
- on the other hand oranges, mandarins and grapefruits.

6.5   Logical Aspects

In this section we wish to address some aspects linked to the conditions of identity and existence, reference, and categorization, with regard to enumeration[18].

6.5.1   Conditions of Identity

We only quote these conditions here for reference, since it is evident that they cannot contain general answers. Such conditions can only be constructed in relation to a definition ('no entity without identity', according to Quine), and we saw that such a definition involves considerations that can vary: if we consider that the presence of a classifier is an obligatory element in an enumeration, but that this is not the case for a quantifier, then:

[16] John likes numerous kinds of fruit: apples, pears and oranges.

can be considered to be carrying out an enumeration identical to:

[17] John likes three kinds of fruit: apples, pears and oranges.

but this would not be the case for:

[2] John likes apples, pears and oranges.

Another aspect of identity between two enumerations is, undoubtably, the order of the items the change of which (when possible) may or may not affect the identity of the enumeration. For instance, this order can be declared as defining of the enumeration or not:

'I list in order of importance: [...]'

'I list in no specific order: [...]'

or its role may be left unspecified. In addition, when the items have links of syntactical dependency other than coordination between them, any permutation of these items destroys the enumeration (what we have here then is a condition of existence).

However, more generally, the conditions of identity of enumerations are the same as for any text, where only the strict alphanumeric identity and that of the formatting device are sufficient conditions of textual identity, but neither one nor the other is a necessary condition (as long as the functional equivalence between developed discursive forms and forms reduced to particular formatting device are respected).

---

[18] On the problems of the ontological status and the conditions of identity of texts see (Gracia 1996, pp. 9–90).

6.5.2 Conditions of Existence

What is studied here is the presupposition of the existence of the domain of reference of an enumeration. We can note two types of relatively separate situations. On one hand, there are those situations where the domain of reference is manifestly pre-existent to the act of enumeration and in which the speaker's activity has nothing to do with its creation (we can thus list the days of the week, the months of the year, etc.). We can consider that in such cases, the enumerative aspect or the enumerative part in the speaker's speech activity is limited, even almost non-existent. We have such a case, when Searle refers to and quotes a well-known (in the context) enumeration: the classification of performative verbs by Austin. There are, on the other hand, situations in which the speaker creates this referential (contemporaneously or prior to the act of enumeration) (I list the kinds of fruit which, to my knowledge, John likes, or the solutions which in my view answer a given problem). There are, naturally, intermediary cases which are difficult to define and, furthermore, how manifest the existence of something one refers to is varies greatly from one person to another (see below). Additionally, enumeration of a pre-existing referential can include in part creation in the different linguistic choices made to realize it. For instance, numerous recipe books include recipes that are not original in any way. However, the author's originality is evident in the way he/she presents the recipes. We find here again, in a different form, the idea that the adherence of an utterance to the class of enumerations is rather gradual.

6.5.3 Some Questions of Categorization

The element mentioned above is important in relation to the kind of categorization which an enumeration creates.

At first sight, utterances such as:

1. John likes the following kinds of fruit apples, pears and oranges
   and
2. Fruit includes the following varieties: stone fruit, fruit with pips, soft fruit, etc.

are very similar and are formally presented as enumerations.

Yet, from the point of view of the problematic of categorization, they perform different speech acts and objectives.

Utterance (1) mentions a group of fruit the justification of which lies with John's taste: the category of fruit created by this enumeration is completely ad-hoc and linked to John (additionally, at this moment, given that tastes can change). There is no justification for the grouping of these three kinds of fruit other than John's taste; neither scientific, economic, technical, nor any other kind. Utterance (2) describes a classification based on encyclopedia-type categories (that these categories are 'popular' or justified by botany is of no importance here). In cognitive psychology, there is a clear distinction between 'natural' categories (animals, trees, drinks, flowers, fruit, insects, vegetable, metals, birds, fish, etc.), 'artefact' categories (arms, buildings, musical instruments, toys, furniture, tools, recipients, utensils, vehicles, clothing, etc.), 'activity' categories (professions, sports, etc.), and incidental or contextual categories, called 'functional' categories: objects on my table; books in John's library; etc. For instance, the 'objects that one must take when camping' certainly are an extremely heterogeneous inventory should we wish to categorize them in terms of general ontology. Yet, they are indubitably a cognitively anchored category, even for subjects who do not go camping! It seems quite natural to identify a classifier of enumeration by the name of a functional category in the sense explained above. A psycholinguistic or even neuropsycholinguistic study of enumerations could, by means of this identification, be based on research concerning these categories. This hypothesis could be tested in reference to the fact that, for instance, subjects perform differently (in terms of quality or speed of comprehension and/or memorization for instance) according to whether the categories of the objects in the experiment are natural, artefactual or functional.

We can also study this hypothesis in the context of the phenomenon of associative or demonstrative anaphora. This type of anaphora is presented, for instance, as follows (Schnedecker 2006):

'A car violently crashed into a plane tree on a bend at the entry to village X on the N.113 road. *This accident* thankfully had no victims'.

As we can see, this form of anaphora allows, following the description of an entity (here an event described thanks to a given story), reference to be made to it by indicating a category (here: an accident). Although not linked in principle to enumeration, a demonstrative anaphora can also play a role in it, by indicating the classifier and thus also the category of the items:

'John likes apples, pears and oranges. These (three) kinds of *fruit* are also those that Mary likes.'

These examples (previously presented in examples [21] and [22]) lead us to broaden our view of enumeration, or at least of the assumed role of the initializer, given that with demonstrative anaphora, the revelation of the classifier comes after the list of items, without jeopardizing, it seems to us, the existence of the enumeration itself. We thus find, at the level of a textual object that can be appreciably longer and more complex than a simple sentence, a possibility which is quite common on the level of the sentence: the existence of syntactical language resources which allow speakers to choose the order of introduction of syntactical agents in the sentence (for instance choosing an active or passive verbal mode), according to a strategy of the management of the introduction of new in contrast to previously mentioned information or information assumed to be known (Prince 1981; Pattabhiraman and Cercone 2001).

The observable data are even more complex: thus, reversing the items-initializer order does not seem to be linked to the presence of a classifier and a demonstrative anaphora, as presented in Aristotle's example [A4] (see above).

6.5.4   Reference to an Enumeration

We mentioned above the importance of the domain of an enumeration. It is possible to refer to some enumerations without needing to quote them directly, but by using what seems to be close to a proper noun: 'days of the week', 'months of the year', '(letters of the) alphabet', etc[19]. We can even evoke the style of an enumeration by saying that it is a 'Prevert-like enumeration' or 'in the style of Prevert' in reference to 'Inventaire' (Inventory), a famous (in France anyway) poem by Prevert, to mean that the inventory is not constructed according to an identifiable classifier, or else that it includes heterogeneous and/or eccentric elements[20].

6.5.5   Reference in an Enumeration

The structure of the initializer allows references to be made in many ways. For instance, let us suppose that the utterance:

[19] John likes the three following kinds of fruit: apples, pears and oranges.

is not alone, but is the initializer of a principal enumeration, the three items of which are devoted to developments related to these three kinds of fruit. There are many ways to refer, in a given item, to the element that concerns it in the initializer; for instance (for the first item):

[19a] As far as the first kind of fruit is concerned, [...]
[19b] As far as the first kind is concerned, [...]
[19c] As far as the first fruit is concerned, [...]
[19d] As far as the kind of apple is concerned, [...]
[19e] As far as apples are concerned, [...]
[19f] As far as the first are concerned, [...]

But, a specific classifier can itself be referenced by a very general substitute (such as, according to the context, 'point' or 'aspect'); and this term can serve in the reference to an item:

[19g] As far as the first point is concerned, [...]

We can see that the reference can be made by the classifier, the term, a combination of the two, or the order of introduction in the text[21].

---

[19] Such nomenclature referred to by proper terms are more numerous than one would think: the four seasons, the countries of the EU, the five continents, army ranks, geological eras, metro stations in a city, chess pieces, the Ten Commandments, the Seven Dwarves, etc.

[20] The fact that we can use a noun or a particular expression to refer to a group, identified thanks to previous enumerations, recalls the case where we give a title to certain utterances or text extracts (which do not have titles that would have been given by the authors): the Poisson law, the Pythagorean theorem, the Hippocratic oath, the motto of the French Republic; in the literary domain: Rodrigue's stanzas, Antigone's farewell speech in Sophocles' tragedy, etc. This phenomenon can also apply to an enumeration: for instance, the Catalogue of Ships in the beginning of the Iliad.

[21] S. Porhiel (2007) made similar observations on examples of so-called two-step enumerations.

6.6   Pragmatic Aspects

The following, more detailed example allows some introductive questions to be addressed, such as:

- What allows a simple inventory to become an enumeration in the true sense?
- What characterises a list of actions more as a directive instruction or more as an assertive series of the events in a narrative?

Let us suppose that the members of a family make a list of all the foodstuffs that are gradually used up by writing down the names on the page of a pad dedicated to that use, with the intention of renewing their stock. Such a list could, for example, at a certain point contain:

- eggs
- tomatoes
- milk
- bacon
- potatoes
- orange jam
- apples
- cheese
- rice
- tea

As we already said, given its mode of creation, this is the list of foodstuffs that have just run out in the house and, implicitly, must be bought. We should note that we speak of a 'list' here but it is not the product of any specific intentional speech act. It could even be better to speak of a chronological order of lexical reminders.

On shopping day, the person who has to do the shopping takes this page and uses it as an instruction to buy. The *same* list changes status here: on one hand, it goes from being a list of objects to buy to a list of purchases (actions) to make; actions noted by the objects required. On the other hand, it would be interesting for the buyer to rewrite this list in order for it to become a true buying-guide. For instance:

```
- Potatoes 2kg
- 6 tomatoes
- Apples 1kg
------------------------------------
- Cheese 300g
- 2l Milk
- Bacon (6 pcs)
------------------------------------
- 12 eggs
------------------------------------
- 500g rice
------------------------------------
- 1 small jar of orange jam
------------------------------------
- 1 box of tea
```

We can, thus, see the change of status between the two lists: the first list records objects to buy. Despite its appearance, its mode of production and its function do not seem to give it the status of enumeration. The second list includes the same objects but with three differences: the quantity details make buying the correct amounts more probable; the objects are grouped according to their place in the store, thus rationalizing the route and allowing a better control over possible forgetfulness; above all, each object refers, by metonymy, to the action necessary to acquire it in the conditions of structure and organization of a given store. We can consider this second list to be a list of actions and thus, indirectly, as instructions to buy; in other words, like a series of directive illocutionary acts. As we have already seen, deciding whether we are dealing with an enumeration depends on the definition we adopt and in particular on the types of ellipses accepted. In addition, we should note that what makes it an enumeration of directive acts is not its content, but the intentionality of the users (writers, readers). We can draw inspiration here from a famous example analyzed by E. Anscombe (Anscombe 1957): if the person who does the shopping using the directive list is followed by a detective who notes down all of the shopper's acts, this detective will eventually produce a list which may have the same content but a different status: it will be the realization of an assertive illocutionary act concerning the buyer's actions. As E. Anscombe also indicates, the difference appears in the case of a mistake: if the buyer gets home and realizes he made a mistake in his purchases (he bought pasta instead of rice for instance), he will have to return to the store to adjust his purchases to the list; while the detective who realizes he made a mistake during the surveillance (he wanted to note down rice but wrote pasta instead), will not need to return to the store but only to correct his list.

This example reveals numerous aspects that concern us here:

• what makes the illocutionary nature of a document does not necessarily come from its form and/or its content; or: the semantic and syntactical identity of a document cannot be sufficient to determine its illocutionary force;

- the enumerative force may not be realized explicitly (i.e.: performed). In the small scenario above what gives the force of enumeration to the list is the 're-working' of the first list in order to obtain an organized and exhaustive list of actions to carry out. The term 'scenario' here is not used accidentally: we wish to say that in such a case knowledge of the history of the document is necessary to interpret its status (yet this is not rare: it is for instance the case of speaking consecutively in a conversation or dialogue).
- enumerating can be secondary to efforts to facilitate extra-textual operations such as aiding the correctness of purchases or optimizing the route in the store.

Addressing the pragmatic aspects of enumeration means asking questions such as: if 'enumerating' is a speech act,

- which type does it belong to with regard to the essential conditions?
- what are the other conditions of success in terms of preliminary conditions, sincerity and propositional content?
- if it is a performative, what does it perform? And, in the case of an act other than a direct illocutionary act, what forms of indirect or implicit enumeration acts are there?

We should firstly note that 'enumerate' is compatible with all the types of speech acts: we can enumerate (items which realize):

- assertions (this is the case for most of the examples presented previously),
- but also commitments: 'I promise I will: 1) come tomorrow by 7 o'clock at the latest and 2) help you as best I can'. In such a case the performative verb of the commitment includes the classifier (as we can see in the form with a verb such as 'make' and the nominalization: 'I'm making the two following promises: to come tomorrow [...]'),
- or directives: 'Would you please do the following three things: [...]?',
- declaratives: 'I declare on one hand that the session is open, and on the other hand that I designate John as the president of the discussion.'; the classifier is included in the performative verb here too,
- or even expressives: 'I would like to congratulate you not only on your success, but also on the technique': we can see that, here as well, we understand there are two (reasons for) congratulations

Thus, 'to enumerate' does not perform a specific speech act, but on the contrary, it allows any type of speech act to be performed in a particular way. From this point of view, this verb is comparable to others which, as Searle notes, do not indicate a specific illocutionary force but some other characteristic; he gives the following examples: 'insist', 'suggest', 'announce', 'confide', 'reply', 'answer', 'interject', 'remark', 'ejaculate', 'interpose' (Searle 1979, pp. 28). We can see that we are indeed able to emphasize the fact that we are performing a particular illocutionary act or that we are able to make announcements or give answers with any speech act, given that what is characterized is a more global mode of performance, or a communicational status. We can consider these speech acts as meta-speech acts. Indeed, they do not perform primary illocutionary acts (promise, advise, etc.) but bear on such acts (emphasize, announce, etc.).

'Enumerate' seems to correspond to such a situation and is not, by far, an isolated exception in language. It is useful to note that Searle, in his taxonomic study of illocutionary acts, also cites 'reply' as well as other verbs such as 'deduce', 'conclude', 'object', which, like those listed above, have the same property of expressing a mode of realization or a communicational status of an illocutionary act. These mentions can be found in the seventh type of 'different types of differences between different types of illocutionary acts' (cf. ex. S1): 'Differences in relations to the rest of the discourse' (Searle 1979, pp. 6). The author says: 'These expressions serve to relate utterances to other utterances and to the surrounding context. [...] In addition to simply stating a proposition, one may state it by way of objecting to what someone else has said, by way of replying to an earlier point, by way of deducing it from certain evidentiary premises, etc. 'However', 'moreover' and 'therefore' also perform these discourse relating functions' (ibid.). We should note that 'moreover' is a typical lexical marker of items. What is important for our argument in what Searle wrote is that these verbs mark a particular mode of realization of the propositional acts themselves as well as marking discursive relationships in discourse. This is again a property we will find in enumerations if we compare expressions that are simply serial and plural with properly enumerative expressions.

What distinguishes 'enumerate' from other verbs given by Searle is that 'enumerate', more than all others, is realizable in many discursively implicit yet typo-dispositionally developed and multi-form ways. Yet, even this is only partially the case, should one consider the case of 'insist', which other than its direct use ('I insist on the fact that you must leave immediately.') can also be realized thanks to typographical traces associated to its deletion: 'You must leave *immediately*', 'You must leave **immediately**', 'You must leave immediately'. Additionally, nothing specifies whether 'insist' refers to the whole of the completive sentence or to one of the elements (what would be possible in oral speech), while the choice on such a feature exists here (cf. 'You **must** leave immediately', 'You must **leave** immediately').

6.6.1   Which Speech Act does an Enumeration Constitute?

We highlighted above: enumeration does not perform a specific illocutionary act having identified illocutionary goals, but it allows any type of illocutionary act *to be performed in a particular way*. Which particular way though?

In order to address this question and to answer it, we will use an aspect of enumeration that we have not yet discussed, except by allusion: contrary to very popular belief, the items of an enumeration do not necessarily have the same syntactical function, if we are dealing with an intra-sentence enumeration, or the same textual function, in the case of a multi-sentence enumeration. Here are two examples, among hundreds, which prove it conclusively:

WHAT IS (NONSOLIPSISTIC) CONCEPTUAL ROLE SEMANTICS?

In this paper I will defend what I shall call '(nonsolipsistic) conceptual role semantics'. This approach involves the following four claims:
(1) The meanings of linguistic expressions are determined by the contents of the concepts and thoughts they can be used to express;
(2) the contents of thoughts are determined by their construction out of concepts; and
(3) the contents of concepts are determined by their 'functional role' in a person's psychology, where
(4) functional role is conceived nonsolipsistically as involving relations to things in the world, including things in the past and future.

M _____

_____

_____ .

(This example is taken from the Internet)

   This enumeration contains four items with numbers indicating the same level in the hierarchy, under the classifier 'claims'. From a syntactical point of view, we can see that items (1), (2) and (3) have a coordination between them: (1) and (2) are implicitly connected; (2) and (3) explicitly (by 'and'); while (4) is subordinate to (3). On another hand, from the point of view of the meaning, we note that (1) is a kind of presentative of (2) and (3) which repeat and specify (in a different order) the terms introduced in (1) of which they are, thus, dependent in a way. We can also note that from the point of view of the punctuation, all the items belong to a single sentence. Thus, we can say that constituents of a sentence which occupy different syntactical positions, and which have links of conceptual dependence, are explicitly organized in terms of typography and position to suggest a strict enumerative identity. This aspect is even more spectacular in the following example:

:

M _____

_____

Est considéré comme "lecture savante", du point de vue fonctionnel, une pratique de lecture répondant aux critères suivants :
-       c'est une lecture "qualifiée",
-       qui se développe sur le temps long de la recherche scientifique,
-       dans un parcours forcément individualisé,
-       où l'écriture se combine à la lecture, souvent dans une perspective de publications.

M _____

_____

:

(...is considered as 'scholarly reading', from the functional point of view, a reading practice which answers to the following criteria:—it is a 'qualified' reading—which developed slowly in the long run of scientific research—in a necessarily individualized way—where writing is combined with reading, often in a view to publication...).

In this enumeration of the criteria of 'scholarly reading', four items are visually presented as of the same rank, while each of the last three depends syntactically on the one preceding it in the enumeration[22].

These facts are syntactical in essence, and we could have presented them in our first section. Yet, we analyze them here because they shed light on what seems to be the specific performative property of enumeration: the declaration of the quality of 'co-enumerability' of items of equal importance in the attention devoted to them. The existence of the syntactical non-parallelism between items compels us to radically question the concept of enumeration as a kind of coordination, and to note that enumerative structures hold primacy over sentence structures, which must, in a way, obey the demands of enumerations. In other words, what is, in this hypothesis, the source of enumeration, is not the intention to assertively reinforce an existent syntactical coordination, but the intention to realize a communicational act which consists in declaratively creating the fact that the enumerated entities are equally co-enumerable. This interpretation also explains the subordinate character of sentence structures in relation to those of the enumeration, as well as the central role of the classifier, which designates the domain of reference in respect to which the equality of importance is performed. It is an aspect which does not appear immediately in the case of syntactical parallelism between items, yet immediately in the cases of non-parallelism, hence their importance as an indicator[23].

---

[22] We should note that there is a textual side to this phenomenon: all the items of an enumeration do not necessarily have the same structural position in the text (for instance, form a paragraph each).

We inserted an example of this case in the abstract of our chapter:

we announce four aspects but there are only three paragraphs, because the second and third aspects are presented in a single paragraph, by contrast to the first and last one for each of which there is a dedicated paragraph. We can, thus, in the text itself devote a section to each of these four "aspects", but in this summary we implicitly indicate that the semantic and logical aspects share very strong links.

Similarly, the textual fragment of the first 'objective' is developed in four paragraphs (of which the three mentioned above), while the second 'objective' is only approached in the frame of the last of the summary.

[23] We have already addressed arguments claiming that the cases of non-parallel enumerations are rather artificial, leave one sceptical about their reality, or are used arbitrarily because they are poorly formed or not very representative. We wish to indicate here on these methodological points that:

• we have observed hundreds of examples of such enumerations attested in many languages;
  the problem of their supposed poor formation is difficult, if not impossible, to evaluate: the constraints which rule the composition of a text are not of the same nature as those of the sentence; a consequence is, for instance, that there is no textbook, to the best of our knowledge, on the 'correct' realisation of enumerations
• whether an example is representative, supposing this can be evaluated, is not always, and not even often, what makes it interesting from the theoretical point of view, but its power to reveal

It should be noted that while this feature appears to us to thoroughly character-ize enumeration, the reasons for which an author uses an enumeration rather than a simple non-indicated plurality can be numerous; as can the exploitation of the fact that an enumeration has been created:—efficiency of the 'first glance' effect that visual properties provide;—the facilitation of references and 'see above/be-low';—the possibility of making ambiguous sequences univocal: cf.: 'There has been a disagreement between John, Jack and Mary' in contrast to 'There has been a disagreement between John on one side and Jack and Mary on the other'. It is also possible to exploit an enumeration-effect as such[24], that is based on the sole fact that an enumeration exists.

Thus, the illocutionary force of an act of enumeration is declarative, and it con-cerns a particular text structure, organized with items with the same importance and the same weight in that structure. 'Enumerate' is thus a meta-speech act that may be realized by lexical markers (enumerate, list, check, …) or by particular formatting devices with embedded text organizers such as 'first', 'second', 'a)', 'b)', 'on the one hand', …. Its effect (its essential condition) is, by its very existence, to create a set of items having the same status and possibly characterized in terms of concep-tual or functional categorization.

6.6.2   The Exhaustiveness of an Enumeration?

There is another aspect of enumeration about which one might wonder whether it belongs to another type of speech act: it is that of completeness. There is a logical component in this question (the meaning that we can or must give for instance to: 'one', 'at least one', 'some', 'all': we shall address certain aspects below); and also an epistemological component on the status and the forms of guaranty of complete-ness (for instance the enumeration of the elements of a set). We shall examine here if the speaker's position can be considered to be a form of commitment in the il-locutionary sense (and which recalls of the ideas of A. Tadros, 1994).

It is possible for a speaker to engage overtly in one way or another: if we refer to the series of examples [0] to [22], we will find cases of admitted incompletion, like for instance:

---

subjacent properties or those masked by more apparent properties, or even intermittent proper-ties, depending on the context.

[24] There is a very explicative illustration in the following anecdote. The poet Baudelaire was put on trial when *Les Fleurs du Mal* (The Flowers of Evil) was published in 1857. This book is made up of exactly one hundred pieces organised according to an 'elaborate architecture', according to Baudelaire himself, who absolutely insisted on this number and structure. The prosecution said that eleven of these pieces were condemnable and demanded the author to remove them from the book. Baudelaire, and the defence, argued that removing these pieces would not only disfigure the original editorial project and provoke the destruction of the book, but would also allow the prosecution to make, using the removed pieces, a group which would be completely different from the poet's intentions *'dans une habile et dangereuse énumération'* (in a ingenious and dangerous enumeration) (Baudelaire 1972, pp. 18).

[15] John likes numerous kinds of fruit such as apples, pears and oranges.

[15a] John likes numerous kinds of fruit, for instance apples, pears and oranges.

[15b] John likes numerous kinds of fruit, such as apples, pears, oranges, etc.

Yet,

[19] John likes the following three kinds of fruit: apples, pears and oranges.

and:

[19a] John only likes the following three kinds of fruit: apples, pears and oranges.

indicate that the list of kinds of fruit that Johns likes is complete

Numerous other cases are rather undetermined from this point of view, and we emphasize 'undetermined' instead of 'ambiguous' or 'polysemous':

[16] John likes numerous kinds of fruit: apples, pears and oranges.

Additionally, it is possible that being complete is not a property of a given set, which could be more or less happily expressed in the frame of one enumeration or another. Completeness can even be the object of the enumerative construction and that of the speaker's commitment. As we have already done on many occasions, let us study some cases taken from specific texts' (here by Aristotle).

6.6.3   The Empirical Status of Completeness

[A5] 'Evidently we have to acquire knowledge of the original causes [...] and causes are spoken of in four senses. In one of these we mean the substance, i.e. the essence (for the 'why' is reducible finally to the definition, and the ultimate 'why' is a cause and principle); in another the matter or substratum, in a third the source of the change, and in a fourth the cause opposed to this, the purpose and the good (for this is the end of all generation and change). We have studied these causes sufficiently in our work in the Physics, but yet let us call to our aid those who have attacked the investigation of being and philosophized about reality before us. For obviously they too speak of certain principles and causes; to go over their views, then, will be of profit to the present inquiry, for we shall either find another kind of cause, or be more convinced of the correctness of those which we now maintain.' (*Metaphysics*, book 1, part 3).

After this examination of previous doctrines, Aristotle concludes on this point:

'Our review of those who have spoken about first principles and reality and of the way in which they have spoken, has been concise and summary; but yet we have learnt this much from them, that of those who speak about 'principle' and 'cause' no one has mentioned any principle except those which have been distinguished in our work on nature, but all evidently have some inkling of them, though only vaguely.' (*Metaphysics*, book 1, part 7)

And to finish:

'All these thinkers then, as they cannot pitch on another cause, seem to testify that we have determined rightly both how many and of what sort the causes are. Besides this it is plain that when the causes are being looked for, either all four must be sought thus or they must be sought in one of these four ways.' (*ibid.*)

At the end of book 1, he summarizes this research:

> 'It is evident, then, even from what we have said before, that all men seem to seek the causes named in the Physics, and that we cannot name any beyond these; but they seek these vaguely; and though in a sense they have all been described before, in a sense they have not been described at all.' (part 10).

We can note that in the beginning Aristotle finds two advantages in soliciting opinions of earlier philosophers: either one will find other kinds of causes in this examination than those already presented, or, if none are found, the enumeration realized will be more credible. And this is effectively what happens, and what he emphasizes, but by moving even further: it is this inability of the other philosophers to discover other kinds of causes which testifies the correctness of the enumeration of kinds of causes according to Aristotle, in relation to their number and nature. In the final conclusion (part 10), he takes a step further in the characterization of the situation: earlier philosophers, because they had not discovered some of the four kinds of causes, did not clearly discover those of which they spoke since they lacked this framework to think of the whole. In a way, what emerges from Aristotle's arguments is that only a complete enumeration confers the status of truth to its content, and conversely, an incomplete enumeration is not really an enumeration, and what it speaks of does not have the same cognitive status.

6.6.4   The Formal Status of Completeness

Thus, Aristotle does not arrive at four kinds of causes following a deduction which would guaranty the completeness of the inventory; on the contrary, it is from the inability to find other causes in other authors' works that he deduces the correctness, by means of a kind of reasoning by default. We can compare this situation of empirically established completeness to another situation where the principle of a formally completed inventory is presented:

> [A6] 'Of things themselves some are predicable of a subject, and are never present in a subject. Thus 'man' is predicable of the individual man, and is never present in a subject.
> By being 'present in a subject' I do not mean present as parts are present in a whole, but being incapable of existence apart from the said subject.
> Some things, again, are present in a subject, but are never predicable of a subject. For instance, a certain point of grammatical knowledge is present in the mind, but is not predicable of any subject; or again, a certain whiteness may be present in the body (for color requires a material basis), yet it is never predicable of anything.
> Other things, again, are both predicable of a subject and present in a subject. Thus while knowledge is present in the human mind, it is predicable of grammar.
> There is, lastly, a class of things which are neither present in a subject nor predicable of a subject, such as the individual man or the individual horse.' (*Categories*, part 2)

As we can see, Aristotle obtains an inventory of four cases by crossing two binary variables: predicable of a subject or not, present in a subject or not; however, only the presence of 'lastly' indicates at the end that we have reached the last case, and it is only retrospectively that we can comprehend the combinatory reasoning that is the basis of the completeness of the inventory of the cases; moreover, nothing indicates in the beginning of the extract that an enumeration is starting.

6.6.5   The Status of Choice

On the question of exhaustiveness and of the role it can play in reasoning, and for the properties of the enumerative textual object deriving from it, we can find in works by the same author a situation which is different from the previous two:

> [A7] 'Quality is a term that is used in many senses. One sort of quality let us call 'habit' or 'disposition'. [...] Another sort of quality is that in virtue of which, for example, we call men good boxers or runners, or healthy or sickly: in fact it includes all those terms which refer to inborn capacity or incapacity. [...] A third class within this category is that of affective qualities and affections. Sweetness, bitterness, sourness, are examples of this sort of quality, together with all that is akin to these; heat, moreover, and cold, whiteness, and blackness are affective qualities. [...] The fourth sort of quality is figure and the shape that belongs to a thing; and besides this, straightness and curvedness and any other qualities of this type; each of these defines a thing as being such and such. [...] There may be other sorts of quality, but those that are most properly so called have, we may safely say, been enumerated.' (*Categories*, part 8).

We can see that the inventory of qualities here, and the enumerative structure which corresponds to it, is somewhat foundational and almost declarative, in SAT terms ('These, then, are qualities, and the things that take their name from them as derivatives, or are in some other way dependent on them, are said to be qualified in some specific way. In most, indeed in almost all cases, the name of that which is qualified is derived from that of the quality' adds Aristotle just after the quotation above), yet we are dealing with here neither the case of a certain empirical guaranty of being exhaustive, nor that of being exhaustive because we have exhausted the examination of all the formally possible cases. What is declared here is the sufficient aspect of the inventory of accepted qualities, which makes further examination unnecessary.

Here is another example of this case:

[A8] 'The term 'to have' is used in various senses. In the first place [...].

Other senses of the word might perhaps be found, but the most ordinary ones have all been enumerated.' (*Categories*, part 15).

6.6.6   The Status of Numbers

When a number is mentioned in the initializer, it plays the role of a guide both in the progression of the reading of the enumeration and in the question of completeness examined above. And one would think that if four cases are announced, we are entitled to expect to read about these four cases, and not about three or five. Nevertheless, Aristotle offers an example of variation of this number during the development of the enumeration, and it is, we believe, interesting to shed light on the role of this device.

[A9] 'There are four senses in which one thing can be said to be '**prior**' to another. Primarily and most properly the term has reference to time: in this sense the word is used to indicate that one thing is older or more ancient than another, for the expressions 'older' and 'more ancient' imply greater length of time.

Secondly, one thing is said to be 'prior' to another when the sequence of their being cannot be reversed. In this sense 'one' is 'prior' to 'two'. For if 'two' exists, it follows directly that 'one' must exist, but if 'one' exists, it does not follow necessarily that 'two' exists: thus the sequence subsisting cannot be reversed. It is agreed, then, that when the sequence of two things cannot be reversed, then that one on which the other depends is called 'prior' to that other.

In the third place, the term 'prior' is used with reference to any order, as in the case of science and of oratory. For in sciences which use demonstration there is that which is prior and that which is posterior in order; in geometry, the elements are prior to the propositions; in reading and writing, the letters of the alphabet are prior to the syllables. Similarly, in the case of speeches, the exordium is prior in order to the narrative.

Besides these senses of the word, there is a fourth. That which is better and more honourable is said to have a natural priority. In common parlance men speak of those whom they honour and love as 'coming first' with them. This sense of the word is perhaps the most far-fetched.

Such, then, are the different senses in which the term 'prior' is used.' (*Categories*, part 12)'

Yet, Aristotle immediately adds:

'Yet it would seem that besides those mentioned there is yet another. For in those things, the being of each of which implies that of the other, that which is in any way the cause may reasonably be said to be by nature 'prior' to the effect. It is plain that there are instances of this. The fact of the being of a man carries with it the truth of the proposition that he is, and the implication is reciprocal: for if a man is, the proposition wherein we allege that he is true, and conversely, if the proposition wherein we allege that he is true, then he is. The true proposition, however, is in no way the cause of the being of the man, but the fact of the man's being does seem somehow to be the cause of the truth of the proposition, for the truth or falsity of the proposition depends on the fact of the man's being or not being.

Thus the word 'prior' may be used in five senses.'

If we exclude a distraction (which is possible but not very likely in this case), we can interpret this extract in four to five senses between the beginning and the end of the enumeration in two different ways:

- Aristotle wishes to explain to the reader using a reasoning based on the remark that there is a very particular sense of 'prior' which the reader would not think of spontaneously; not even Aristotle could, since he pretends in a way to not have been able to notice it in the beginning. And this fifth sense, linked to the presupposition of existence, is effectively of a completely different nature from the four others. This entire system would be a correction *in praesentia*, where the corrected element is maintained, it is the difference between crossing out and deleting (In other words it could be: 'It is not four senses that we need to count but five' in contrast to 'We must count five senses').

- Aristotle wants to make the readers notice that there are effectively five senses and not four as he announced initially, but there are two classes, one of four senses and one of one sense, the fifth: there are indeed 4 + 1 senses; in Fregean terms, we could say that the expressions '5' and '4 + 1' have the same referent (number 5) but not the same sense (not the same mode of presentation of this referent)[25].
- We will only note as a reminder here that nothing seems to us to be making these two interpretations incompatible. On the contrary, the reader should note that while we announced 'two ways' in the initializer of this enumeration, it really contains three items, without provoking inconsistency in the enumeration or incoherence of its sense[26].

6.7 Perspectives

The study presented here is quite diverse and yet, at the same time, incomplete. We cannot yet fully apprehend all the aspects of the enumerative textual object. Some points are discussed here by means of an illustrative conclusion.

Like all language phenomena, enumeration overlaps with many others. Other aspects pertain to the relations of this textual object to others. For instance, E. Pascual and M.P. Péry-Woodley (Pascual and Péry-Woodley 1997; Péry-Woodley 1998) demonstrate that a detailed analysis of the 'definition' type textual object (e.g. the definitions of operators or tasks in a software manual) overlaps with that of enumeration in an original way. Similarly, M. Ho-Dac (Ho-Dac et al. 2001) presents an analysis of 'titles' where interferences with enumerations of titled items play an important role.

We have, on many occasions, indicated that, to go beyond the structural appearance of the enumeration in the comprehension of its actual operation, we must make assumptions on the nature of operations subjacent to its realization and its exploitation during reading. This is what we have attempted, in a provisional manner, to do in particular in Sect. 6.5. But, such preoccupations should be considerably widened. Let us indicate, now, the existence of research in cognitive sciences on enumeration anchored in research by Goody 1977; Olson 1994; Ong 1982; Chemla 1990; Leroi-Gourhan 1993. For instance, in the ACTE (Acquisition of Textual Competences) project (Grandaty et al. 2000), the enumerations under study were present in the

---

[25] Another aspect of this question is that Aristotle seems to put the four first senses in a specific order, from 'Primarily and most properly' to 'perhaps the most far-fetched'. From this point of view, and yet again, the fifth sense does not seem to have a designated place.

[26] More generally, we were careful to introduce in the writing of this chapter (and its abstract) a number of enumerations corresponding to cases indicated, and even one case of non-parallel enumeration. We leave the reader the pleasure of discovering them…

rules of (strategy type) games presented to children of different ages (9 to 12). The experiment showed that the performances of the subjects vary not only according to age, but also in relation to the cognitive process (comprehension, memorization, judgment on the legality of moves, transmission of information to a neophyte) demanded, in relation to the version of rules (discursively developed or with visual markers), and to the nature of entities listed in the rules (elements of the equipment needed for the game, alternatives which a player may come across at some point in the game, and the action he/she then can or needs to realize).

In a different vein, the applicative context of the written-oral conversion of documents with enumerations for blind or visually-disabled people fully reveals the linguistic role of visual markers (Maurel et al. 2003, 2006; Bouraoui and Vigouroux 2003). In yet another direction, the SARA project aims to link the Theory of Text Signaling developed in psycholinguistics with the TAM, conceptually and methodologically overhauling its ambitions and approach (Lemarié et al. 2008; Lemarié 2006). In a similar direction, Ch. Luc demonstrated the feasibility of a composition between the TAM and RST (Luc 2000).

Finally, there is a point which intrigued us in our research: a naïve acceptance of communicational efficiency could lead to the belief that it is more efficient to introduce the initializer of an enumeration first, then the items; there are tested cognitive arguments to justify this 'naivety'[27]. And yet, observable reality compels us to consider that authors, for good or bad reasons, are lead to do the opposite. Furthermore, as shown in the example from Aristotle [A6], the very fact that an enumerations is being developed in the text we are reading can be kept a 'secret' up to the end. This type of phenomenon can lead to the discovery of an as yet overlooked property of enumeration[28]. We could refer to it as follows: there are strategies of introduction of the information in a discursive universe (textual in our case), where the classical distinction 'known information in contrast to new information' plays a significant, but not unique role. Even if we do not know how to explain it satisfactorily in terms of efficient treatment of information, the order 'propositional content—illocutionary status' seems to have a degree of efficacy, at least in certain circumstances. Additionally, this characteristic is not unique to enumeration; we can find it in performative realization such as, for instance: 'Starting tomorrow I don't want you to […] anymore. Or else, […]. This is a warning.' There is much to suggest that these questions must be addressed with regard to the concept of salience, in connection with that of relevance (cf. for instance Landragin 2004), and as the implementation of "draw attention to", and not only for enumeration taken in isolation, but in conjunction with 'stress/emphasize', 'entitle', 'cite/quote', at the very least.

---

[27] J. Lemarié (Lemarié 2008) has recently emphasised a different kind of 'naivety' related to the author's supposed good-will to facilitate the reader's access to information.

[28] Just as the syntactical non-parallelism is, to our view, a revealing factor of the performative function of enumeration: the co-enumerability of items.

6.8 ANNEX

In this Annex we give the examples of Aristotle's enumerations used in the text with a particular formatting device that allows the readers to perceive their structure at first glance.

[A1] "**Forms of speech** are either

- simple
- or composite.

Examples

- of the latter are such **expressions** as
  - 'the man runs',
  - 'the man wins';
- of the former
  - 'man',
  - 'ox',
  - 'runs',
  - 'wins'." (*Categories*, part 2)

[A2] "**Quantity** is either discrete or continuous. Moreover, *some* quantities are such that each part of the whole has a relative position to the other parts: *others* have within them no such relation of part to part." (*Categories*, part 6)

[A3] "There are *six sorts* of **movement**:

- generation,
- destruction,
- increase,
- diminution,
- alteration, and
- change of place." (*Categories*, part 14)

[A4] "Since definition is said to be the statement of a thing's nature, obviously
- one *kind* of **definition** will be a statement of the meaning of the name, or of an equivalent nominal formula. A definition in this sense tells you, e.g. the meaning of the phrase 'triangular character'. [...] That then is one way of defining definition.
- Another *kind* of **definition** is a formula exhibiting the cause of a thing's existence. Thus the former signifies without proving, but the latter will clearly be a quasi-demonstration of essential nature, differing from demonstration in the arrangement of its terms. [...] Again, thunder can be defined as noise in the clouds, which is the conclusion of the demonstration embodying essential nature.
- On the other hand the **definition** of immediates is an indemonstrable positing of essential nature.

We conclude then that definition is
a. an indemonstrable statement of essential nature, or
b. a syllogism of essential nature differing from demonstration in grammatical form, or
c. the conclusion of a demonstration giving essential nature." (*Posterior Analytics*, II, 10)

[A5] "Evidently we have to acquire knowledge of the original causes [...] and **causes** are spoken of in *four senses*.
- In one of these we mean the substance, i.e. the essence (for the 'why' is reducible finally to the definition, and the ultimate 'why' is a cause and principle);
- in another the matter or substratum,
- in a third the source of the change,
- and in a fourth the cause opposed to this, the purpose and the good (for this is the end of all generation and change)

We have studied these causes sufficiently in our work in nature [i.e.: the *Physics*], but yet let us call to our aid those who have attacked the investigation of being and philosophized about reality before us. For obviously they too speak of certain principles and causes; to go over their views, then, will be of profit to the present inquiry, for we shall either find another kind of cause, or be more convinced of the correctness of those which we now maintain." (*Metaphysics*, book 1, part 3)

---

[29] In this example, and numerous others below, the markers of items such as (a) or (i) are translators' notes and cannot be attributed to Aristotle.

After this examination of previous doctrines, Aristotle concludes on this point:

"Our review of those who have spoken about first principles and reality and of the way in which they have spoken, has been concise and summary; but yet we have learnt this much from them, that of those who speak about 'principle' and 'cause' no one has mentioned any principle except those which have been distinguished in our work on nature, but all evidently have some inkling of them, though only vaguely." (*Metaphysics*, book 1, part 7)

And to finish:"

"All these thinkers then, as they cannot pitch on another cause, seem to testify that we have determined rightly both how many and of what sort the causes are. Besides this it is plain that when the causes are being looked for, either all four must be sought thus or they must be sought in one of these four ways." (*ibid.*)

At the end of book 1, he summarizes this research:

"It is evident, then, even from what we have said before, that all men seem to seek the causes named in the Physics, and that we cannot name any beyond these; but they seek these vaguely; and though in a sense they have all been described before, in a sense they have not been described at all." (*Metaphysics,* book 1, part 10)

[A6]
1. "Of **things** themselves some are predicable of a subject, and are never present in a subject. Thus 'man' is predicable of the individual man, and is never present in a subject.
   By being 'present in a subject' I do not mean present as parts are present in a whole, but being incapable of existence apart from the said subject.
2. Some **things**, again, are present in a subject, but are never predicable of a subject. For instance, a certain point of grammatical knowledge is present in the mind, but is not predicable of any subject; or again, a certain whiteness may be present in the body (for colour requires a material basis), yet it is never predicable of anything.
3. Other **things**, again, are both predicable of a subject and present in a subject. Thus while knowledge is present in the human mind, it is predicable of grammar.
4. There is, lastly, a class of **things** which are neither present in a subject nor predicable of a subject, such as the individual man or the individual horse." (*Categories*, part 2)

[A7] "**Quality** is a term that is used in many *senses*.

- One *sort* of quality let us call 'habit' or 'disposition'. [...]
- Another *sort* of quality is that in virtue of which, for example, we call men good boxers or runners, or healthy or sickly: in fact it includes all those terms which refer to inborn capacity or incapacity. [...]
- A third *class* within this category is that of affective qualities and affections. Sweetness, bitterness, sourness, are examples of this sort of quality, together with all that is akin to these; heat, moreover, and cold, whiteness, and blackness are affective qualities. [...]

  The fourth sort of quality is figure and the shape that belongs to a thing; and besides this, straightness and curvedness and any other qualities of this type; each of these defines a thing as being such and such. [...]

  There may be other sorts of quality, but those that are most properly so called have, we may safely say, been enumerated." (*Categories*, part 8).

[A8] "The *term* 'to have' is used in **various senses**.

- In the first place [...]
- Other senses of the word might perhaps be found, but the most ordinary ones have all been enumerated." (*Categories*, part 15)

[A9] "There are *four senses* in which one **thing can be said to be 'prior' to another**.

- Primarily and most properly the term has reference to time: in this sense the word is used to indicate that one thing is older or more ancient than another, for the expressions 'older' and 'more ancient' imply greater length of time.
- Secondly, one thing is said to be 'prior' to another when the sequence of their being cannot be reversed. In this sense 'one' is 'prior' to 'two'. For if 'two' exists, it follows directly that 'one' must exist, but if 'one' exists, it does not follow necessarily that 'two' exists: thus the sequence subsisting cannot be reversed. It is agreed, then, that when the sequence of two things cannot be reversed, then that one on which the other depends is called 'prior' to that other.
- In the third place, the term 'prior' is used with reference to any order, as in the case of science and of oratory. For in sciences which use demonstration there is that which is prior and that which is posterior in order; in

geometry, the elements are prior to the propositions; in reading and writing, the letters of the alphabet are prior to the syllables. Similarly, in the case of speeches, the exordium is prior in order to the narrative.

- Besides these senses of the word, there is a fourth. That which is better and more honourable is said to have a natural priority. In common parlance men speak of those whom they honour and love as 'coming first' with them. This sense of the word is perhaps the most far-fetched.

Such, then, are the different senses in which the term 'prior' is used." (*Categories*, part 12)

Yet, Aristotle immediately adds:

"Yet it would seem that besides those mentioned there is yet another. For in those things, the being of each of which implies that of the other, that which is in any way the cause may reasonably be said to be by nature 'prior' to the effect. It is plain that there are instances of this. The fact of the being of a man carries with it the truth of the proposition that he is, and the implication is reciprocal: for if a man is, the proposition wherein we allege that he is true, and conversely, if the proposition wherein we allege that he is true, then he is. The true proposition, however, is in no way the cause of the being of the man, but the fact of the man's being does seem somehow to be the cause of the truth of the proposition, for the truth or falsity of the proposition depends on the fact of the man's being or not being.

Thus the word 'prior' may be used in five senses."

[A10] "In establishing a definition by division one should keep *three* **objects** in view:

1. the admission only of elements in the definable form,
2. the arrangement of these in the right order,
3. the omission of no such elements.

- The first is feasible because one can establish [...]
- The right order will be achieved if the right term is assumed as primary, [...]
- Our procedure makes it clear that no elements in the definable form have been omitted [...]" (*Posterior Analytics*, II,13)

[A11] "We must next explain the various senses in which the term 'opposite' is used. **Things** are said to be opposed in *four senses*:
1. as correlatives to one another,
2. as contraries to one another,
3. as privatives to positives,
4. as affirmatives to negatives.

Let me sketch my meaning in outline.
  An *instance of the use* of **the word 'opposite'**

- with reference to correlatives is afforded by the expressions 'double' and 'half';
- with reference to contraries by 'bad' and 'good'.
- Opposites in the sense of 'privatives' and 'positives' are 'blindness' and 'sight';
- in the sense of affirmatives and negatives, the propositions 'he sits', 'he does not sit'.

1. Pairs of opposites which fall under the category of relation are explained by a reference of the one to the other [...]
2. Pairs of opposites which are contraries are not in any way interdependent, but are contrary the one to the other. [...]
3. 'privatives' and 'Positives' have reference to the same subject. [...]
4. Statements opposed as affirmation and negation belong manifestly to a class which is distinct [...]." *Categories*, part 10).

[A12] "The pre-existent **knowledge** required is of *two kinds*.

- In some cases admission of the fact must be assumed,
- in others comprehension of the meaning of the term used,
- and sometimes both assumptions are essential." (Posterior Analytics, I, 1)

References

André, J. 1989. Can structured formatters prevent train crashes?. *Electronic Publishing* 2 (3): 169–173.
Anscombe, G.E.M. 1957. *Intention*. Oxford: Blackwell.
Aristotle. *Categories*. Trans. E. M. Edghill. The Internet Classics Archive. http://classics.mit.edu/Aristotle/categories.html. Accessed April 2015

Aristotle. *Metaphysics*. Trans. W. D. Ross. The Internet Classics Archive http://classics.mit.edu/Aristotle/metaphysics.html. Accessed April 2015.

Aristotle. *Posterior analytics*. Trans. G. R. G. Mure. The Internet Classics Archive. http://classics.mit.edu/Aristotle/posterior.html. Accessed April 2015.

Ballmer, Th., and W. Brennenstuhl. 1981. *Speech act classification. A study in the lexical analysis of English speech activity verbs*. Springer series in language and communication, vol. 8. Berlin: Springer-Verlag.

Baudelaire, C. 1972. *Les Fleurs du mal*. Paris: Le Livre de Poche.

Bouraoui, J.-L., and N. Vigouroux. 2003, November 24–26. Les marqueurs des structures énumératives sur le Web: analyse pour la transmodalité. *CIDE (Conférence internationale sur le document électronique) 6*, 199–217.

Bouffier, A. 2006. Segmentation de textes procéduraux pour l'aide à la modélisation de connaissances: le rôle de la structure visuelle. *Schedae* 10:1, 79–84.

Bush, C. 2003. Des déclencheurs des énumérations d'entités nommées sur le Web. *Revue Québécoise de Linguistique* 32:2, 47–81.

Charolles, M. 1997. L'encadrement du discours—univers, champs, domaines et espaces. *Cahiers de Recherche Linguistique* 6:1–73.

Chemla, K. 1990. De l'algorithme comme liste d'opérations. In *L'Art de la liste, Extrême-Orient Extrême-Occident*, 12.

Damamme-Gilbert, B. 1989. *La Série énumérative: Étude linguistique et stylistique s'appuyant sur dix romans français publiés entre 1945 et 1975*. Langue et Cultures, 19. Genève: Droz

Goody, J. 1977. *The domestication of the savage mind*. Cambridge: Cambridge University Press.

Gracia, J. J. E. 1996. *Texts. Ontological status, identity, author, audience*. Albany: State University of New York Press.

Grandaty, M., Cl. Garcia-Debanc, and J. Virbel. 2000. Evaluer les effets de la mise en page sur la compréhension et la mémorisation des textes procéduraux (règles de jeux) par des adultes et des enfants de 9 à 12 ans. *PAroles* 13:3–38

Greenberg, M., and Harman, G. 2006. Conceptual role semantics. In E. Lepore and B. Smith, 295–322. *The Oxford handbook of philosophy of language*. Oxford: Clarendon Press.

Grosz, B., A. Joshi, and S. Weinstein. 1995. Centering: A framework for modelling the local coherence of discourse. *Computational Linguistics* 21 (2): 203–225.

Harris, Z. S. 1991. *A theory of language and information. A mathematical approach*. Oxford: Clarendon Press.

Ho-Dac, M., A. Le Draoulec, and M.-P. Péry-Woodley. 2001. Cohabitation des dimensions temps, espace et 'phénomènes' dans un texte géographique. *Cahiers de Grammaire* 26:125–142.

Jackiewicz, A. 2002, October 20–23. Repérage et délimitation des cadres organisationnels pour la segmentation automatique des textes [Identification and delimitation of organizational frames for automatic text segmentation]. *Colloque International sur la Fouille de Textes*, Hammamet, Tunisia.

Jacquemin, Ch., and C. Bush. 2000. Combining lexical and formatting cues for named entity acquisition from the web. In *Proceedings, Joint Sigdat Conference On Empirical Methods In Natural Language Processing And Very Large Corpora (EMNLP/VLC-2000)*, ed. H. Schutze, 181–189. Hong Kong: Association for Computational Linguistics.

Landragin, F. 2004. Saillance physique et saillance cognitive. *Corela*, 2 (2), (electronic journal). http://corela.edel.univ-poitiers.fr/index.php?id=603.

Lemarié, J. 2006, September. *La compréhension des textes visuellement structurés. Le cas des énumérations*. Ph. D. Thesis, Université de Toulouse-Le Mirail.

Lemarié, J. July 2008. La signalisation de la structure des textes comme élément facilitateur des traitements de l'information. Examen critique de cette conception et contre-propositions. In *JETCSIC—14ᵉ Journée d'étude sur le traitement cognitif des systèmes d'information complexes*, Toulouse.

Lemarié, J., R. Lorch, H. Eyrolle, and J. Virbel. 2008. SARA: A text-based and reader-based theory of signaling. *Educational Psychologist* 43 (1): 27–48.

Leroi-Gourhan, A. 1993. *Gesture and speech*. Cambridge Mass.: The MIT Press.

Luc, Ch. 2000. *Représentation et composition de structures visuelles et rhétoriques du texte*. Ph. D. Thesis, Université Paul-Sabatier (Toulouse 3).

Luc, Ch., and J. Virbel 2001. Le modèle d'architecture textuelle: fondements et expérimentation. *Verbum* XXIII (1): 103–123.

Luc, Ch., M. Mojahid, M.-P. Péry-Woodley, and J. Virbel. 4–6 juillet 2000. Les énumérations: structures visuelles, syntaxiques et rhétoriques. In *Conférence Internationale sur le Document Electronique (CIDE III)*, eds. M. Gaio and E. Trupin, 21–40. Lyon: Europia Productions.

Luc, C., M. Mojahid, J. Virbel, C. Garcia-Debanc, and M-P Péry-Woodley. 1999, November. A linguistic approach to some parameters of layout: A study of enumerations. In *Using layout for the generation, understanding or retrieval of documents, AAAI 1999 fall symposium*, ed. R. Power and D. Scott, 20–29. North Falmouth (Mass.). http://w3.erss.univ-tlse2.fr/textes/pages-persos/pery/articles/AAAI99.pdf. Accessed April 2015.

Mann, W. C., and S. A. Thompson. eds. 1992. *Discourse description: Diverse linguistic analyses of a fund-raising text*. Pragmatics & beyond, new series. Amsterdam: John Benjamins.

Maurel, F., N. Vigouroux, M. Raynal, and B. Oriola. 2003, September 24–26. Contribution of the transmodality concept to improve web accessibility. In Assistive Technology Research Series (V.12): *Proceedings of 1st International Conference On Smart Homes & Health Telematics (ICOST'2003) "Independent Living for Persons with Disabilities and Elderly People"*, ed. Mounir Mokhtari, 186–193. Marne-la-vallée IOS Press Ohmsha, Amsterdam.

Maurel, F., M. Mojahid, N. Vigouroux, and J. Virbel. 2006. Documents numériques et transmodalité. Transposition automatique ã l'oral des structures visuelles de texte. *Document numérique* (Hermès) 9 (1): 25–42.

Olson, D. 1994. *The world on paper: The conceptual and cognitive implications of writing and reading*. Cambridge: Cambridge University Press.

Ong, W. 1982. *Orality and literacy: The technologizing of the word*. London & NewYork: Routledge.

Pascual, E. 1991. *Représentation de l'architecture textuelle et génération de texte*. Ph. D. Thesis, Université Paul Sabatier, Toulouse.

Pascual E. 1996. Integrating text formatting and text generation. In *Trends in natural language generation: An artificial intelligence perspective*, ed. M. Adorni and M. Zock, 205–221. Berlin: Springer-Verlag.

Pascual, E., and J. Virbel. 1996, June 23–28. Semantic and layout properties of text punctuation. In *34th Annual Meeting of the Association for Computational Linguistics*, International Workshop on Punctuation in Computational Linguistics, 41–48. University of California at Santa Cruz. http://www.hcrc.ed.ac.uk/publications/wp-2.html. Accessed April 2015.

Pascual, E., and M.-P Péry-Woodley. 1997. Modèles de texte pour la définition. *Proceedings of 1ères Journées Scientifiques et techniques du Réseau Francophone de l'Ingénierie de la Langue*, AUPELF-UREF, Avignon, 137–146.

Pattabhiraman, T., and N. Cercone. 2001. Selection: Salience, relevance and the coupling between domain-level tasks and text-planning. *Proceedings of the Fifth Workshop on Cognitively Plausible Models of Semantic Processing*, Edinburgh, 79–86.

Péry-Woodley, M.-P. 1998. Textual signaling in written text: A corpus-based approach. *Workshop on discourse relations and discourse markers. COLING-ACL'98*, 79–85. http://w3.erss.univ-tlse2.fr:8080/index.jsp?perso=pery&subURL=articles/coling98_.pdf.

Porhiel, S. 2007. Les structures énumératives à deux temps. *Revue Romane* 42:106–138.

Prince, E. 1981. Towards a taxonomy of given-new information. In *Radical pragmatics*, ed. P. Cole, 223–255. NewYork: Academic Press.

Searle, J. 1979. *Expression and meaning. Studies in the theory of speech acts*. Cambridge: Cambridge University Press.

Schnedecker, C. 2006. Anaphores prédicatives démonstratives: de la cohésion syntagmatique à la cohérence textuelle. *Corela, Organisation des textes et cohérence des discours*, special issue. http://corela.edel.univ-poitiers.fr/index.php?id=1437. Accessed April 2015.

Tadros, A. 1994. Predictive categories in expository texts. In *Advances in written text analysis*, ed. M. Coulthard, 69–82. Oxon: Routledge.

Virbel, J. 1985. Langage et métalangage dans le texte du point de vue de l'édition en informatique textuelle. *Cahiers de Grammaire* 10:5–7.

Virbel, J. 1989. The contribution of linguistic knowledge to the interpretation of text structures. In *Structured documents*, ed. J. André, V. Quint, and R. Furuta, 161–181. Cambridge: Cambridge University Press.

Virbel, J., S. Schmid, L. Carrio, C. Dominguez, M.-P. Péry-Woodley, Ch. Jacquemin, M. Mojahid, Th. Baccino, and Cl. Garcia-Debanc. 2005. Approches cognitives de la spatialisation du langage. De la modélisation des structures spatio-linguistiques des textes à l'expérimentation psycholinguistique: le cas d'un objet textuel, l'énumération. In *Agir dans l'espace*, ed. C. Tinus-Blanc and J. Bulier, 233–244. Paris: Editions de la Maison des sciences de l'homme.

Wierzbicka, A. 1987. *English speech acts verbs. A semantic dictionary.* NewYork: Academic Press.

Chapter 7
The Enumeration Structure of 爾雅 *Ěryǎ's* "Semantic Lists"

Michel Teboul (戴明德)

Abstract Modern linguistic enumeration theory is applied to a study of 爾雅 *Ěryǎ's Semantic Lists*, leading to an in-depth analysis of the work's first three sections without any recourse to the traditional methods of Chinese classical philology. It is hoped that an extension of the same method can lead to a better understanding of the remaining 16 sections.

7.1 The Content of the *Ěryǎ*

The 爾雅 *Ěryǎ* is the first compilation of Chinese characters[1] which aims to elucidate them not through a graphic analysis as is the case with the 說文解字 *Shuōwén jiězì* of 許慎 Xǔ Shèn, but through contrast with other characters which the various commentaries of the Classics show to have related meanings.

Although this is a somewhat simplified definition, it does underline the work's two main characteristics:

- It is not an original work, but a compilation of characters the literary meaning of which had become all but lost, and which Hàn dynasty scholars tried to retrieve by relating them, by means of the sketchiest possible gloss, to other characters of similar (or supposedly similar) use context in the Classics.
- The work being meant to assist scholars with their reading of the Classics, characters were distributed according to certain principles pointed at through formatting devices that help us clarify its underlying structure.[2]

---

[1] The commentaries we have tell us, most of the time, whether we have a monomial of the form x or a binomial of the form xy. Difficult cases such as 權輿 *quányú* (v. *infra*) must be studied separately.

[2] I use here the term "formatting" in the meaning given in (Luc Ch. et al. 1999.)

M. Teboul (✉)
SPHERE (ex-REHSEIS; CNRS & University Paris Diderot), Université Paris 7,
UMR 7219, 75205 Paris Cedex 13, France
e-mail: michel.teboul@univ-paris-diderot.fr

© Springer International Publishing Switzerland 2015   267
K. Chemla, J. Virbel (eds.), *Texts, Textual Acts and the History of Science*,
Archimedes 42, DOI 10.1007/978-3-319-16444-1_7

The work as we have it now is divided into 19 sections 篇 *piān*, but we know from the *Treatise on Arts and Literature of the Hanshu* 漢書藝文志 that at the time it comprised 20 sections 篇 *piān* distributed through three chapters 卷 *juàn*.[3] Some scholars suppose the missing section 篇 *piān* to have been a general Introduction, now lost, to the work.[4]

The oldest and most important commentaries we have are those by 郭璞 Guō Pú[5] from the 晉 Jìn dynasty and 邢昺 Xíng Bǐng[6] from the 宋 Sòng dynasty. The *commentaries*[7] 注 *zhù* of Guō Pú and their *sub-commentaries*[8] 疏 *shū* by Xíng Bǐng are still essential to our understanding of the *Ěryǎ*. They have spawned a huge literature aimed at its elucidation in order to recover the original meaning of the characters contained in the *Ěryǎ*.[9]

Each of the 19 sections has a title describing its content. Because the last 16 contain concrete terms, they have such titles as 釋親 *Shìqīn On Family Terms* (Chap. 4) or 釋畜 *Shìchù On Livestock* (Chap. 19). The first three sections containing abstract terms (verbs mainly), their precise characterization is still unclear.

It is generally thought that these first three sections are the oldest[10] because they display an elaborate formatting, and because their present titles appear to have been added afterwards on the model of the following ones, since titles for Sect 1 釋詁 *Shìgǔ On On the Interpretations of Ancient Terms* and for Sect 3 釋訓 *Shìxùn On On Explanations of Words* are mere tautologies.[11]

It is best to give the names of the 19 sections of the *Ěryǎ* in tabular form (Table 7.1):[12]

---

[3] Hs₄(30)1718. The number of *juàn* varies with the editions.

[4] See for instance (Zhū Xīng 朱星 1996, p. 33.)

[5] (276–324).

[6] (932–1010).

[7] These commentaries present, in a first part, illustrations on the actual use in context of the characters in the Classics. When this is not possible, Guō Pú gives, in a second part of the same commentary, his own definition of some of the characters.

[8] Xíng Bǐng's sub-commentaries are followed by his own *notes* 注 *zhù* where he gives precise references to the passages of the Classics alluded to by Guō Pú and where he comments on Guō Pú's definitions of characters. There is a very critical review on Xíng Bǐng's work in *A Sung Bibliography* (Balazs (ed.) 1978, p. 54a), which does not give him justice.

[9] The combined labours of the best Chinese scholars in the field have been collected in five volumes edited by 朱祖延 Zhū Zǔyán (1996), under the collective title 爾雅詁林 *Ěryǎ gǔlín*. Following each list of the the *Ěryǎ* one finds, very conveniently arranged in chronological order, textual annotations by authors from Guō Pú to our times, gathered from 94 different works, the latest published in 1984.

[10] For a possible reconstruction of the compilation of the *Ěryǎ*, v. (Zhū Xīng 1996, p. 35s.)

[11] The title for chapter 2 釋言 *Shìyán On Phrases* however is perfectly constructed.

[12] For an excellent introduction to the study of 爾雅 *Ěryǎ* v. (Coblin 1993, p. 94.) On the web one can refer to http://encyclopedia.thefreedictionary.com/Erya. Both give the Chinese title of the 19 sections of the work with an explanation of their meaning.

Table 7.1 The names of the 19 sections of the *Ěryǎ*

Section *piān* 篇	1	2	3	4	5
Chinese title	釋詁	釋言	釋訓	釋親	釋宮
pīnyīn	*Shìgǔ*	*Shìyán*	*Shìxùn*	*Shìqīn*	*Shìgōng*
English translation	On (On) Old terms	On phrases	On (On) words	On family terms	On buildings

Chapter 篇	6	7	8	9	10	11	12
Chinese title	釋器	釋樂	釋天	釋地	釋丘	釋山	釋水
pīnyīn	*Shìqì*	*Shìyuè*	*Shìtiān*	*Shìdì*	*Shìqiū*	*Shìshān*	*Shìshuǐ*
English translation	On utensils	On music	On heaven	On earth	On hills	On mountains	On waters

Chapter 篇	13	14	15	16	17	18	19
Chinese title	釋草	釋木	釋蟲	釋魚	釋鳥	釋獸	釋畜
pīnyīn	*Shìcǎo*	*Shìmù*	*Shìchóng*	*Shìyú*	*Shìniǎo*	*Shìshòu*	*Shìchù*
English translation	On plants	On trees	On insects	On fishes	On birds	On wild animals	On domestic animals

7.2   The First Semantic List of the *Ěryǎ*

Scholars studying the *Ěryǎ* and other dictionaries compiled on its model usually describe them as consisting of lists of words arranged by semantic categories, sometimes known as "Semantic Lists". However correct, this definition cannot help us elucidate the complex underlying structure of the work.

As this is a rather complicated problem, let us first give an overview of the paradigmatic list of the *Ěryǎ*, traditionally the one with which it begins.

This very first list is made up of 14 different characters which do not make a meaningful sentence. It is therefore advisable to present it in tabular form, numbering the characters in descending order from 11 to 0, the last one 也 *yě* not being numbered, for reasons that will soon become apparent:

Rank	11	10	9	8	7	6	5
Character	初	哉	首	基	肇	祖	元
pīnyīn	*chū*	*zāi*	*shǒu*	*jī*	*zhào*	*zǔ*	*yuán*
Modern meaning	At the beginning of	Exclamation point	Head	Foundation	To initiate	Ancestor	Initial

Rank	4	3	2	1.1	1.2	0	*
Character	胎	俶	落	權	輿	始	也
pīnyīn	*tāi*	*chù*	*luò*	*quán*	*yú*	*shǐ*	*yě*
Modern meaning	Foetus	To commence	To fall	To weigh	Carriage	To begin	Also

Traditionally, the two characters 1.1＝權 *quán* and 1.2＝輿 *yú* are considered to form a binome 權輿 *quányú* meaning "commencement". It is still used with this meaning in modern Chinese, and we shall accept this interpretation here.[13]

The remarkable thing about that list is that neither Guō Pú nor Xíng Bǐng gives any comment on the meaning and use of the character 始 *shǐ* which serves to define each of the first ten characters of the present list and the final binome 權輿 *quányú*.

In fact, each of these items is quoted by the various commentators as being used in one sentence from the Classics inside which it is always glossed as having a meaning related to that of the character 始 *shǐ* through the semantic formula:

z_n 始也 The character z_n (or the binome 權輿 *quányú*) "equals" the character 始 "to begin, beginning". (n=11, ..., 2).

Traditionally, this is taken to mean that each item is used as the character 始 *shǐ*, and therefore must coincide with one of its meanings, which is rather strange considering that these characters appear frequently in, for example, the 詩經 *Shījīng* or *Classic of Poetry*, famous for the richness of its vocabulary and the refinement of its shades of meaning.

In fact, a closer look at the commentaries shows that their authors satisfy themselves with *ad hoc* explanations in order to justify a meaning that seems to fit with the general context of the Classic quoted, within the traditional school of interpretation they belong to. However, one must never forget that after the nearly complete destruction of all the anterior literature by the first Emperor of China, the scholars of the Hàn dynasty could not fully restitute the ancient meaning of the texts they strived hard to retrieve, and that it is still a hotly debated problem among scholars of the field. The difficulty is compounded by the fact that even the *Hàn* school of interpretation is still not fully understood.

It therefore seems advisable to try to turn to a linguistic study of this first list of the 爾雅 *Ěryǎ*.

---

[13] It can be shown that originally both characters referred *separately* to the character 始 shǐ "beginning". This has no incidence on the present study.

7.3 The First Semantic List of the *Ěryǎ* as an Enumeration

In the case of the first list of the 爾雅 *Ěryǎ*, the ancient meaning "to begin, beginning" attributed by the commentaries to items 11–1. is still apparent in their modern Chinese meanings, except for items 10. 哉 *zāi* "exclamation point", 2. 落 *luò* "to fall" and 1. 權輿 *quányú* whose two component characters mean "to weigh" and "carriage" respectively. Therefore, we can easily admit the fact that each of these 11 items primitively referred to some "beginning" concept, as the philological analyses of all the Chinese commentators tend to show.

As these 11 items, put together, do not make a meaningful sentence, as also underlined by Chinese commentators who have always analysed them separately, it is quite clear that in this particular *Semantic List* they are put on an equal footing, either as nouns or verbs relating to some "beginning" concept.

This leads us to suspect that what we have here is actually a *Classical Enumeration* according to the following definition:

> D_0. To enumerate is to attribute an equal level of importance to entities and to classify these entities according to various criteria.[14]

It is not clear according to which criteria the entities making up the semantic lists[15] of the *Ěryǎ* are classified. The example of the first list under consideration here seems to suggest that the further the meaning of one item deviates from the concept of beginning, the nearer it is placed to $z_0 = 始$ *shǐ*. This hypothesis may explain why, apparently, items belonging to a Chinese enumeration cannot be interchanged.

The classification criterium is dropped in the general definition for enumeration which simply states that:

> D1. An enumeration is a set of items preceded by an introducer or by an initial organizer which serve to announce the enumeration without being a part of it.[16]

Conversely, if we accept that the semantic lists of the *Ěryǎ* are in fact enumerations, the theory of classical or general enumeration can help us understand the roles of item $z_0 (= 始$ *shǐ*), and of the particle 也 *yě* which ends all the enumerations found in the *Ěryǎ*.

In what follows, whenever quoting the text of the *Ěryǎ* it is convenient to give its location by section and paragraph in the *Index to Erh Ya* compiled by the Harvard-Yenching Institute,[17] as well as its usual reference in the 十三經 *Shisanjing* edition. Thus, its first semantic list will simply be referred to as 1A.1.

---

[14] (Pascual 1991; quoted in Luc Ch. et al. 1999).

[15] The word "list" describes here any sequence of items found in the text of the *Ěryǎ*. It is only when these can be shown to satisfy the definition of an enumeration that the word "enumeration" will be used.

[16] Luc Ch. et al. (1999, p. 37.)

[17] Sinological Index Series, Supplement Series (number 18).

7.4   The Semantic Lists of the *Ěryǎ* as Enumerations

Each semantic list of the *Ěryǎ* is made up of a finite number of named entities which are functionnally equivalent and whose formatting defines that of the main text in which they are embedded. These entities are themselves semantic units of one or two characters, as instanced in 1A.1.

In European languages an enumeration of named entities equivalent from a syntactic and semantic point of view is always preceded by an introducer, as stated in definition D_1. Any introducer must contain a NP making explicit the nature of all the named entities making up the enumeration, since otherwise one could not guess their nature just by looking at the enumeration. For instance, the sentence

The following is a list of Chinese scholars who recently published papers on the 爾雅 *Ěryǎ*.

could be the introducer of a list of names of Chinese scholars who all contributed recently to the field of *Ěryǎ* studies. Given the list of their Chinese (or non-Chinese) names, it would be impossible to guess that they all had recently published in this particular field of study.

Here the NP "Chinese scholars who recently published papers on the 爾雅 *Ěryǎ*" is the *kernel* of the introducer, "Chinese scholars" is hypernymic to the *kernel*, which itself is hypernymic to each item of the enumeration. Or one can say that each item of the enumeration, i.e. each Chinese name, is a hyponym of the *kernel*, which itself is a hyponym of "Chinese scholars".

Equivalently, we can consider that the seme "Sinologue" is common to all the items of the list under consideration.

If we try to map these notions onto the case of *Ěryǎ*'s enumerations as instanced in 1A.1, we see at once that in classical Chinese there is no equivalent—meaning that there is no need—for such markers or parts of the introducer as "The following" and "is a list of". The *kernel* is reduced to the Chinese verb $z_0 = $ 始 *shǐ* "to begin", since characters z_n ($n = 11, \ldots, 2$) and $z_1 = $ 權輿 *quányú* are all defined as hyponymic to z_0. Being the *kernel* of the enumeration, it is numbered z_0.

As for the Chinese particle 也 *yě* which ends nearly all the enumerations found in the *Ěryǎ*,[18] it appears as a special case of organizer as defined in D_1, with the special feature that it always occurs at the end of the enumeration.[19] This position at the end of the enumeration fulfills two needs at the same time:

---

[18] The particle 也 *yě* is used predominantly in the first three sections 篇 *piān* of the *Ěryǎ*, much less in the remaining 16 sections. Indeed, it disappears completely in Sect 7 釋樂 *Shìyuè*, 10 釋丘 *Shìqiū*, 14 釋木 *Shìmù*, 15 釋蟲 *Shìchóng* and 18 釋獸 *Shìshòu*. In the first three sections the particle 也 *yě* is always used as a marker of enumeration, except in a few sentences all located at the end of Sect 3 釋訓 *Shìxùn* in which it seems to have a different function. In some other sentences, also located at the end of Sect 3, the particle 也 *yě* is not used. It has already been noted that these last exceptions were out of context and that they should belong to other sections of the work.

[19] Properly speaking, it functions as an anaphora. One could translate "Such characters as z_n ($n = 11, \ldots, 2$) and $z_1 = $ 權輿 *quányú* are hyponyms of $z_0 = $ 始".

- It defines the *kernel* of the enumeration as the Chinese character which immediately precedes it, and
- It solves at once the problem of the final boundary of enumerations which in Chinese are always clearly delimitated (contrary to the case of European languages), because otherwise no one could tell which string of characters was meant to be hyponymic to a given z_0.

Since the Chinese particle 也 *yě* is both the organizer and the boundary of (nearly all the) enumerations found in the 爾雅 *Ěryǎ*, it is not a part of the enumeration. Therefore, it is better not to number it.

One fundamental consequence of this analysis of the Chinese particle 也 *yě* is that one cannot translate the semantic formula

$$z_n \text{ 始也 as}$$

"The character z_n equals the character 始 shǐ to begin",

but should translate it as

"The character z_n is hyponymic to the character 始 shǐ "to begin"".

However, because semantic formulas of the kind z_n 始也[20] came to be used in all Chinese dictionaries (starting with 許慎 Xǔ Shèn's 說文解字 *Shuōwén jiězì*) without specific reference to embedding enumerations, it is probably best to translate it as:

"The character z_n *refers* to the character 始 shǐ "to begin"".

where the English verb "to refer to" is a technical expression meaning that z_n is hyponymic to the verb 始 *shǐ* "to begin".

7.5 Some Examples of *Ěryǎ's* Enumerations

7.5.1 *Embedded Enumerations*

Let us consider the four enumerations found under 1A.9 and 1B.94:

1A.9   舒, 業, 順, 敘 也. 舒, 業, 順, 敘, 緒 也.[21]

  shū, yè, shùn, xù yě. shū, yè, shùn, xù, xù yě.

1B.94   俾, 拼, 抨, 使 也. 俾, 拼, 抨, 使, 從 也.[22]

  bǐ, pīn, pēng, shǐ yě. bǐ, pīn, pēng, shǐ, cóng yě.

---

[20] Of course, z_n始也 is an enumeration with one item ($n = 1$).

[21] ssj$_{8\cdot 3}$ (釋詁) 8.11b$_4$, EYGl$_1$110b.

[22] ssj$_{8\cdot 3}$ (釋詁) 28.17a$_{10}$–17b$_1$, EYGl$_1$774a.

The first enumeration found under 1A.9, U=(舒, 業, 順, 敘), is included in the second one U'=(舒, 業, 順, 敘, 緒). Similarly, the first enumeration found under 1B.94, V=(俾, 拼, 抨, 使), is included in the second one V'= (俾, 拼, 抨, 使, 從).[23] Both U' and V' keep the order of the items in U and V.

In his commentaries, Guō Pú writes that the three characters 舒, 業, 順 all mean 次敘 *cìxù* "order", and that the four characters 舒, 業, 順, 敘 *moreover* (又) mean 端緒 *duānxù* "main thread, clue", and similarly that the three characters 俾, 拼, 抨 all mean 使令 *shǐlìng* "to order about", while the four characters 俾, 拼, 抨, 使 *moreover* (又) mean 隨從 *suícóng* "to accompany (one's superior)".

In other words Guō Pú specifies the meaning of the *kernel* z_0 by a binome c_0 of which z_0 is one component, and then explains that each z_n (n=i, i−1, ..., 1) means c_0.

This amounts to saying that c_0 is hypernymic to each z_n (n=i, i−1, ..., 1), but it does not explain the link between the two different hypernyms given. 陳晉 Chén Jìn in his 爾雅學 *Ěryǎxué* puts forward the traditional Chinese explanation of 1A.9:

敘緒聲義同, 故同用. 事之有次敘者, 自有耑緒可尋, 故舒[24]業順敘, 皆訓緒.

> The characters 敘 *xù* and 緒 *xù* having the same pronunciation and the same meaning can be used one for the other. Ordered things have by themselves some beginning that can be sought, which is why the four characters 舒, 業, 順 and 敘 are all glossed by the character 緒.[25]

His argument is not so compelling in the case of 1B.94:

使與從其義相通, 故連舉以釋之.
> The meanings of the two characters 使 and 從 are interchangeable, which is why they are quoted in succession to explain (these two enumerations).[26]

Anyway, the result of the above is that all commentators of a list of characters which is in fact an enumeration, try to philologically prove that the *kernel* z_0 is the hypernym of the characters preceding it.

This is not always easy. Let us, as an example of how this can be done from the point of view and with the methods of Chinese traditional philology, treat the case of the character 俾 *bǐ* whose common meaning is that of the character 使 *shǐ* "make, cause", which justifies its belonging to V. In the 尚書 *Shàngshū* or *Book of Documents* we find the sentence 罔不率俾 twice,[27] and the sentence 罔不率從

---

[23] For another similar cases, see 1B.33, 1B.115, 2.48.

[24] Printed, by error, 敘.

[25] Quoted in EYGl₁115a.

[26] Quoted in EYGl₁778a.

[27] ssj₁.₂ (武成) 162.23a₁₀ (Legge 1861–1872, p. 313) and ssj₁.₂ (君奭) 249.27b₅ (Legge 1861–1872, p. 485).

once [28] with clearly the same meaning of "There was no one who did not obediently follow", which shows that 俾 can also have the meaning of 從.

However, this can be greatly simplified if, as suggested by the concept of enumeration, we can *semantically* show that one character is the hypernym of another.

In the case of the character 俾 *bǐ* we know that there is no difference in the bronze script 金文 *Jīnwén* between 俾 *bǐ* and 卑 which, when pronounced bēi, means 執事 zhíshì "attendant" according to the *Shuōwén jiězì*.[29] The character 俾 is therefore a specialized form of the character 卑, with the added radical for man 亻 *rén* and a variant pronounciation in order to differentiate the meaning of "attendant" from the other shade of meaning given by the *Shuōwén jiězì*, namely that of 賤 jiàn "of low status". Now an attendant is a person which can be defined through the two semes [to be ordered about (by one's superior) 使令; to follow (one's superior) 隨從], which explains Guō Pú's commentary nicely. Further, as it is clear that you cannot be ordered about if you do not follow (your superior), the first seme is dependent upon the second one which shows that the character 從 *cóng* is the hypernym of the character 使 *shǐ*.

We can also treat semantically the case of U and U′ where we have to show that the character 緒 *xù* is the hypernym of the character 敘 *xù*. The common signification of the character 敘 *xù*, now rather written 序 *xù*, is "to arrange in order". In the *Shuōwén jiězì* the character 緒 is defined as:

緒,絲耑也.
The character 緒 refers to the extremity of a silk thread.[30]

which is explained by 段玉裁 *Duàn Yùcái*, in clear reference to the first step in silk manufacturing, namely the unwinding of the cocoons, by the following paraphrase:

抽絲者得緒而可引. 引申之, 凡事皆有緒可纘.
When unwinding raw silk, one has to get hold of the extremity of a silk thread 緒 before he can pull (the whole thread). Figuratively speaking, it is said of anything with a beginning which can be followed up 纘.

Now anything arranged in order 敘 *xù* reaches this state according to a certain sequence with a given beginning, which expresses the fact that it possesses a 緒 *xù*. In other words, the character 緒 *xù* is the hypernym of the character 敘 *xù*.

The same metaphor appears in reverse use in the following sentence of the 尚書 *Shàngshū* or *Book of Documents*:

荒墜厥緒, 覆宗絕祀.
In our neglect 荒 (of the principles of our forefathers) we dropped their thread 緒, overturned the ancestral shrine and left it without sacrifices.[31]

Here the character 緒 *xù* refers to the guiding principles of the imperial institution as underlined by the gloss which explains the characters 墜厥緒 "we dropped their

---

[28] ssj$_{1·2}$ (文侯之命) 309.2a$_7$ (Legge 1861–1872, p. 613).

[29] 卑 sw (卷 3下, 十部) 65.10a$_6$=Dsw116.20b$_5$.

[30] sw (卷 13上, 糸部) 271.1b$_2$=Dsw643.1b$_1$.

[31] ssj$_{1·2}$ (五子之歌) 101.7a$_8$ (Legge 1861–1872, p. 160).

thread" by the characters 失其業以取亡 "We lost their imperial institution picking up doom instead", contrasting the character 失 *shī* "to lose" with the character 取 *qǔ* "to pick up" and explaining the character 緒 *xù* by the character 業 *yè* as in 祖業 "property handed down from our forefathers".[32]

This use of the character 緒 *xù* is duly reflected in enumeration 1A.28 which links together the two characters 緒 *xù* and 業 *yè* as in 1A.9 but with 事 *shì* as hypernym instead of 緒 *xù*:

1A.28   績, 緒, ..., 業, ..., 事也.[33]
              jì, xù, ..., yè, ..., shì yě.

This is allowed by the fact that, as just seen, in Chinese the character 緒 *xù* has the seme 業 *yè* which in turn has the seme 事 *shì* as in 事業 *shìyè* "cause, enterprise". One can also note that the first item in 1A.28, the character 績 *jì*, has for primary signification "to twist hempen thread" and is therefore very much akin the character 緒 *xù*. Just like the character 緒 *xù*, the character 績 *jì* has the seme 業 *yè* as explained in the following gloss of Ode 244 of the 詩經 *Shījīng* or *Classic of Poetry*:

績, 業皇大也.
The character 績 *jì* refers to a grand cause.[34]

and as evidenced in modern Chinese by the compound 業績 *yèjì* "outstanding accomplishment", which explains its appearing in enumeration 1A.28.

As we have seen, for any Chinese character to belong to a given *Ěryǎ* enumeration defined by its kernel z_0, it is necessary that it has at least one seme in common with z_0. This explains why one Chinese character can appear in several *Ěryǎ*'s enumerations.[35]

There are many variants of the case instanced by 1A.9 and 1B.94. Most of the time, only one part (which can be reduced to one item) is repeated in the second enumeration appearing in the same paragraph of the *Index to Erh Ya*, as in

1A.11   通,遵,率,循,由,從,自也. 通,遵,率,循也.
              yù, zūn, shuài, xún, yóu, cóng, zì yě. yù, zūn, shuài, xún yě.

Since the three characters 通 *yù*, 遵 *zūn*, 率 *shuài* all have a seme in common with the character 循 *xún*, and that these four characters all have a seme in common with the character 自 *zì*, they also appear in the first enumeration.

---

[32] ssj$_{1·2}$ (五子之歌) 101.7a$_9$.

[33] ssj$_{8·3}$ (釋詁) 10.16b$_7$.

[34] ssj$_2$ (文王有聲) 584.13a$_7$ (Legge 1861–1872, p. 462).

[35] For one good example, see 2.269.

7.5.2 *"Mirror" Enumerations*

These enumerations are defined by the fact that the kernel $z_{0,m}$ of the first enumeration M belongs also to the second enumeration M whose kernel $z_{0,m'}$ is an item of M. Such a case appears in the two enumerations found under 1A.24:

1A.24 永, 悠, 迥, 違, 遐, 邈, 闊, 遠也. 永, 悠, 迥, 遠, 遐也.[36]

yǒng, yōu, jiǒng, wéi, xiá, tì, kuò, yuǎn yě. yǒng, yōu, jiǒng, yuǎn, xiá yě.

from which we may imply that the two characters 遐 *xiá* and 遠 *yuǎn* are at least partial synonyms. In modern Chinese both mean "far, distant", but the first one belongs only to the written language. In his commentary to the second enumeration Guō Pú writes:

遐亦遠也.轉相訓.
The character 遐 also refers to the character 遠. Their glosses alternate.[37]

meaning that here we have the two semantic formulas 遐, 遠也 and 遠, 遐也. In his sub-commentary Xíng Bǐng writes:

皆謂遼遠也. … 永, 悠, 迥, 遠四者, 又遠遐也. 遐亦遠也. 轉相訓爾.
All these characters mean far away. … The four characters 永, 悠, 迥 and 遠 moreover (又) refer to 遠遐 "in the far away distance". The character 遐 also refers to the character 遠. Their glosses merely alternate.[38]

In other words he equates each item of these two enumerations with the character 遠 *yuǎn*, going so far as to create the new compound 遠遐 *yuǎnxiá*.[39]
Both Guō Pú and Xíng Bǐng thus make it very clear that there is no difference between the two characters 遠 *yuǎn* and 遐 *xiá*, a fact immediately apparent from 1A.24.[40]

7.5.3 *Uncertain Membership*

A case of "uncertain membership" upon which the present study can throw some light is given by enumeration 2.31:

---

[36] $ssj_{8\cdot3}$ (釋詁) $10.15b_{9s}$.
[37] $ssj_{8\cdot3}$ (釋詁) $10.16a_1$.
[38] $ssj_{8\cdot3}$ (釋詁) $10.16a_{1-3}$.
[39] 遠遐, see 遠 M1139047–41 which equates it with 遐遠.
[40] For other similar cases, see 1B.3, 1B.15.

2.31   撫,敉,[41]撫也.[42]
fǔ, mǐ, fǔ yě.

Since the two distinct characters 撫 *fǔ* and 敉 *mǐ* are both hyponymic to $z_0 =$ 撫 *fǔ*, we can suspect that the first character 撫 *fǔ* is a copy error. This is confirmed in the collation notes of the 爾雅詁林 *Ěryǎ gǔlín* which prove that 2.31 should read

2.31′   憮, 敉, 撫也.[43]
wǔ, mǐ, fǔ yě.

Indeed, in the *Ěryǎ gǔlín* edition, enumeration 2.31 is printed throughout as in 2.31′.

7.6   Conclusion

The use of the modern linguistic enumeration theory has allowed us to give a fine-grained characterization of the various items making up *Ěryǎ's* enumerations, without any recourse to the traditional methods of Chinese classical philology. Indeed, some of the results thus obtained can be arrived at by these traditional methods, and are thus justified, but this necessitates much longer technical developments. The most important results, concerning the characterization of the *kernel* z_0 of the enumeration, or the role of the Chinese particle 也 *yě*, are completely new.

The importance of these results lies in the fact that as the *Ěryǎ* is the oldest compilation of Chinese characters to have been handed down to us, all subsequent dictionaries, be they compiled on its model or along new lines like the 說文解字 *Shuōwén jiězì* of 許慎 Xǔ Shèn, take their cues from it. In particular, they all use the Chinese particle 也 *yě* in its meaning first instanced in the *Ěryǎ*, though, as we have shown, this has been largely misunderstood until now.

The next line of research opened up by this study would be an extension of the same method in order to get a better understanding of the 16 last sections of the *Ěryǎ* and a characterization of Xǔ Shèn's *Six Scripts* theory which he must have arrived at through his long study of the *Ěryǎ* which he quotes 28 times in his 說文解字 *Shuōwén jiězì* though, curiously enough, not in his famous preface.

---

[41] The collation notes of the *Shàngshū* 尚書, under the sentence 敉寧武圖功 ($ssj_{1\cdot2}$ (大誥) 191.17b_6, Legge 1861–1872, p. 367 and his note p. 366b_6), indicate (Legge 1861–1872, ib., p. 198b.4a_2) that in ancient editions of the *Shàngshū* 尚書 the character 敉 was written 撫. This cannot apply to 2.31 since then we would have an enumeration made up of three identical characters.

[42] $ssj_{8\cdot3}$ (釋言) 38.3b_7, EYGl$_2$922a.

[43] EYGl$_2$924a. The fact that in 2.31 the first character should read 憮 is only hinted at in collation note $ssj_{8\cdot3}$ 47b_{10}.

7.7   Bibliography

7.7.1   *Acronyms Used*

ssj	Shīsānjīng zhùshū
EYGl	Ěryǎ Gǔlín 爾雅詁林
sw	Shuōwén jiězì
Dsw	Shuōwén jiězì zhù
Hs	Hànshū
M	Dai Kan-Wa Jiten 大漢和辭典

7.7.2   *Sources and Chinese Titles*

Shīsānjīng zhùshū 十三經注疏, **Ruǎn Yuán** 阮元 ed., 1815. 8 vol. Taibei reprint, Taibei: Yìwén yìnshūguǎn 藝文印書館, 1976.

ssj$_{1.1}$: *Zhōuyì*	周易
ssj$_{1.2}$: *Shàngshū*	尚書
ssj$_{2}$: *Shījīng*	詩經
ssj$_{3}$: *Zhōulǐ*	周禮
ssj$_{4}$: *Yílǐ*	儀禮
ssj$_{5}$: *Lǐjì*	禮記
ssj$_{6}$: *Zuǒzhuàn*	左傳
ssj$_{7.1}$: *Gōngyángzhuàn*	公羊傳
ssj$_{7.2}$: *Gǔliángzhuàn*	穀梁傳
ssj$_{8.1}$: *Lúnyǔ*	論語
ssj$_{8.2}$: *Xiàojīng*	孝經
ssj$_{8.3}$: *Ěryǎ*	爾雅
ssj$_{8.4}$: *Mèngzǐ*	孟子

Ěryǎ Gǔlín 爾雅詁林, 朱祖延 **Zhū Zǔyán** ed. 1996. Húběi: Húběi jiàoyù chūbǎnshè 湖北教育出版社.

Ěryǎ yǐndé 爾雅引得, Index to Erh Ya. 1941. Cambridge (Mass.): Harvard-Yenching Institute, Sinological Index Series, Supplement Series (number 18). Taiwan reprint, nd.

Hànshū 漢書. Běijīng: Zhōnghuá shūjú 中華書局, 1975.

Morohashi Tetsuji 諸橋 轍次 1955–1960. *Dai Kan-Wa Jiten* 大漢和辭典. Tōkyō: Taishūkan shoten 大修館書店

Shuōwén jiězì 說文解字. **Xǔ Shèn** 許慎. Běijīng: Zhōnghuá shūjú 中華書局, 1978.

Shuōwén jiězì zhù 說文解字注. **Duàn Yùcái** 段玉裁, Shànghǎi: Shànghǎi gǔjí chūbǎnshè 上海古籍出版社, 1981.

Zhū Xīng 朱星, *Zhū Xīng gǔhànyǔ lùnwén xuǎnjí* 朱星古漢語論文選集. Taibei: Hóngyè wénhuà 洪葉文化, 1996.

References

Balazs, E., (initiator) and Y. Hervouet. 1978. *A Sung Bibliography* (Bibliographie des Sung). Hong Kong: The Chinese University Press.

Bottéro, F. 1996. *Sémantisme et Classification dans l'Ecriture Chinoise*. Paris: Collège de France, Institut des Hautes Etudes Chinoises.

Bush, C. 1999. *Initialisateurs report*, LIMSI (Draft version of Bush 2003).

Bush, C. 2003. Des déclencheurs des énumérations d'entités nommées sur le Web. *Revue québécoise de linguistique* 32 (2): 47–81.

Coblin, W. South. 1993. Erh ya 爾雅. In *Early Chinese texts: A bibliographical guide*, ed. Michael Loewe, 94–99. Berkeley: The Society for the Study of Early China.

Karlgren, Bernhard. 1931. The early history of the *Chou Li* and *Tso Chuan* texts. *Bulletin of the Museum of Far Eastern Antiquities* 3:1–59.

Legge, James. 1861–1872. *The Chinese classics*. 5 vols. London: Trubner. (This work is cited according to the reprint in 4 vol., Taibei reprint, nd.)

Luc, Ch., C. Garcia-Debanc, M. Mojahid, M.-P. Péry-Woodley, and J. Virbel. 1999. A linguistic approach to some parameters of layout: A study of enumerations. In *The AAAI fall symposium technical report FS-99–04*, ed. R. Power and D. Scott, 35–44. North Falmouth. http://www. aaai.org/Papers/Symposia/Fall/1999/FS-99-04/FS99-04-005.pdf.

Pascual, E. 1991. *Représentation de l'Architecture Textuelle et Génération de Texte*. (Thèse de doctorat). Toulouse: Université Paul Sabatier.

Xue, Shiqi. 1982. Chinese lexicography past and present. *Dictionaries: Journal of the Dictionary Society of North America* 4:151–169.

Yong, H., and J. Peng. 2008. *Chinese lexicography: A history from 1046 BC to AD 1911*. New York: Oxford University Press.

Chapter 8
A Tree-Structured List in a Mathematical Series Text from Mesopotamia

Christine Proust

Abstract The written culture of the Ancient Near East, whose history covers more than three millennia (from the beginning of the third millennium to the end of the first millennium BCE), underwent profound transformations over the centuries and showed many faces according to the region of the vast territory in which it developed. Yet despite the diversity of contexts in which they worked, the scholars of Mesopotamia and neighboring regions maintained and consistently cultivated a true 'art of lists', in the fields of mathematics, lexicography, astrology, astronomy, medicine, law and accounting. The study of the writing techniques particular to lists represents therefore an important issue for the understanding of the intellectual history of the Ancient Near East. In this chapter, I consider extreme cases of list structures, and to do this I have chosen very long lists, most items of which are not semantically autonomous. More specifically, I shall study one of the most abstract and concise lists that have come down to us. It belongs to a series, of which one tablet is kept in the Oriental Institute in Chicago (no. A 24194). The study of this case will allow to set forth some of the writing techniques that were particularly developed in the series. Such a study of the structures of the mathematical texts could benefit other areas in Assyriology.

8.1 Introduction

The written culture of the Ancient Near East, whose history covers more than three millennia (from the beginning of the third millennium to the end of the first millennium BCE), underwent profound transformations over the centuries and showed many faces according to the region of the vast territory in which it

Translation Theodora Seal

C. Proust (✉)
SPHERE (ex-REHSEIS; CNRS & University Paris Diderot), Université Paris 7, UMR 7219, 75205 Paris Cedex 13, France
e-mail: christine.proust@univ-paris-diderot.fr

© Springer International Publishing Switzerland 2015
K. Chemla, J. Virbel (eds.), *Texts, Textual Acts and the History of Science*, Archimedes 42, DOI 10.1007/978-3-319-16444-1_8

developed. Yet despite the diversity of contexts in which they worked, the scholars of Mesopotamia and neighboring regions maintained and consistently cultivated a true 'art of lists', in the fields of mathematics, lexicography astrology, astronomy, medicine, law and accounting. The study of the writing techniques particular to lists represents therefore an important issue for the understanding of the intellectual history of the Ancient Near East.

The cuneiform mathematical texts—most of which date from the Old Babylonian period (twentieth–seventeenth centuries BCE)—are usually given in the format of a succession of problems. Each problem is written in a section delimited by vertical and horizontal lines on the tablet. Thus many mathematical texts can be considered as lists, where each item is defined by a section containing a problem. In the *procedure texts*[1], the problems consist of a statement accompanied by a detailed resolution. But, it happens, in particular in the catalogues, that the statement might only be followed by an answer without any explanation. Sometimes, the problems are limited to a statement with neither question nor answer. Some of these lists of statements can reach considerable proportions and cover several successive tablets; they are named *series*. In the case of the lists of solved problems, the content of each section is generally rich and informative enough to be of interest in itself. For this reason, the *procedure texts* have rarely been studied as lists. However, in the case of *catalogues* or *series*, a section considered individually might show to be poor in content or even incomprehensible. It is then absolutely necessary to identify the structure of the list in order to understand the meaning of the texts. Between the two extremes, namely the list of problems with detailed solutions and the series, all intermediate cases are found. The existence of this *continuum* stresses the importance of paying attention to the particular meanings conveyed by the list structures, even in the cases where the items are few and self-sufficient.

In this article, I shall consider extreme cases of list structures, and to do this I have chosen very long lists, most items of which are not semantically autonomous. More specifically, I shall study one of the most abstract and concise lists that have come down to us. It belongs to a series, of which one tablet is kept in the Oriental Institute in Chicago (no. A 24194). The study of this case will allow to set forth some of the writing techniques that were particularly developed in the series. Insofar as these techniques are found in other types of mathematical lists, although to a lesser extent, this analysis could lead to a better understanding of a wider corpus of cuneiform mathematical texts. Further, the strong internal logic of the mathematical texts makes the structure of the lists of problems clearer than that of texts from other genres, for example certain divinatory lists. Thus the study of the structures of the mathematical texts could benefit other areas in Assyriology.

---

[1] The categories of mathematical texts referred to here (*procedure texts, catalogue texts, series texts*) are those defined by J. Friberg and J. Høyrup; these authors completed the categories previously defined by Neugebauer (Høyrup 2002, p. 9). For more discussion on these classifications, see (Proust 2012.)

8.2   Mathematical Series

As indicated above, a series is a succession of numbered tablets containing a list that runs from one tablet to another. Mathematical series[2] have very particular characteristics. The tablets end with a colophon[3] giving the number of sections and the place of the tablet in the series (by its number). The script is logographic: the cuneiform signs represent Sumerian words[4]. The style is extremely concise. The sections only contain the statements, sometimes with a question and an answer, but without any indication of a solution. This set of properties clearly distinguishes the series from other mathematical texts.

Twenty tablets are known to belong to a mathematical series. Today most of them are kept at Yale University, two are in Berlin, two in the Louvre, and two in the Oriental Institute in Chicago[5]. The mathematical series have been studied in depth by Neugebauer in his first publications[6]. But later, they were of little interest to the historians of science, probably because these texts do not provide direct evidence on the methods of solving mathematical problems. The series, however, raise many issues of major interest, for instance questions concerning the language in which

---

[2] Neugebauer named them *Serientexte* in his first publications in German (Neugebauer 1934–1936), then *series texts* in his following publications. Thureau-Dangin used the name *textes de séries* (Thureau-Dangin 1938, p. 214).

[3] A colophon is a small additional text usually written at the end of the text on the reverse of the tablet, sometimes on the edges, that gives information on the tablet or its context (number of lines or sections, author, date, name of the text, praise to a good, etc.). In the Old Babylonian period, mathematical texts rarely have colophons; when they do, the colophons are brief (giving one or two pieces of information, not more).

[4] A detailed discussion on the use of Sumerian logograms in mathematical texts is given in (Høyrup 2002). Further, it might be useful to recall some pieces of information concerning the written languages in the Ancient Near East. The Sumerian language was written and spoken in southern Mesopotamia during the entire third millennium BCE. Later, Sumerian was supplanted by Semitic languages, in particular Akkadian in the Old Babylonian period. Nevertheless, Sumerian continued to be used in the scribal schools and for scholarly activities during a major part of the second millennium, and was maintained within certain erudite circles until the disappearance of cuneiform writing at the beginning of our era. In their great majority, the mathematical texts are written in Akkadian. Akkadian writing is syllabic, therefore the cuneiform signs represent sounds. However, Sumerian logograms are frequently inserted into this phonetic writing. Although they originally represented words of the Sumerian language, they were probably read in Akkadian. With respect to the mathematical series, the connection between writing and language is more complex (see below). Let us end this note with some points concerning Sumerian. Sumerian words are formed of an invariable root, usually monosyllabic, to which grammatical particles are added: suffixes (which, for example, give the cases for the nouns), prefixes and infixes (for the verbs). For example, in the text examined in this article, the root 'zi' (to subtract) is found alone or in a conjugated form ('ba-zi', I have subtracted).

[5] Yale: tablets YBC 4668, YBC 4669, YBC 4673, YBC 4695, YBC 4696, YBC 4697, YBC 4698, YBC 4708, YBC 4709, YBC 4710, YBC 4711, YBC 4712, YBC 4713, YBC 4714, YBC 4715 (Neugebauer 1935, Chap. 7 and Neugebauer 1935–1937) ; Berlin: tablets VAT 7528 et VAT 7537 (Neugebauer 1935, Chap. 7); Louvre: tablets AO 9071 and AO 9071 (Proust 2009); Chicago: tablets A 24194 and A 24195 (Neugebauer and Sachs 1945, texts T and U).

[6] Neugebauer (1934–1936).

the texts were written, their function, their relation to the Old Babylonian scholarly tradition. Neugebauer questioned the link between the writing used in the series and the Sumerian and Akkadian languages. For him, the writing in the *Serientexte* has nothing to do with a spoken language, and the cuneiform signs could be considered as mathematical symbols[7]. For the translation of such texts, he therefore chose a word for word style that follows the order of the terms in the cuneiform text; this makes the translation difficult to understand. Moreover, he thought, in a way, that the best translation was that of mathematical formulae:

> Many such texts are virtually untranslatable but could best be represented by mathematical formulas. [Neugebauer and Sachs, p. 3].

As for Thureau-Dangin, he thought that the Sumerian logograms in the series texts as well as those in other mathematical texts ought to be read in Akkadian. He therefore adopted an Akkadian transcription and a more literary translation compared to that of Neugebauer. Another question, also raised by Neugebauer, concerns the function of the text. For him, the mathematical texts are clearly intended for teaching and this is, in his view, particularly clear in tablets A 24194 and A 24195:

> This text, like the following one [i.e., A 24194 and A 24195 (C.P.)], can best be compared to an extensive collection of problems from a chapter of a textbook. It is obvious that a collection of this sort was used in teaching mathematical methods. They constitute a large reservoir of problems from which individual problems of any required type (say, speaking from a modern point of view, of a certain category of quadratic equations) could be selected. (Neugebauer and Sachs 1945, p. 116).

This interpretation will be discussed in the conclusion of this article.

Specialists usually consider that the mathematical series texts date back to the end of the Old Babylonian period (seventeenth century BCE), and possibly, with respect to the two exemplars in Chicago, to the Kassite period (sixteenth–twelfth centuries BCE). The hypotheses on their origin are varied: Neugebauer and Høyrup are inclined to believe that they come from the northern parts of Mesopotamia; Friberg thinks they come from the southern regions. For my part, I have suggested that the structure of the colophons might speak in favor of a connection between the mathematical series texts and a tradition which developed in Sippar at the end of the dynasty of Hammurabi[8].

The two tablets kept in Chicago—tablet no. A 24194 and A 24195– have sides measuring approximately 10 cm and contain several hundred of statements. Their square shape and extremely dense text distinguishes them from the other mathematical series tablets. The colophon at the end of tablet A 24194 indicates that the tablet contains 240 sections and is the tenth of a series:

> 4 sixties of sections, 10th tablet (4 šu-ši im-šu dub-10-kam-ma)

Tablet A 24195 does not have a colophon; however, the last column being empty and the penultimate column being only partially inscribed, it seems as if the tablet is unfinished; this could explain the absence of colophon. Because of the numerous similarities between the two tablets in Chicago, including the fact that they have

---

[7] Neugebauer (1934–1936, p. 107).

[8] Reign of Ammi-ṣaduqa (1711–1684); Sippar lies north of the Mesopotamian plain. For a more detailed presentation of the different hypotheses concerning the date and provenance, along with the corresponding bibliography, see (Proust 2009).

consecutive museum inventory numbers (and therefore must have been purchased at the same time), it is likely that they have the same origin. It is even possible that they belong to the same series.

We shall now concentrate on tablet A 24194. A text extract, which seems representative of the whole, will first be given. Some particular aspects of series will then be discussed: the relation between the statements and the sections; the exact meaning of the Sumerian word 'im-šu' that is given in the colophon; the nature of the statements of the list; the status of the calculations. The article will then be mainly devoted to a study of the text on two scales: that of the entire tablet and that of the sections. An analysis of the relation between these two scales will shed light on the organization and the distribution of information. To conclude I shall tackle again the question of the function of the text. A complete copy of the cuneiform text (appendix 1), a diagram representing the organization and distribution of information within the chosen extract (appendix 2) and a glossary (appendix 3) are given at the end of the article. Given the complexity of the text structure, it will probably be useful to the reader to refer frequently to appendix 2.

Before going further, some information on cuneiform numerical notations and transcription will be given. In cuneiform texts, the numbers used in measure notations (length, area, volume, weight) belong to numerical systems using an additive principle, generally with a sexagesimal structure; these texts can be mathematical, administrative, commercial, legal etc. It is of no help here to go into the details of these systems[9], because the only metrological notation in our text is that of area (cf *infra*). In addition to these metrological notations, the mathematical texts use numbers written in a sexagesimal place value notation with 'floating value': the units of a given rank represent 60 units of the preceding one (i.e. to its right); the orders of magnitude are not given (1, 60, 1/60 are written in the same manner). The transcription of the number in place value notation chosen here follows Thureau-Dangin's system: the sexagesimal digits are transcribed into a modern Indo-Arab notation; the digits inside a number are separated by dots. Further, in the transcription, the translation and the commentaries, I have kept the floating value notation found in the cuneiform texts. For example, the notation 'uš × sag = 10' means that the product of the length (uš) and the width (sag) is 10, but it does not indicate how this number 10 is placed in relation to other sexagesimal numbers of the text; in particular, "10" does not necessarily mean "10 units" (see Sect. 8.3 for more details).

Moreover, to facilitate the switching back and forth between the cuneiform text and the modern algebraic representations used to describe the statements, I applied the following principles: in the formulae the terms are written in the same order as the one used in the cuneiform text; the length and the width are denoted by 'uš' and 'sag' respectively (vs. the usual x and y). For example, the text transliterated as 'a-ra$_2$ 3 uš a-ra$_2$ 2 sag' meaning '3 times the length, two times the width' is represented by the formula: $3 \times$ uš $+ 2 \times$ sag. It is more difficult to always follow the order of the signs of operation $(+, -, \times)$, because in the ancient text the operators are placed either before or after the arguments, whereas we usually infix them. For example, in the case of addition the ancient order is "A B dah" and the order of the expressions

---

[9] The interested reader will find the list of numerical graphemes and the systems to which they belong on the CDLI site (http://cdli.ucla.edu/).

is "A + B". For subtraction the ancient order of terms leads to fairly unusual notations: I chose to represent "A B zi" by "− A + B" (vs. the usual "B − A").

8.3   Extract of Tablet A 24194

An extract is given below; it represents only approximately a quarter of the text inscribed on tablet A 24194 (i.e. two columns out of eight). This excerpt is nevertheless
sufficient to give a good idea of the main properties of the text. The translation is
close to that of Neugebauer: it follows as far as possible the word order of the ancient
text, but it is somewhat obscure. For ease of reading, indents and line breaks that
highlight the text structure have been added; of course this layout is not found on the
original tablet. However, the horizontal lines represented in the table below correspond to the section lines inscribed on the tablet. The first column contains the section
numbering (#), to which I shall refer in the following; the second column contains the
line numbering (l.). Neither of these is on the original tablet (Table 8.1, 8.2).

The list of statements has apparently a linear structure, since the statements are
written one after the other in parallel with no visible hierarchy. However, a closer look
reveals a more complex structure. On the one hand, the text is divided into sections
of different sizes: long sections (with respect to extract: #1, unfortunately partly destroyed, and #42), medium-sized sections (#8, 14, 21, 28, 35, 49) and short sections
(all others). On the other hand the same statements recur—not considering the specific numerical values—cyclically. This applies to the statements of the short sections
as well as to those of the medium-sized and long sections. In fact the list is built on
cycles of various lengths that are inserted in each other; this will be seen below.

Let us now examine the excerpt formed by sections 42–53, which is relatively
well preserved. Section 42 provides some information:

The area is $1(\text{eše}_3)$ GAN_2[10].

The same information is found at the beginning of each long section of the tablet. This
area datum gives the product of the length (uš) and the width (sag). According to the
metrological tables that were used during the Old Babylonian period, an area of $1(\text{eše}_3)$
GAN_2 corresponds to the number 10. Thus the following "equation" is obtained:

$$\text{uš} \times \text{sag} = 10$$

(let us recall that this equality is defined up to a power of 60)[11].

---

[10] An area of $1(\text{eše}_3)$ GAN_2 (or 6 GAN_2) approximately represents 21 600 m². Indeed, 1 GAN_2 is
equal to 100 sar; a sar is the area of a square of side 1 ninda; a ninda is a unit of length approximately
equal to 6 m.

[11] This relation can be verified by a very simple calculation: as will be seen below, the length uš
corresponds to 30, the width sag to 20, therefore the area corresponds to: $30 \times 20 = 600 = 60 \times 10$,
which is written 10 in the cuneiform 'floating value' system. For more details on the relations
between measures and place value notations, see (Proust 2008).

Table 8.1 Obverse, column 1

#	l.	Transliteration[a]	Structured translation (indents added by the author)
1	1.	[...]	[...]
	2.	a-[...]	[...]
	3.	igi [...]	1/x [...]
	4.	20 dah [...]	20 I added [...]
	5.	igi 7 [...]	one seventh [...]
	6.	uš-še$_3$ dah-ma [32]	To the length I added: [32]
2	7.	a-ra$_2$ 2-e tab dah 34	2 times I repeated, I added: 34
3	8.	ba-zi-ma 28	I subtracted: 28
4	9.	a-ra$_2$ 2-e tab zi 26	2 times I repeated, I subtracted: 26
5	10.	a-ra$_2$ 15-e tab-ma uš	15 times I repeated: the length.
6	11.	a-ra$_2$ 20-e tab 10 diri	20 times I repeated: it exceeded by 10.
7	12.	a-ra$_2$ 10-e tab-ma 10 [ba]-la$_2$	10 times I repeated: it was less by 10.
8	13.	sag-še$_3$ dah 22	To the width I added: 22
9	14.	a-ra$_2$ 2-e <tab> dah 24	2 times I repeated, I added: 24
10	15.	zi-ma 18	I subtracted: 18
11	16.	a-ra$_2$ 2-e <tab> zi-ma [16]	2 times I repeated, I subtracted: 16
12	17.	a-ra$_2$ 10-e tab sag	10 times I repeated: the width
13	18.	a-ra$_2$ 15-e <tab> 10 diri	15 times I repeated: it exceeded by 10
14	19.	uš sag-še$_3$ dah 52	To the length and width I added: 52
15	20.	a-ra$_2$ 2-e <tab> dah 54	2 times I repeated, I added: 54
16	21.	zi 48	I subtracted: 48
17	22.	a-ra$_2$ 2-e <tab> zi 46	2 times I repeated, I subtracted 46
18	23.	a-[ra$_2$] 2[5]-e <tab> uš sag	25 times I repeated: the length/width
19	24.	a-ra$_2$ 30-e tab 10 diri	30 times I repeated: it exceeded by 10
20	25.	a-ra$_2$ 20-e <tab> 10 ba-la$_2$	20 times I repeated: it was less by 10
21	26.	a-ra$_2$ 3 uš a-ra$_2$ 2 sag	To 3 times the length, 2 times the width,
	27.	dah 2.12	I added: 2.12
22	28.	a-ra$_2$ 2-e <tab> dah 2.14	2 times I repeated, I added: 2.14
23	29.	zi-ma 2.8	I subtracted: 2.8
24	30.	a-ra$_2$ 2-e <tab> zi 2.6	2 times I repeated, I subtracted: 2.6
25	31.	[a-ra$_2$] 1.5-e <tab> uš sag	1.5 times I repeated: the length/width
26	32.	a-ra$_2$ 1-e <tab> 10 ba-la$_2$	1 times I repeated: it was less by 10

Table 8.1 (continued)

#	*l.*	Transliteration[a]	Structured translation (indents added by the author)
27	33.	a-ra$_2$ 1.10-e <tab> 10 diri	I multiplied by 1.10: it exceeded by 10

[a] Transliteration and translation are based on Neugebauer and Sachs (1945, p. 107 ss.,) as well as on an examination of the tablet I made in November 2010. I warmly thank Walter Farber for the authorization he gave me to work on the mathematical tablets kept at the Oriental Institute of Chicago, and for his kind help. In the translation, I followed exactly the order of the words in the cuneiform text, and chose terms closer to the original meaning (as "repeated" instead of "multiplied"). The importance of the original order will appear later in the chapter. Meaning of transliteration symbols are the following: [x] means that the sign x is destroyed, but that it can be reconstituted thanks to the context; <x> means that the sign x was omitted by the scribe. In order to simplify the reading for non-specialists, some information has been omitted (for example, half square brackets that designate partially destroyed signs). For a complete transliteration, see Neugebauer and Sachs 1945, p. 107 ss.

Table 8.2 Obverse, column 2

28	1.	[uš u$_3$ a-na uš ugu sag diri]	[To the length and that by which the length exceeded the width
	2.	[dah-ma 4]2	I added: 42]
29	3.	[a-ra$_2$ 2-e tab dah-ma 4]4	[2 times I repeated, I added: 4]4
30	4.	[zi 38]	[I subtracted: 38]
31	5.	[a-ra$_2$ 2-e tab zi-ma 3]6	[2 times I repeated, I subtracted: 3]6
32	6.	[a-ra$_2$ 20-e tab-ma uš sa]g	[20 times I repeated: the length]/width
33	7.	[a-ra$_2$ 25-e tab 10 di]ri	[25 times I repeated:] it exceeded by 10]
34	8	[a-ra$_2$ 15-e tab-ma 10 ba]- la$_2$	[15 times I repeated:] it was less by [10]
35	9.	uš [sag u$_3$ a-ra$_2$ 2]	To the length, [the width and 2 times]
	10.	a-na [uš ugu sag]	[that by which the length exceeded the width]
	11.	[diri dah-ma 1.12]	[I added: 1.12.]
36	12.	[a-ra$_2$ 2-e tab dah 1.14]	[2 times I repeated, I added: 1.14.]
37	13.	zi-ma [1.8]	I subtracted: [1.8]
38	14.	a-ra$_2$ 2-e <tab> zi 1.6	2 times I repeated, I subtracted: 1.6
39	15.	a-ra$_2$ 35-e <tab> uš [sag]	35 times I repeated: the length/width
40	16.	a-ra$_2$ [40]-e <tab> 10 diri	40 times I repeated: it exceeded by 10
41	17.	a-ra$_2$ 30-e tab 10 ba-la$_2$	30 times I repeated: it was less by 10

Table 8.2 (continued)

42	18.	a-ša₃ 1(eše₃) GAN₂	The area is 1(eše₃) GAN₂
	19.	igi-14- gal₂ uš [sag]	To one-fourteenth of the length, the width
	20.	ù a-ra₂ 2 [a-na uš ugu]	and 2 times [that by which the length exceeded]
	21.	sag diri [2.29 dah]	the width, [2.29 I added],
	22.	igi-7-gal₂ igi-11- gal₂	one-seventh of one-eleventh
	23.	uš-še₃ dah 32	to the length I added: 32
43	24.	a-ra₂ 2-e \<tab> dah 34	2 times I repeated, I added: 34
44	25.	zi 28	I subtracted: 28
45	26.	a-ra₂ 2-[e tab zi 26]	2 times I repeated, [I subtracted: 26]
46	27.	a-ra₂ [15-e tab uš]	[15] times [I repeated: the length]
47	28.	a-ra₂ 20-e \<tab> 10 diri	20 times I repeated: it exceeded by 10
48	29.	a-ra₂ 10-e \<tab> 10 ba-la₂	10 times I repeated: it was less by 10
49	30.	sag-še₃ dah 22	To the width I added: 22
50	31.	a-ra₂ 2-e \<tab> dah 24	2 times I repeated, I added: 24
51	32.	zi 18	I subtracted: 18
52	33.	a-ra₂ 2-e \<tab> zi 16	2 times I repeated, I subtracted: 16
53	34.	a-ra₂ 10-e tab sag	10 times I repeated: the width
54	35.	a-ša₃ 1(eše₃) GAN₂ (end of statement in the following column)	The area is 1(eše₃) GAN₂

The following three lines define a fairly complicated linear combination of the length (uš) and the width (sag) of a rectangle that will be referred to below as 'main expression'[12], denoted P:

$$P=\left\{\frac{1}{14}\times\left[uš+sag+2\times\left(uš-sag\right)\right]+2.29\right\}\times\frac{1}{7}\times\frac{1}{11}.$$

Then comes a very simple second expression, "uš", which I shall designate as 'secondary expression', denoted S. Here:

$$S = uš.$$

---

[12] This term is inspired from Neugebauer's work; Neugebauer designated the content of the long sections by "main problems" and the content of the others by "variants". Nevertheless, I noted the name of these expressions as "P" (*principal*) and "S" (*secondaire*) in order to be coherent with the notations I used in my French publications, as (Proust 2009).

Finally, the content of the section ends with a verb, to add (dah), and a result, 32, indicating a relation between P and S. In modern terms, this relation could be translated by the following formula:

$$P + S = 32$$

or, if P and S are written explicitly:

$$\left\{\frac{1}{14} \times \left[u\check{s} + sag + 2 \times (u\check{s} - sag)\right] + 2.29\right\} \times \frac{1}{7} \times \frac{1}{11} + u\check{s} = 32.$$

Therefore, the statement of Sect. 42 is formed of four segments: the area, the definition of the main expression P, the definition of the secondary expression S, and a relation between P and S. The Table 8.3 below summarizes this segmentation.

Section 43 is short: "2 times I repeated, I added: 34". What is repeated 2 times (i.e. multiplied by 2)? To what is the result added? As the text is extremely regular and repetitive, it is easy to identify these expressions. The first is the complex linear combination of uš and sag (or main expression P) defined in Sect. 42. The second is uš (or secondary expression S), also defined in Sect. 42. The statement 43 can therefore be represented by the following formula:

$$P \times 2 + S = 34$$

where P and S are defined above, in Sect. 42, and are implicitly used in Sect. 43. This brief statement is thus a reduced formulation of a full statement, which could be expressed in modern language in the following way:

$$\left\{\frac{1}{14} \times \left[u\check{s} + sag + 2 \times (u\check{s} - sag)\right] + 2.29\right\} \times \frac{1}{7} \times \frac{1}{11} \times 2 + u\check{s} = 34$$

The next section (#44) gives another relation between the same expressions P and S, here too implicit: "I subtracted: 28":

Table 8.3 Section 42 (#42)

a-ša₃ 1(eše₃) GAN₂	The area is 1(eše₃) GAN₂	Area
igi-14- gal₂ uš [sag] ù a-ra₂ 2 [a-na uš ugu] sag diri [2.29 dah] igi-7-gal₂ igi-11- gal₂	To one-fourteenth of the length, the width and two times [that by which the length exceeded] the width, [2.29 I added], one-seventh of one-eleventh	Main expression P
uš-še₃ dah 32	To the length	Secondary expression S (S = uš)
	I added: 32	Relation between P and S (P + S = 32)

$$-P + S = 28$$

The following sections (#45–48) in turn give other relations between these same expressions.

Section 49 is a middle-sized section that introduces a new value for S: 'To the width I added: 22'. P is no longer added to uš, but to sag:

$$P + sag = 22$$

The following sections give in turn relations between P and this new value of S. Let us note that this new cycle of relations almost identically reproduces the one of Sect. 42–48.

Let us now consider all the sections of the extract. Their content can be represented by the formulae provided in Table 8.4 (when the expressions P and S are explicitly given in the cuneiform text, they are given in underlined bold type in the formulae in Table 8.4; see also appendix 2).

Section 54 contains the beginning of a statement (continued in column 3) which gives the area and defines another variant of P; the cycles described above for sections 1–41 and 42–53 then recurs in column 3 and following.

The above list of formulae shows several important properties of the text: the regularity of the cycles of statements; the structure of the information (definition of the product uš × sag, then P, S, and the relation between P and S); the system of elision of information. Concerning this last point, we note that P and S are only given explicitly in a section if these expressions take a new value. In the following sections, they are implicit. However, some exceptions are found (#5, 12, 18), where S is given explicitly, probably for grammatical reasons[13]. In sections 25 and 39, expression S is named without being fully developed (see Sect. 8.4, 'level 2', for some remarks on the names given to the expressions).

8.4   Sections, Statements and Calculations

After this first partial examination of the text, it is possible to clarify some aspects that were briefly introduced at the beginning of this article.

The first aspect is the relation between the sections and their content. In principle, each section of a series contains a statement and the number of sections (N), which is equal to the number of statements, is given in the colophon according to the expression 'N im-šu'. However, there are cases where the one-to-one mapping between sections and statements is disrupted. For instance, it may happen that some sections

---

[13] In section 5, for example, if uš had been omitted, the sentence would end with '-ma'; this seems impossible within the syntax of the text. This is quite comprehensible, since it is difficult to imagine a sentence in modern languages that would end with ':'. The problem is the same in sections 12, 18 and 39. Let us note that in other series texts the difficulty is circumvented by using the verb 'sa₂', which means 'is equal' (Proust 2009).

Table 8.4 Formulae representing the sections of the extract

Section	Area	P	S	Relation
#1	uš × sag = 10	?	uš	P + S = [32]
#2	uš × sag = 10	?	uš	P × 2 + S = 34
#3	uš × sag = 10	?	uš	− P + S = 28
#4	uš × sag = 10	?	uš	− P × 2 + S = 26
#5	uš × sag = 10	?	uš	P × 15 = S
#6	uš × sag = 10	?	uš	P × 20 = S + 10
#7	uš × sag = 10	?	uš	P × 10 = S − 10
#8	uš × sag = 10	?	sag	P + S = 22
#9	uš × sag = 10	?	sag	P × 2 + S = 24
#10	uš × sag = 10	?	sag	− P + S = 18
#11	uš × sag = 10	?	sag	− P × 2 + S = 16
#12	uš × sag = 10	?	sag	P × 10 = S
#13	uš × sag = 10	?	sag	P × 15 = S + 10
#14	uš × sag = 10	?	uš + sag	P + S = 52
#15	uš × sag = 10	?	uš + sag	P × 2 + S = 54
#16	uš × sag = 10	?	uš + sag	− P + S = 48
#17	uš × sag = 10	?	uš + sag	− P × 2 + S = 46
#18	uš × sag = 10	?	uš + sag	P × 25 = S
#19	uš × sag = 10	?	uš + sag	P × 30 = S + 10
#20	uš × sag = 10	?	uš + sag	P × 20 = S − 10
#21	uš × sag = 10	?	3 × uš + 2 × sag	P + S = 2.12
#22	uš × sag = 10	?	3 × uš + 2 × sag	P × 2 + S = 2.14
#23	uš × sag = 10	?	3 × uš + 2 × sag	− P + S = 2.8
#24	uš × sag = 10	?	3 × uš + 2 × sag	− P × 2 + S = 2.6

Table 8.4 (continued)

Section	Area	P	S	Relation
#25	uš × sag = 10	?	3 × uš + 2 × sag	P × 1.5 = uš/sag
#26	uš × sag = 10	?	3 × uš + 2 × sag	P × 1 = S − 10
#27	uš × sag = 10	?	3 × uš + 2 × sag	P × 1.10 = S + 10
#28	uš × sag = 10	?	uš + (uš − sag)	P + S = 42
#29	uš × sag = 10	?	uš + (uš − sag)	P × 2 + S = 44
#30	uš × sag = 10	?	uš + (uš − sag)	− P + S = 38
#31	uš × sag = 10	?	uš + (uš − sag)	− P × 2 + S = 36
#32	uš × sag = 10	?	uš + (uš − sag)	P × 20 = S
#33	uš × sag = 10	?	uš + (uš − sag)	P × 25 = S + 10
#34	uš × sag = 10	?	uš + (uš − sag)	P × 15 = S − 10
#35	uš × sag = 10	?	uš + sag + 2 × (uš − sag)	P + S = 1.12
#36	uš × sag = 10	?	uš + sag + 2 × (uš − sag)	P × 2 + S = 1.14
#37	uš × sag = 10	?	uš + sag + 2 × (uš − sag)	− P + S = 1.8
#38	uš × sag = 10	?	uš + sag + 2 × (uš − sag)	− 2 × P + S = 1.6
#39	uš × sag = 10	?	uš + sag + 2 × (uš − sag)	P × 35 = S
#40	uš × sag = 10	?	uš + sag + 2 × (uš − sag)	P × 40 = S + 10
#41	uš × sag = 10	?	uš + sag + 2 × (uš − sag)	P × 30 = S − 10

Table 8.4 (continued)

Section	Area	P	S	Relation
#42	uš × sag = 10	$\left\{\dfrac{1}{14}\times[uš+sag+2\times(uš-sag)]+2.29\right\}\times\dfrac{1}{7}\times\dfrac{1}{11}$	uš	$P+S=32$
#43	uš × sag = 10	$\left\{\dfrac{1}{14}\times[uš+sag+2\times(uš-sag)]+2.29\right\}\times\dfrac{1}{7}\times\dfrac{1}{11}$	uš	$P\times2+S=34$
#44	uš × sag = 10	$\left\{\dfrac{1}{14}\times[uš+sag+2\times(uš-sag)]+2.29\right\}\times\dfrac{1}{7}\times\dfrac{1}{11}$	uš	$-P+S=28$
#45	uš × sag = 10	$\left\{\dfrac{1}{14}\times[uš+sag+2\times(uš-sag)]+2.29\right\}\times\dfrac{1}{7}\times\dfrac{1}{11}$	uš	$-P\times2+S=26$
#46	uš × sag = 10	$\left\{\dfrac{1}{14}\times[uš+sag+2\times(uš-sag)]+2.29\right\}\times\dfrac{1}{7}\times\dfrac{1}{11}$	uš	$P\times15=S$
#47	uš × sag = 10	$\left\{\dfrac{1}{14}\times[uš+sag+2\times(uš-sag)]+2.29\right\}\times\dfrac{1}{7}\times\dfrac{1}{11}$	uš	$P\times20=S+10$
#48	uš × sag = 10	$\left\{\dfrac{1}{14}\times[uš+sag+2\times(uš-sag)]+2.29\right\}\times\dfrac{1}{7}\times\dfrac{1}{11}$	uš	$P\times10=S-10$
#49	uš × sag = 10	$\left\{\dfrac{1}{14}\times[uš+sag+2\times(uš-sag)]+2.29\right\}\times\dfrac{1}{7}\times\dfrac{1}{11}$	sag	$P+S=22$
#50	uš × sag = 10	$\left\{\dfrac{1}{14}\times[uš+sag+2\times(uš-sag)]+2.29\right\}\times\dfrac{1}{7}\times\dfrac{1}{11}$	sag	$P\times2+S=24$
#51	uš × sag = 10	$\left\{\dfrac{1}{14}\times[uš+sag+2\times(uš-sag)]+2.29\right\}\times\dfrac{1}{7}\times\dfrac{1}{11}$	sag	$-P+S=18$

Table 8.4 (continued)

Section	Area	P	S	Relation
#52	uš × sag = 10	$\left\{\dfrac{1}{14} \times [u\check{s} + sag + 2 \times (u\check{s} - sag)] + 2.29\right\} \times \dfrac{1}{7} \times \dfrac{1}{11}$	sag	$-P \times 2 + S = 16$
#53	uš × sag = 10	$\left\{\dfrac{1}{14} \times [u\check{s} + sag + 2 \times (u\check{s} - sag)] + 2.29\right\} \times \dfrac{1}{7} \times \dfrac{1}{11}$	sag	$P \times 10 = S$

contain two statements, or that some statements begin in a section and end in another, notably in the case of a change of column[14]. When the number of sections and the number of statements on a tablet differ, it is usually the number of statements that is recorded in the colophon. Yet, literally, the term 'im-šu' refers to a physical reality, namely a box bounded by vertical and horizontal lines. Thus, the term 'im-šu' is ambivalent: it sometimes designates the container (the box) and sometimes the content (the statement), and it probably most often refers to both[15]. On this subject, let us remark that the only cuneiform tablets in which the expression 'im-šu' is found in a colophon are the mathematical tablets. In the latter, the section is clearly associated with the text unit represented by the problem (possibly reduced to its statement). In other corpuses, for instance certain divinatory texts, the colophons give the number of 'lines' (mu); but, just as it is the case for the mathematical series, the term 'mu' designates both the container (the line) and the content (the sentence)[16]. This comparison highlights the link between the physical unit (section or line), and the textual unit (problem or sentence)[17]. It is difficult to find a translation of the term 'im- šu' that covers these various meanings. Following Neugebauer and Thureau-Dangin, the literal translation as 'section' has been retained here and in some instances the word 'section' has been used to refer to both the boxes and their content.

However in the tablet that is considered here, the number expressed in the colophon is not quite clear, because it indicates 4 times 60 of 'im-šu' that neither corresponds to the number of sections, nor to the number of statements. This discrepancy cannot be explained.

The second aspect that must be clarified is the nature of the content of the sections. All the specialists who have studied this type of text acknowledge that the contents of the sections are problem statements, however in my opinion this deserves a justification. Indeed, nowhere in tablet A 24194 is there the slightest allusion to a request to solve a problem: neither questions nor answers are given with the statements. Further, the grammatical structure is so poor that it is doubtful that the text is written in the first person, as it is usually the case for problem statements in cuneiform texts. But, if we turn to other mathematical series texts, one can see that many of them consist of similar statements, more obviously written in the first person singular, and followed by a question ('What are the length and the width?') and sometimes by an answer, which is always the same ('The length is 30 and the width is 20'). In these cases the statements

---

[14] The first case is attested to in tablet AO 9071 (*Ibid.*), the second case in tablet YBC 4712 (Neugebauer 1935, p. 433, n. 12a) and in the present text (as indicated above, section 52 begins at the end of column 2 of the obverse and ends at the beginning of column 3).

[15] This double meaning can be compared to that of a word such as 'book': it can denote an object such as 'a library of 300 books', or a text such as '300 copies of the book were printed'.

[16] On this subject, see the typology of the Old Babylonian divination texts developed by J.J. Glassner, in particular that of the series from Sippar, which date from the end of Hammurabi's dynasty (seventeenth century BCE) (Glassner 2009). Let us note that a 'line' is sometimes more a theoretical entity than a practical one: when a sentence is long, it can be written on two lines

[17] This comparison is also of historical interest. Indeed, careful observation of the colophons suggests that the tradition of the mathematical series is not so different than that of the divination texts studied by J.-J. Glassner, see note above (for more details, see Proust 2009).

are clearly problems that are to be solved. Is it the same for our text? In fact, tablets A 24194 and A 24195 are quite different from those belonging to the other known mathematical series and it is doubtful that the function or even the nature of the texts is the same. It is possible that the two Chicago tablets might be more recent and that their authors developed techniques to elaborate statements on the basis of an existing tradition, but for different purposes. In order to provide a more reliable answer, let us consider the question of the nature of the statements from a mathematical point of view. All the information provided explicitly or implicitly in each section consists in two relations between the length and the width of a rectangle. The first gives the area of the rectangle, thus the product of the length and the width; the second gives a linear relation between the two dimensions. Considering the length and the width as unknowns, the necessary and sufficient information is available for the calculation of these unknowns. It is difficult to ascribe such a strong mathematical property to chance. For this reason, the statements of tablet A 24194 are indeed problem statements, whose unknowns are the length and the width, each consisting of two 'equations'[18]. In modern mathematical language, one would say that the content of each section describes (explicitly or implicitly) a quadratic system of two equations with two unknowns.

Do these statements refer to calculations? The text does not explicitly mention calculations since the problems are given, but are not solved. However, as the solution is always the same (i.e. uš $= 30$ and sag $= 20$), the elaboration of the statements requires calculations ensuring that the chosen data lead to this immutable solution decided in advance. The calculations are not intended to solve the problems, but to produce problems.

What exactly are these calculations? The answer is probably not the same for the reader and for the writer of the text. As has been seen, the reader is not asked explicitly to solve the problems, maybe not even implicitly. But he is encouraged, if only to control his reading of the text, to verify that the statement leads to the correct solution. The calculations thus consist in replacing uš by 30 and sag by 20 and then performing the operations given in the statements in their order of occurrence. It appears as if, when the statements were read, they functioned as instructions. J. Virbel, in his contribution to this book, has given other examples of lists that vary in nature depending on the person using it and his intentions[19]. The process seems to have been different for the writer of our text, since it consisted in adjusting the data in order to produce problems with solution 30 and 20. It is possible to identify certain values in the statements that were more specifically subject to these adjustments; Table 8.5 below identifies them for the first cycle of statements, but it could be reproduced almost identically for the other cycles:

---

[18] The word equation is understood here in a very broad sense: a numerical relation between two unknown magnitudes.

[19] See the example of the list of food lacking in a house which, in the hands of the person shopping, can be used to carry out the act of giving instructions (Chap. 6, Sect. 6.6). Other examples have been suggested by J. Virbel: '*a very large number of texts have both the status of an assertion and that of an order (and often also of commitment): the agenda of a meeting, the menu of a restaurant*' (REHSEIS seminar, September 2008; see Chap. 6, Sect. 6.4).

Table 8.5 Adjustments

#	The adjustment concerns
1	The result: 32
2	The result: 34
3	The result: 28
4	The result: 26
5	The coefficient of P: 15
6	The coefficient of P: 20
7	The coefficient of P: 10

Table 8.6 'Abacus'

uš			30
sag			20
$2 \times (uš - sag)$			20
add		1	10
1/14			5
		2	29
add		2	34
1/7			22
1/11			2
add uš			32

Further let us remark that the main expressions are built so as to take particular values (2, 3 and 4 in this tablet; 5 and 10 in others).

How were the calculations done in practice? There is no trace of calculation techniques in cuneiform texts, probably because the operations were made on physical instruments. On this matter, the modern reader remains without resources. For instance, when performing an addition or a subtraction, the indetermination of the orders of magnitude poses a problem: the positioning of the numbers with respect to each other. Here, we shall limit ourselves to an empirical approach by noting that, if the numbers of our text are placed one under each other as they would be on an abacus, the data are usually lined up to the right. For example, the calculations that enable to verify the text of Sect. 42 are the following (Table 8.6):

However this approach does not solve all the positioning problems of the text. For example, in section 26, the number 1 is not lined up to the right. Further, one might wonder where to place the number 10; the latter corresponds to the area whereas the numbers 30 and 20 correspond to the length and the width. I shall leave this last question aside for it does not specifically arise in this text[20].

---

[20] It does arise however in other series, in which the statements give sums of lengths and areas, in particular in the twin tablet A 24195 and in one of the Louvre tablets. On this subject, see (Proust 2009).

8.5   Structure and Distribution of the Information

Let us return to the analysis of the text as a whole: How is the information organized and distributed among the different sections? The presentation of an excerpt has shown that the information contained in each statement can be decomposed into four segments:

- The area of a rectangle;
- A complex linear combination of the length (uš) and width (sag) of rectangle (P);
- A simple linear combination of uš and sag (S);
- A relation between the two expressions P and S.

The area of the rectangle is the same for all statements, whether explicit or not. The expression P takes several different values, each of which initiates a cycle of variants for expressions S; each new value of S initiates a new cycle of variants for the relations between P and S. Thus the text is built on a system of linked variants of the four components. The information is structured in the form of a four level tree: the highest (level 4) gives a first equation (E_1); level 3 defines the main expressions P; level 2 defines the secondary expressions S; finally level 1, the lowest, defines the relations between P and S, i.e. equations (E_2). This structuring is shown in the diagram of appendix 2. Let us now specify the content of these different levels of information and the way the latter is distributed among the various sections.

Level 4

As we have seen, the tablet A 24194 is the tenth of a series. But the other tablets in the series are not known with certainty, therefore it is difficult to specify any differences between the tablets. The existence of tablet A 24195, very similar to tablet A 24194, allows us to suggest several hypotheses. The statements of tablet A 24194 form a homogenous group, each defining a system of two relations. The first (equation E_1) is the same in all the statements of the text: 'the area is $1(\text{eše}_3)$ GAN_2'. The second (equation E_2) is a linear relation between uš and sag, which is subject to variations, following however a single model throughout the text. The statements of tablet A 24195 also form a homogenous group, but follow a different model than that of tablet A 24194. In A 24195, we have two quadratic relations between uš and sag. The variations concern both relations. For the rest, both tablets from Chicago are identical in every respect: same square format, same dimensions, same tiny writing, same way of structuring the information in a tree form, same reduction processes. Therefore, the two twin tablets A 24194 and A 24195 could represent two level 4 variants of the same series.

Level 3

Eight variants of expressions P occur in tablet A 24194. Given the state of the text, only three are identifiable:

#1	$P = ?$ (text partly destroyed)
#42	$P = \left\{ \dfrac{1}{14} \times [\text{uš} + \text{sag} + 2 \times (\text{uš} - \text{sag})] + 2.29 \right\} \times \dfrac{1}{7} \times \dfrac{1}{11}$
#54	$P = ?$ (text partly destroyed)

#81	$P = ?$ (text partly destroyed)
#119	$P = ?$ (text partly destroyed)
#139	$$P=\left\{\left[\left(\frac{1}{2}u\check{s}\right)-\frac{1}{3}\times\left(\frac{1}{2}u\check{s}\right)+25\right]\times\frac{1}{7}+17\right\}\times\frac{1}{11}$$
#197	$$P=\left\{\left[\left(\left(u\check{s}+\left(u\check{s}-sag\right)\right)\times\frac{1}{4}+6\right)\times\frac{1}{8}+15\right]\times\frac{1}{11}+u\check{s}\right\}\frac{1}{8}$$
#242	$P = ?$ (text partly destroyed)

Each new value of P initiates a cycle of variants of the other components of the statement. All the statements of the cycle use the same value P without mentioning it explicitly. For example, the value:

$$P=\left\{\frac{1}{14}\times\left[u\check{s}+sag+2\times\left(u\check{s}-sag\right)\right]+2.29\right\}\times\frac{1}{7}\times\frac{1}{11},$$

defined in Sect. 42, is used implicitly in all the sections that depend on it (#43–53).

The values of P are defined in the long sections.

Level 2

To each of the variants of expression P corresponds a cycle of variants of expression S. In sections 1–41, S takes in turn the following values:

S = uš
S = sag
S = uš + sag
S = 3uš + 2sag
S = uš + (uš − sag)
S = uš + sag + 2(uš − sag)

This succession of values of S is more or less found in all the cycles initiated by the variants of P, sometimes with a few omissions or with some new combinations. For example, in the cycle of sections 42–53, expression S has only two variants: uš and sag. But in other cycles, the variations are more developed: in addition to the six expressions given above, there are the expressions S = 1/3 uš + 1/4 sag (#113), S = 1/3 uš + 1/4 sag + (uš − sag) (#180, #231), S = uš + 1/3 uš + sag + 1/4 sag (#186). Thus, the variations of S form relatively regular patterns. Note the significant fact that expression S is sometimes designated in the text as "uš/sag"[21]:

#25	'uš/sag' designates the expression 3uš + 2sag
#39	'uš/sag' designates the expression uš + sag +2(uš − sag)

[21] In the cuneiform text, this concerns the juxtaposition of the signs uš and sag ([Image]). I represent this sequence by 'uš/sag' and not 'uš-sag' such as the usual transliteration norms would require; this, in order to avoid the graphical similarity with the subtraction uš—sag.

#163	'uš/sag' designates the expression 3uš +2 sag
#170	'uš/sag' designates the expression uš +(uš − sag)
#177	'uš/sag' designates the expression uš +sag +2(uš − sag)
#221	'uš/sag' designates the expression uš +(uš − sag)

Each new value of S initiates a cycle of variants for the relations between P and S. For each new value of S, all the following variants implicitly use this same value of S. For example, the value S = uš defined in Sect. 42 is implicitly used in all the following sections (#43–48) until a new value of S is defined (#49).

The values of S are defined in the long and medium-sized sections. However, S is given explicitly in certain short statements, probably for grammatical reasons (see the above excerpt commentary).

Level 1

The relations between P and S also form quite regular patterns more or less composed of the following variants (N designates a specified number):

$$P + S = N$$
$$P \times 2 + S = N$$
$$-P + S = N$$
$$-P \times 2 + S = N$$
$$P \times N = S + 10$$
$$P \times N = S - 10$$

The only element that changes from one cycle to another is the number N; the writer of the text is able to produce a statement, whose solution is uš = 30 and sag = 20, mainly by adjusting this number.

In each cycle of relations between P and S, three types of relations alternate regularly:

A-ma B	A is B (translated A: B)	$A = B$
A B C diri	A exceeds B by C	$A = B + C$
A B C ba-la$_2$	A is less than B by C	$A = B - C$

The relations between P and S (thus the equation E_2) are defined at the end of the long and middle-size sections and in the short sections. It is at this level, i.e. at the leaf of the tree, that the main clause of each sentence is given: is (nominal clause), exceeds (diri), is less (la$_2$).

The structuring of the text is based on two combined processes: (1) a hierarchical organization of the information in four levels (2) a distribution of the information based on a system of elision: a piece of information common to several consecutive statements is given in the first statement and then is generally omitted in the following ones. The diagram in appendix 2 highlights these two processes: on the one hand the tree shows how the information is structured; on the other hand, the column to the right of the diagram gives the segments of information found in each section. Level 4 information corresponds to the definition of E_2 (root of tree); level 3 information to the definitions of P (main nodes), level 2 information to the definitions of S (nodes); the level 1 information to equations E1 (leaves). The distribution of the information is the following: the long sections contain information of level 1, 2, 3 and 4; the middle-sized sections contain information of levels 1 and 2; the short sections contain level 1 information (and sometimes level 2 information for the grammatical reasons mentioned above). Let us insist on the fact that the structure of the information and its distribution are two independent aspects: for example, there are texts in which the information is given in the form of a tree structure, similar to that of A 24194, but with each section containing all the information[22].

These writing techniques, plus the fact that the grammatical elements are mostly absent, lead to an extremely concise text. For instance, some sentences are so reduced that they are only formed of two signs (see for example Sect. 16: 'zi 48').

The organization of the text is thus very rigorous. It is the reflection of a systematic search for concision and regularity. Is it pure virtuosity in the art of lists or does this structure have a significance of a different nature? In what follows, I shall show how tree structures increase the expressive possibilities of mathematical language.

8.6   The Writing of Operations

The expressions P and S are linear combinations of uš and sag. They are composed of arguments (numbers, uš, sag, linear combinations of uš and sag) on which operations act (additions, subtractions, repetitions, taking the Nth part). Does the expression of these operations follow regular rules? Is it identical on all levels of the tree? How do they relate to the structure of the text? Before answering these questions, let us further examine the different operations.

Addition is generally built with the verbal stem dah (to add), which is always placed after the arguments[23]:

A B-še₃ dah	A to B I have added	A + B

The suffix -še₃ (terminative case suffix, translated in English by 'to') is often omitted. The arguments can be of all type (numbers or combinations of uš and sag), explicit or implicit. In the great majority of cases, the first argument is a main expression and the second is a secondary expression: For example:

#49	sag-še₃ dah	(P) to the width I added	P + sag = 22

---

[22] See for example tablets YBC 4673, VAT 7528, YBC 4698, only to mention the tablets belonging to series (Neugebauer 1935–1937).

[23] In the following, the ordinary arguments are denoted by the letters A, B etc. In the particular cases where they represent numbers, they are denoted N, as mentioned above.

In this example, the first argument is implicit; it is an expression P defined previously #42.

Addition can also be expressed by means of simple juxtaposition or coordination "and" (u_3), such as in section 42, line 20.

Subtraction is generally built with the verbal stem zi (to subtract), which is always placed after the arguments[24]:

A B ba-zi	A from B I have subtracted	$-A + B$

Except for one case (#139), both arguments are implicit. For example:

#44	zi	(P from S) I have subtracted	$-P + S$

There also is another way of expressing subtraction:

a-na A ugu B diri	That by which A exceeds B	$A - B$

It is a frozen formula in which the arguments are almost always uš and sag. For example:

#42	a-na uš ugu sag diri	That by which uš exceeds sag	$uš - sag$

It is used in the secondary statements and in some of the main statements. In all cases, the arguments are explicitly given and placed immediately before the operator.

Repetition is the addition of an argument to itself a certain number of times:

A a-ra₂ N-e tab[a]	A repeated N times	$A \times N$

[a] About the construction N-e tab, see (Proust 2009, pp. 183–186)

The argument A is often an implicit expression P. For example:

#53	a-ra₂ 10-e tab	(P) repeated 10 times	$P \times 10$

Repetition is equally used in the main statements and in the secondary ones. The form is then often abbreviated and the operator can be placed before:

#21	a-ra₂ 3 uš	3 times uš	$3 \times uš$
#42	a-ra₂ 2 a-na ugu sag diri	2 times (uš − sag)	$2 \times (uš - sag)$

---

[24] Let us remark that the complete grammatical form would be A B-ta ba-zi, but the suffix –ta (ablative, translated by 'from') never appears in this text and the verbal prefix ba- is often omitted.

Let us note that the form "A a-ra$_2$ N-e tab" is rarely found outside the mathematical series[25]. It constitutes a kind of neologism, which appeared well after the demise of Sumerian as a living language.

Operation of taking the Nth part (igi-N-gal$_2$, written 1/N in the formulae) appears in all the main expressions. The number of parts N is often a non regular sexagesimal number[26] (for example 7, 11, 14). Generally, the operator is placed after the argument and carries a possessive suffix (-bi).

A igi-N-gal$_2$-bi	A, its Nth part	A × 1/N

The argument A is then formed of everything preceding the operator igi-N-gal$_2$ within the section.

The operator igi-N-gal$_2$ is sometimes placed before the argument, as has been seen at the beginning of section 42. This construction is also found in certain secondary statements (#113, 180, 231, 186):

igi-3-gal$_2$ uš	The third of uš	1/3 × uš
igi-4 gal$_2$ sag	The quarter of sag	1/4 × sag

In the cases where the operator igi-N-gal$_2$is placed before the argument, the possessive suffix (-bi) disappears.

The construction of igi-N-gal$_2$ is thus relatively regular: if the operator is placed after the argument, it carries the possessive suffix -bi and its argument is formed of everything preceding it in the statement; if it is placed before the argument, it does not take a possessive suffix and its argument is the term following it (generally uš or sag). However some ambiguities remain, mainly due to the frequent omission of grammatical suffixes. Sect. 42, which begins with a fraction 1/14 of everything following it, has a singularity that I cannot explain.

Let us now examine how the arguments and the operators are distributed in the sections. The constructions can be divided into two classes that will be called class I and II.

Class I: the arguments are explicit;

Class II: the arguments are implicit.

The constructions using the operators diri (to exceed), igi (Nth part of) and the additions by juxtaposition are mainly of class I. The constructions using the operators zi (to subtract) and tab (to repeat) are mainly of class II. This classification can be summarized in the following diagram (in which only the most frequent construc-

---

[25] I have only found this exact form in tablet Str 366. It is found in slightly different forms in some tablets dating from the end of the Old Babylonian period (for example: A a-ra$_2$ N tab-ba in tablet BM 85194), and in some tablets of the Schøyen Collection (for example: A a-na N-e tab in Tablets MS 2792, MS 3052, MS 5112– see (Friberg 2007)). Let us note that according to J. Friberg, Tablet MS 5112 could date back to the Kassite period (sixteenth–twelfth century BCE)

[26] This expression designates a number whose inverse cannot be written in base 60 with a finite number of digits. A number is regular in a given base if it can be written as the product of divisors of the base. In base 60, this means that its decomposition into prime factors only contains the factors 2, 3 and 5.

tions for each operator are given). The constructions are placed in grids formed of three boxes: the first corresponds to the long sections, the second to the middle-sized sections and the third to the short sections. As previously, letters A, B, C refer to any argument, letters P and S denote the main and secondary arguments.

8.6.1   Classe I constructions

dah (to add)

A B dah		

u$_3$ (and)

A B (u$_3$) C		

Or

	A B (u$_3$) C	

diri (to exceed)

a-na A ugu B diri		

Or

	a-na A ugu B diri	

igi (to take an Nth part)

A igi-N-gal$_2$-bi		

Or

	igi-N-gal$_2$ A	

8.6.2   Class II constructions

dah (to add)

P	S-še$_3$ dah	

Or

P	S-$še_3$	dah

zi (to subtract)

P	S	zi

tab (to repeat)

P		a-ra_2 N-e tab

This diagram shows several important phenomena with respect to class II constructions. As mentioned above, one has to go back up in the text in order to find the arguments on which the operators dah (to add), zi (to subtract) and tab (repeat) act. Further, this system of elision is only possible because the operator is placed after the arguments, thus placed at the end of the expression. Finally, due to the fact that the arguments are defined in previous sections, they form units enabling to express expressions with several levels of calculation. In some ways, these units play the same role as expressions inside brackets in modern algebraic notations. Let us note that the units S and P were recognized by the scribes as particular objects since a specific name was given to them. In this text, the unit S is denoted in several cases by the expression 'uš/sag' (see above). In other texts belonging to mathematical series texts, the unit P is denoted by a kind of 'keyword', i.e. an 'igi-N-gal_2' expression that occurs in its definition (Proust 2009). Therefore, the names 'P' and 'S' used in this article are not completely artificial in the sense that they refer to objects created and named by the scribes who practiced writing the series.

8.7 Enumerative Structure

In connection with the questions raised in the other contributions of this book, one may wonder whether a text such as that of tablet A 24194 is an exotic specimen or whether it has its place within the system developed by J. Virbel to describe enumerations (Chap. 6). Let us first remark that the text of tablet A 24194 presents in a particular acute manner one of the remarkable aspects of enumerations: the exploration of the utmost limits of the possibilities of writing. No oral discourse could completely reproduce the embedded system of elisions—on three or four levels—on which the text is built. The historians' difficulty today to see the link between this text and a spoken language (Sumerian or Akkadian) is probably a symptom of the fact that we are dealing with an elaboration based on writing, rather than speech.

But, is the list of statements of our tablet an enumeration? The colophon states that the tablet contains 240 sections ('im- šu'), and thus provides the nature and the number of components of the list, the items of which are formed of the sections and their content. Therefore the indications given in the colophon may be considered as some kind of an initializer to the enumerations; let us note however that the latter is placed at the end of the text and that the number 240 is not quite exact. The boxes delimited by horizontal and vertical lines represent the characteristic typographical features of enumerations (indents, bullets, etc.). The tree structure of the information corresponds to the model of embedded enumerations on different levels. And even though these different levels are not indicated by visual marks (such as the indentations used in the translated extract given at the beginning of this article), they are recognizable by the fact that each level contains items with a specific function in the sentence. The example of John's fruits (cf. Chap. 6, Sect. 6.3) might shed some light on this phenomenon. The enumeration structure of our tablet is of the same type as the one obtained by making a list of the possible variants of the sentence 'John likes apples':

John	Apples	Likes
		Appreciates
		Is able to eat
	Pears	Likes
		Appreciates
		Is able to eat
	Oranges	Likes
		Appreciates
		Is able to eat
Lea	Apples	Likes
		Appreciates
		Is able to eat
	Pears	Likes
		Appreciates
		Is able to eat
	Plums	Likes
		Appreciates
		Is able to eat
	Oranges	Likes
		Appreciates
		Is able to eat
		etc.

Level 1 information consists of verbs (likes, appreciates, is able to eat); it corresponds to the relations of our tablet. Level 2 information consists of fruits (apples, pears, prunes, oranges); it corresponds to the secondary expressions of our tablet. Level 3 information is only composed of people (John, Lea etc.); it corresponds to the main expressions of our tablet. It is therefore easy to distinguish the different levels. If the indentations were suppressed, it would nevertheless be possible to reconstitute the various items of this list:

John apples likes
appreciates
is able to eat
pears likes
appreciates
is able to eat
oranges likes
appreciates
is able to eat
Lea apples likes
appreciates
is able to eat
pears likes
appreciates
is able to eat
plums likes
appreciates
is able to eat
oranges likes
appreciates
is able to eat

A comparison between the two enumeration presentations given here shows that the layout may constitute an aid for reading without providing any supplementary information. The lack of indentation makes the circulation within the text slower and more difficult, but does not introduce any reading ambiguities.

This illustration reveals other interesting aspects of our tablet. The system of elision is based on the fact that the verbs of the main clauses are placed at the end of the sentence or, in other words, on the fact that the corresponding operators are placed after the arguments. Moreover, it is easier to see how the partial initializer works at each enumeration level: the first item is at the same time the first item of the list and the initializer. The combinatorial nature of enumerations can also be seen: the system, ideally, could enable to consider all possible cases. And although the list of our tablet is not exhaustive, it aims at exhaustiveness.

8.8   Circulation within the Text

As has been seen, an abridged statement in a short or middle-sized section represents a full statement which is formed of elements belonging to several related sections. How can these elements be identified? Let us first note that the main and the secondary expressions can be located by the size of the sections in which they are defined:

the main expressions are found in the sections of approximately 6 lines, some of the secondary expressions are to be found in the 2 line sections, and some of the relations between P and S in the one line sections. Further, several textual or physical marks can be observed such as vocabulary, syntax, layout, lines between sections that help guiding the eye and, often, permit to grasp the content of a section at a glance.

Grammatical and Lexical Marks As indicated above in the description of the enumerative structure, each level of information (main expression, secondary expression, relation) corresponds to a particular segment of the sentence (arguments, operator). These levels can thus easily be identified by both their vocabulary and syntax. The forms "igi-N-gal$_2$," (1/N) where N is a non regular number, therefore remarkable, are only found in the main expressions (it is the case of the suffix—še$_3$). Similarly, the form "a-ra$_2$ N-e tab" (N times I repeated) is only found in the relations between P and S; it is the same for the suffix -ma. These specific terms and constructions thus enable to distinguish the levels of information in the sections.

Change of Column Does the layout of the text in the different columns take into account the information levels of the tree? At first glance, it seems not, as can be seen in the following Table 8.7.

For example, at the end of the second column of the obverse there is a change of column in the middle of a statement. The scribe prefers to write the first line of a new section at the end of a column, where there remains little room, rather than moving to the next column for the section to be all in one block. In this case, the change of column could have been used to indicate a change of level. The chosen layout shows that the main constraint is the density of the writing: the text is packed as much as possible into the available space. However there are three places where the text moves up a level in the case of a change in column (end of columns obv. I, III, IV). A fourth location is at the end of the last column, which corresponds to a

Table 8.7 Layout of the text on the tablet

End of column	Movement in the tree	Information of the text
obv. I	1 –>3	Goes up two levels
obv. II	3 –>3	Stays on level 3
obv. III	1 –>3	Goes up two levels
obv. IV	1 –>2	Goes up one level
obv. V	1 –>1	Stays on level 1
rev. I	1 –>1	Stays on level 1
rev. II	1 –>1	Stays on level 1
rev. III	1 –>1	Stays on level 1
rev. IV	1 –>1	Stays on level 1
rev. V	1 –>?	Goes to the following tablet

Table 8.8 Double lines in A 24194

Between short sections	Between short and middle-sized sections	Between short and long sections	Between short sections and ?
15, 43, 77, 78, 139, 143, 144, 146, 147, 152, 153, 160, 173, 174, 175, 187, 192, 195, 198, 237	48, 23, 145, 151, 172, 223	53, 80, 118, 138	247 (end of the tablet)

tablet change: the tablet is complete, as indicated by the presence of the colophon, and the following tablet probably starts with a long section[27].

In conclusion, the layout of the text does not entirely follow the distribution of the information: the changes of column do not always correspond with the changes of level. This dissociation between the physical and textual elements is sometimes also observed on a different scale: the tablets of the series[28].

Section Lines The statements are placed in sections delineated by simple or double lines. Are these double lines related to the structure of the text? It is difficult to analyze this kind of detail on the sole basis of the copy of the text. However, an examination of the original shows that, in his copy, Neugebauer carefully kept the distinction between simple and double lines, even though the condition of the tablet does not always allow such a clear distinction. The positions of the double lines that are clearly identifiable are indicated in Table 8.8.

According to this table, there seems to be no strict rules. However, we can see that out of the six visible long sections, four begin with a double line. It is therefore possible that the distribution of the double lines might have helped to locate pieces of information.

Although these various marks are useful, they do not permit comfortable circulation within the text; other signs might have existed that cannot be seen today. In any case it is likely that to skim through such a text required much expertise. This poses the problem of the reading mode and use of such types of texts.

8.9   Conclusion

Let us go back to the questions concerning the function of the text that were raised in the introduction. Does tablet A 24194 contain a collection of problems for teaching, as Neugebauer thought? This question raises two others, larger in scope: what was teaching at the time the text was written? What were the relations between

---

[27] It is rare for a statement (or a cycle of statements) to begin on one tablet and end on another. Nevertheless, this case is documented (see tablet AO 9072, Proust 2009).

[28] For example, the thematic groupings of statements may not correspond to the groupings by tablets (*Ibid*).

the scribes that produced the scholarly texts and those who taught? It is beyond the scope of this article to tackle these questions in detail. Let us just underline some important data in this regard. Old Babylonian elementary instruction is well documented thanks to the thousands of pupils' rough copies found near the scribal schools. Further, the Old Babylonian scholarly mathematical texts of known origin (Ur, Nippur, Tell Harmal, Susa…) have almost always been found associated with school remains. This suggests that the authors of the erudite mathematical texts had close connections with the scribal school milieus, or at least with teaching activities. Unfortunately, as stated in the introduction, the discovery context of tablet A 24194 is not known, neither is that of the other mathematical series, therefore no archeological evidence informs us about their possible date[29], origin or the relation between these series and the scribal schools. Further, to my knowledge, most of the statements of our tablet have no parallel in the mathematical cuneiform documentation available today, in particular in that of the Old Babylonian period. Indeed, tablets containing solved problems, of which statements are similar to the ones found at the beginning of some series. But none of them reach the level of complexity found in the Chicago tablets or in many mathematical series texts[30].

Of course, these arguments *ex silentio* are not evidence and the only fact is that there are no sources—neither archeological nor textual—that might shed some light on a possible teaching context. Yet, let us add another argument. Some mathematical series have statements leading to equations of the third or fifth degree that the scribes obviously were unable to solve. In this case, Neugebauer's interpretation, cited at the beginning of this article (*It is obvious that a collection of this sort was used in teaching mathematical methods*), has to be ruled out. Although we remain cautious concerning the Chicago tablets, it is nevertheless possible to state that the series are not all 'reservoirs of problems for teaching'. Further, the characteristics of the text and its probable late date might indicate that it belongs to the emerging tradition of scholarly compilations i.e. a possibly different intellectual milieu than that of the Old Babylonian schools.

Let us try to clarify some of the intentions reflected by the content of the text itself. The structure of the text shows how the statements are constructed, in a systematic way, by the interplay of linked variants acting on the different segments of an initial statement. The selection criteria and the classification of the problems are due to this process of statement construction, and thus are not based on the resolution methods which are not addressed in the text. This distinguishes our text—and series in general—from a list of problems such as that of tablet BM 13901 mentioned

---

[29] As mentioned above, it is generally believed that the series date back to the end of the Old Babylonian period (seventeenth century BCE).

[30] See in this regard the comparative study of a list of solved problems (BM 13901) and a list of statements from a series text (YBC 4714), in which J. Høyrup shows a connection between the problems in terms of the topics (concentric squares) and the resolution methods considered, so far as one can reconstruct them as regards the tablet YBC 4714. He notes however that the variants found in the series are much more sophisticated than those found in the list of solved problems. He suggests that one of the first tablets of the series to which YBC 4714 belongs might contain statements similar to those of tablet BM 13901 (Høyrup 2001, p. 199).

above, the organization of which is mainly based on the resolution methods. In tablet A 24194, the scribe's efforts concentrate on the elaboration of the statements, both with respect to their form (writing style) and substance (mathematical content).

The list of statements is generated by specific writing techniques: tree structure of the information, combinatory method of exploration of the possible, rationalized distribution of the information by means of elision, logographic writing, reduced and specialized vocabulary including neologisms. In particular, these techniques allow dealing with difficulties created by the need to determine the hierarchy of operations that are dealt with today by using parentheses. Therefore the resources of ordinary writing are increased; these techniques permit writing highly complex operations. The result is an extremely concise style. Thus, this conciseness seems more a consequence of the writing constraints of the statements than an objective in itself. Nevertheless brevity is clearly valued. Indeed, we have seen that the tablet filling strategies favored the compactness of the text rather than its readability, and that the omission of grammatical particles was a general phenomenon, even though it could lead to ambiguities. Let us remark that the search for conciseness is much greater in the two Chicago tablets than in the other mathematical series. Therefore, the writing techniques are directly related to the purpose of the text, which is the production of statements, but they also seem to have been developed for their own sake.

If the mathematical content does not concern the methods of problem resolution, than what is it? The statements obey two strong mathematical constraints. The first is that the statements contain the necessary and sufficient information for their resolution, i.e. two independent relations between two unknowns. The second is that the problems can be reduced to quadratic equations and are therefore solvable, at least theoretically, by means of mathematical procedures known at the end of the Old Babylonian period. We might wonder whether the scribes were aware of the fact that their method of statement generation—which essentially consisted in modifying a linear relation between unknowns starting from a standard initial problem—always led to problems they could solve, *even if the resolution in question was not effective*. More precisely, we can hypothesize that the purpose of the text is precisely the relation between modes of producing problem statements and the possibility of solving them *in theory*.

Appendix 1: Copy of Tablet A 24194 (Fig. 8.1)

Fig. 8.1 Copy Neugebauer
and Sachs 1945, pl. 15

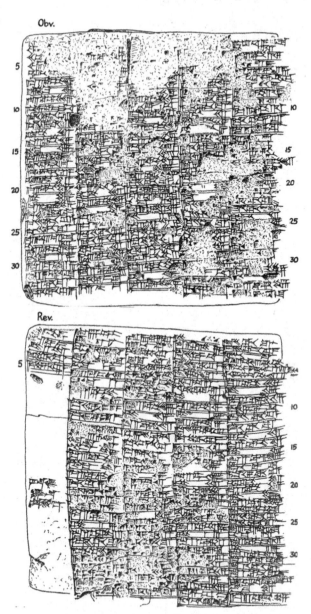

Appendix 2: Structure and Distribution of the Information in Tablet A 24194 (Fig. 8.2)

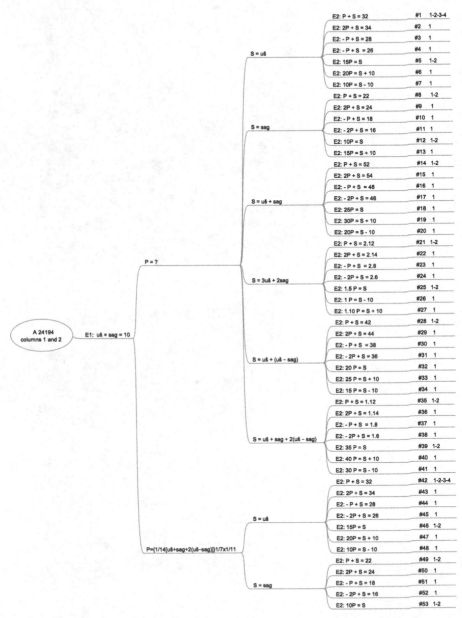

Fig. 8.2 Tree structure in tablet A 24194, obverse, columns 1 and 2. The branches of the tree correspond to the different levels of information: root = level 4; main nodes = level 3; secondary nodes = level 2; extremities = level 1. The column located to the right gives the number and content of the sections that correspond to each path in the tree

Appendix 3: Glossary

This glossary is a list of forms attested in the extract analyzed in this article. The notations are the same as before: N denotes specified numbers and the letters A, B and C denote any type of expression (specified numbers, uš, sag, simple or complex combinations of uš and sag, explicit or implicit). The grammatical suffixes and prefixes frequently omitted in the cuneiform text are given in parentheses.

Arguments

uš	length
sag	width
a-ša$_3$	area

Relations

A-ma B	A: B	$A = B$
A-ma ugu B C diri	A: exceeds B by C	$A = B + C$
A-ma B C ba-la$_2$	A: is less than B by C	$A = B - C$

Operations

A B-(še$_3$) (bi$_2$)-dah	A to B I added	$A + B$
A B GAR.GAR	A B I accumulated	$A + B$
A u$_3$ B	A and B	$A + B$
A B	A and B	$A + B$
A B (ba)-zi	A, from B I subtracted	$-A + B$
a-na A ugu B diri	That by which A exceeds B	$A - B$
A a-ra$_2$ N-e tab	A, N times I repeated	$A \times N$
A igi-N-gal$_2$-(bi)	A, its Nth part	$A \dfrac{1}{N}$
igi-N-gal$_2$	A la Nième partie de A	$\dfrac{1}{N} A$

A	Tablet of the Oriental Institute, Chicago
AO	Tablet from Antiquités Orientales, the Louvre, Paris
BM	Tablet from British Museum, London
MS	Manuscript of the Schøyen Collection
VAT	Tablet of the Vorderasiatisches Museum, Berlin
YBC	Tablet from Yale Babylonian Collection, New Haven

References

Friberg, Jöran. 2007. *A remarkable collection of Babylonian mathematical texts*. Manuscripts in the Schøyen collection: Cuneiform Texts Vol. I. New York: Springer.

Glassner, Jean-Jacques. 2009. Ecrire des livres à l'époque paléo-babylonienne: le traité d'extispicine. *Zeitschrift für Assyriologie und Vorderasiatische Archäologie* 99:1–81.

Høyrup, Jens. 2001. The Old Babylonian square texts—BM 13901 and YBC 4714. Retranslation and analysis. In *Changing Views on Ancient Near Eastern Mathematics*, ed. J. Høyrup and P. Damerow, 155–218. Berlin: Dietrich Reimer Verlag.

Høyrup, Jens. 2002. *Lengths, Widths, Surfaces. A Portrait of Old Babylonian Algebra and its Kin*. Berlin & Londres: Springer.

Høyrup, Jens. 2006. Artificial language in ancient mesopotamia—a dubious and less dubious case. *Journal of Indian Philosophy* 34:57–88.

Neugebauer, Otto. 1935. *Mathematische Keilschrifttexte I*. Berlin: Springer.

Neugebauer, Otto. 1935–1937. *Mathematische Keilschrifttexte II–III*. Berlin: Springer.

Neugebauer, Otto, and Abraham J. Sachs. 1945. *Mathematical Cuneiform texts*. American Oriental Studies Vol. 29. New Haven: AOS & ASOR.

Proust, Christine. 2008. Quantifier et calculer: usages des nombres à Nippur. *Revue d'histoire des mathematiques* 14:143–209.

Proust, Christine. 2009. Deux nouvelles tablettes mathématiques du Louvre: AO 9071 et AO 9072. *Zeitschrift für Assyriologie* 99:1–67.

Proust, Christine. 2012. Reading colophons from Mesopotamian clay-tablets dealing with mathematics. *NTM Zeitschrift für Geschichte der Wissenschaften, Technik und Medizin* 20:123–56.

Thureau-Dangin, François. 1938. *Textes Mathématiques Babyloniens*. Leiden: Ex Oriente Lux.

Chapter 9
Describing Texts for Algorithms: How They Prescribe Operations and Integrate Cases. Reflections Based on Ancient Chinese Mathematical Sources

Karine Chemla (林力娜)

Abstract The texts of algorithms fall under the general rubric of instructional texts, discussed by J. Virbel in this book. An algorithm has two facets. It has a text—a written text—, which usually appears to be an enumerated list of operations. In addition, whenever an algorithm is applied to a specific set of numerical values, practitioners derive from its text a sequence of actions, or operations, to be carried out. In the execution of the algorithm, these actions generate events that constitute a flow of computations eventually yielding numerical results. This chapter aims mainly to develop some reflections on the relationship between these two facets: the text and the different sequences of actions that practitioners derive from it. I use two tools in my argumentation. Firstly, I use the description of textual enumerations, as developed by Jacques Virbel, to find out how enumerations of operations were carried out in the text of algorithms and how these enumerations were used. Then I focus on the language acts carried out in some of the sentences composing the texts, since, when prescribing operations, the texts of the algorithms differ in that they use distinct ways of carrying out directives. The conclusion highlights different ways in which the text of an algorithm can be general and convey meanings that go beyond simply prescribing operations.

An algorithm has two facets. It has a text—a written text—, which usually appears to be an enumerated list of operations. In addition, whenever an algorithm is applied to a specific set of numerical values, practitioners derive from its text a sequence of actions, or operations, to be carried out. In the execution of the algorithm, these actions generate events that constitute a flow of computations eventually yielding numerical results. This chapter aims mainly to develop some reflections on the relationship between these two facets: the text and the different sequences of actions that practitioners derive from it.

K. Chemla (林力娜) (✉)
SPHERE (ex-REHSEIS; CNRS & University Paris Diderot), Université Paris 7, UMR 7219,
75205 Paris Cedex 13, France
e-mail: chemla@univ-paris-diderot.fr

© Springer International Publishing Switzerland 2015
K. Chemla, J. Virbel (eds.), *Texts, Textual Acts and the History of Science,*
Archimedes 42, DOI 10.1007/978-3-319-16444-1_9

317

Why is the question of how these facets relate to each other a problematic and an interesting one? I will limit myself here to considering two reasons for this, on which I will focus in this chapter.

First, an algorithm usually has a certain generality. In particular, it provides a means for carrying out a task—e.g., multiplying—for a whole range of numerical values.[1] Although these values may fall into different cases and thus require different sequences of actions in order to be dealt with, a general algorithm allows the reader to deal with them all. What are the means employed in writing a text for an algorithm in such a way that a single text can relate to the different flows of computation corresponding to all the various cases? This is one of the issues behind the question of the relationship between the two facets, and we will see that our sources attest to the shaping of different solutions for this problem.

Second, we will see that the text of an algorithm can designate the same action in different ways. More generally, our sources attest to the fact that the same flow of computations can correspond to different texts. How can we describe, and account for, this variety? What are the different meanings conveyed by these different texts? This constitutes the second set of issues we will address.

Two tools will prove useful in developing my argumentation and I will use them systematically. Firstly, I will use the description of textual enumerations, as developed by Jacques Virbel, to find out how enumerations of operations were carried out in the text of algorithms and how these enumerations were used.[2] This approach deals with the text of the algorithm as text. Then I will focus on the language acts carried out in some of the sentences composing the texts, since we will see how, when prescribing operations, the texts of the algorithms differ in that they use distinct ways of carrying out directives. We will in particular be led to discussing several modes of carrying out indirect acts to which they bear witness. The texts of algorithms fall under the general rubric of instructional texts, discussed by J. Virbel in this book.[3] In this second approach, I will call on Virbel's work on Speech Act Theory to analyze the types of directives occurring in the sentences in these kinds of texts. From this analysis, we will be in a position to derive conclusions regarding the texts themselves.

Addressing the issues raised earlier with these tools will enable us, in our conclusion, to highlight different ways in which the text of an algorithm can be general and convey meanings that go beyond simply prescribing operations.

---

[1] In fact, an algorithm can even integrate means for carrying out different tasks. An example from ancient Chinese mathematical sources is described in Chemla 1997a. However, I'll leave this aspect aside here. The completion of this article was supported by the Chinese Academy of Sciences Visiting Professorship for Senior Foreign Scientists 外国专家特聘研究员, grant number 2009S1–34 (Beijing, 2010).

[2] See J. Virbel, "Textual enumerations," Chap. 6, in this book. This chapter could not have been written without the joint research work carried out over the last few years with the authors of the other chapters in the book and the various colleagues who took part in the seminar "History of science, history of text." It is my pleasure to acknowledge my intellectual debt to this collective endeavor as well as to Ramon Guardans for his stimulating reactions. I am also glad to express my gratitude to Richard Kennedy for his help in preparing the final version of this text.

[3] See J. Virbel, "Speech Act Theory and Instructional Texts," Chap. 2, in this book.

In this chapter, I will illustrate and examine these questions on the basis of a small corpus composed of texts found in mathematical sources from ancient China—see the overview of the sources used in the Appendix. The core of my corpus will consist of texts from different time periods devoted to the task of dividing quantities composed of integers and fractions by similar quantities. To make the exposition easier, I will also include algorithms given for multiplying quantities of the same type.

The writings from which these texts are extracted are of different natures. One text is taken from a manuscript recently excavated from a tomb sealed around 186 BCE: the *Book of Mathematical Procedures* 算數書 (*Suanshushu*).[4] In contrast to this document, which is available to us thanks to archeological excavations, all the other texts are taken from books that were handed down through the written tradition. This is the case for the Han book *The Nine Chapters on Mathematical Procedures* 九章算術 (*Jiuzhang suanshu*), which probably reached its present form in the first century CE, as well as the *Mathematical Classic by Zhang Qiujian* 張邱 建算經 (*Zhang Qiujian suanjing*). Qian Baocong 錢寶琮 dates the composition of the latter to the second half of the fifth century, whereas he attributes the "detailed procedures 草 *cao*" that it contains to Liu Xiaosun 劉孝孫, who composed them in the sixth century.[5] Both books were to be considered as classics and were, as such, the objects of commentary. From Liu Hui 劉徽's commentary on *The Nine Chapters*, completed in 263, we will consider the section bearing on the text of the algorithm for dividing.[6] In 656, Li Chunfeng 李淳風 et al. presented to the throne an annotated edition of *Ten Classics of Mathematics* 算經十書 (*Suanjing shishu*), which included both *The Nine Chapters*, with Liu Hui's commentary, and the *Mathematical Classic by Zhang Qiujian*. Within this context, Li Chunfeng and his associates composed a commentary on each classic—which I refer to as "Li Chunfeng's commentary". We will also use the commentary written in this context on the texts of algorithms for multiplying and dividing included in the *Mathematical Classic by Zhang Qiujian*.

These sources provide an abundant resource for the analysis, on a general level, of the different ways in which the text of an algorithm relates to the sequences of computations it gives rise to. In this way they will help us reflect on the general questions raised. When using our sources with this aim in mind, we will not bring them into play in chronological order, but instead privilege an order that introduces the phenomena to be observed according to a scale of increasing complexity.

Once we have brought to light the diversity of techniques that practitioners developed to textualize algorithms, we will be led to a second set of issues, essential to plead for the interest of furthering such lines of inquiry: how can History of

---

[4] I will rely on the critical edition given in Peng Hao 彭浩 (2001). It is still difficult to determine the status of this text.

[5] See the discussion of the attribution and the critical edition of the book, respectively, in Qian Baocong 錢寶琮 (1963, pp. 326–327, 329–405). I will also refer to another recent critical edition: Guo Shuchun 郭書春 and Liu Dun 劉鈍 (2001).

[6] For both *The Nine Chapters* and Liu Hui's commentary, I refer the reader to the critical edition in Chemla and Guo Shuchun (2004). Chapter B and the introduction to Chap. 6 of the latter book present the various views held regarding the dating of *The Nine Chapters*.

Science benefit from such research? What use can be made of such considerations for more purely historical purposes? Addressing these issues require that we change our point of view with respect to our sources. So far, I have suggested using a set of Chinese sources to *illustrate* the treatment of general questions with respect to algorithms. These sources have allowed us to show that the text of an algorithm was something problematic and that its description could reveal diversity, where we may have spontaneously assumed that it would display homogeneity. Now, to address the second set of issues just raised, we will have to consider our Chinese sources in and of themselves and ask ourselves whether our analyses allow them to be seen in a new light. I suggest, in conclusion, two perspectives for this.

The analyses carried out will have provided methods for identifying systematic differences between various kinds of texts for algorithms. The discovery of such a phenomenon raises key questions for History of Science, and in particular, that of determining how we can account for these differences. I will suggest that they may offer clues to the existence of different "professional cultures," which elaborated different ways of writing down texts for algorithms. In other words, the benefit for History of Science here would be that we should be able to identify scholarly cultures on the basis of the kind of texts specific to them.

In addition, our analyses will identify, among the various texts examined, a class of texts sharing unique features, which differentiate them not only from the other texts analyzed, but also from modern texts for algorithms. Moreover, the singular features in question present remarkable stability, since they can be attested on a time span that goes from the second century B.C.E. to the seventh century C.E. Again, this phenomenon could only be brought to light by an analysis of the kind we have developed. The type of historical continuity that this analysis discloses, which goes beyond the differences between the texts, calls for a historical explanation. I will suggest that such phenomena are a powerful tool in defining and capturing the development of a tradition.

Before raising such questions, however, let us first turn to concrete algorithms and describe their texts.

9.1 Two Ways of Writing down Texts for Algorithms and a First Type of General Algorithm

The opening section of the *Mathematical Classic by Zhang Qiujian* is composed of six problems, each followed by a text referring to an algorithm that solves it—this preliminary and coarse description is given more precision below. The first three problems deal with multiplying quantities that combine an integer and fractions, whereas the following three deal with dividing quantities of the same type. The procedures and Li Chunfeng's commentary on them provide us with a clear-cut contrast between two ways of writing down a text for an algorithm. It is therefore convenient to start our analysis from here and I will consider each of them one by one.[7]

We begin with the first three problems in the *Mathematical Classic by Zhang Qiujian*, which offer excellent source material when addressing the issues on which we focus in this chapter. In Sect. 9.2, it will be instructive to address the contrast between the texts of procedures recorded for multiplication and those given for division.

9.1.1 A Simple Text for an Algorithm—First Remarks

The first problem of the Classic reads as follows:[8]

(1.1) "以九乘二十一、五分之三。問得幾何。
　　答曰:一百九十四、五分之二。
　　草曰:置二十一,以分母五乘之,内子三,得一百八。然以九乘之,得九百七十二。却以分母五而一,得合所問。

One multiplies 21 and 3/5 by 9. One asks how much it yields.

Answer: 194 and 2/5

Detailed procedure (*cao*):

(a) One puts 21, one multiplies it by the denominator 5, and incorporates (the result) into the numerator 3, hence yielding 108.

(f) Then, multiplying this by 9 yields 972.

(h) If, in return, one divides by the denominator, 5, what is yielded conforms to what was asked."

The text is quite simple. After the statement of a problem—identified by the key expression "One asks...how much..."—and the answer, the text provides the "detailed procedure (*cao*)." We come back to this term below, which qualifies the kind of text provided here for the algorithm. This section starts with the operation:

---

[7] See Qian Baocong 錢寶琮 (1963, pp. 331–334), Guo Shuchun 郭書春 and Liu Dun 劉鈍 (2001, pp. 297–299).

[8] In what follows, I refer to any problem from any book by a pair of numbers, the first one indicating the chapter in which the problem was included and the second one giving its rank in the chapter. We will make an exception for the *Book of Mathematical Procedures*, which was found in separate bamboo strips and the organization of which is still the subject of some debate. Moreover, to allow us to refer to the text of an algorithm in a convenient way, I distinguish steps in the text by introducing letters "(a), (b)..." The reader must keep in mind that these signs do not belong to the original text and are merely a tool for the sake of the analysis. Further, note that I sometimes use letters that do not follow each other alphabetically, in order to allow comparison between the texts of different procedures. Finally, note that in this way, I set up an enumeration that groups together operations that were listed as distinct actions in the sources. More generally, the original texts list operations in a certain way. I group them into larger sets in such a way as to allow comparison between the texts. The beginnings of the groups of operations I form regularly correspond to markers such as "one puts," "moreover," "then," whereas their endings regularly correspond to the assertion of results or to declaratives, giving a name to a value obtained (see below).

Table 9.1 The flow of computations for Problem (1.1)

Upper row	21	105			194
Middle row	3	3	108	972	2
Lower row	5	5	5	5	5

"one puts…," which is quite common in all Chinese mathematical sources, from the oldest known. The term refers to the fact that the values, to which the sequence of operations was to be applied, were put on a surface on which the computations were carried out.[9] On this surface, of whose physical features we have no direct evidence, numbers were represented with counting rods and their values could thus be transformed through the process of computation.[10]

In modern symbols, we would represent the computation as follows:

$$\left(21+\frac{3}{5}\right).9 = \frac{21.5+3}{5}.9 = \frac{108.9}{5} = \frac{972}{5} = 194+\frac{2}{5}$$

On the surface, the list of operations carried out yielded something like the sequence of states represented in Table 9.1.

We see that, in this example, the relationship between the text and the actions carried out on the surface is straightforward: each verb corresponds to either an event or an assertion about a result, the sequence of sentences corresponds to the sequence of actions and their outcomes. It illustrates a case that, we may spontaneously be tempted to believe, holds in general for the relationship between the text of an algorithm and the corresponding flow of computations. However, we will see that this first representation fails to account for the actual variety of relationships between the text of an algorithm and the actions derived from it, to which the

---

[9] For want of a better word, throughout this chapter I will use the term "surface" to designate the surface on which one computed in ancient China. Probably, the calculations were not done on a surface specific to that use, but on an ordinary flat area (Martzloff 1987, p. 170).

[10] Below, I will systematically represent by columns the configurations of numbers that, during a computation, succeed to each other on the surface, as I suggest recreating them. Moreover, in order not to increase the burden placed on the reader, I replace the counting rods by the Arabic digits that are now common. While each state of the surface corresponds to a column of values, the global computation is represented by the sequence of columns arranged in a table. The reader has to imagine that the columns are states that follow each other, one replacing the other. For the sake of clarity, I divide the columns into zones separated by lines. However, there is no reason to believe that the original surface bore any marks. Let me stress that for the operations discussed, the *Mathematical Classic by Zhang Qiujian* contained only texts for algorithms, but no further information regarding how the surface was used. In particular, no extant writing composed before the thirteenth century includes illustrations showing the computations on the surface. I do not present any arguments regarding how I recreate the layout of the computations, referring the reader to another publication in which this aspect is addressed in greater detail: Chemla (1996). In particular, note that I interpret "one puts 21" as corresponding to the action of placing 21 on the surface together with the fraction attached to it. Since I am mainly interested in the relationship between the text of the algorithm and the actions to be taken, the way in which the concrete computations are recreated is of no crucial importance here.

ancient sources bear witness. Also note that already, in this case—the simplest one we will encounter—, we must qualify the common assumption regarding texts of algorithms: such texts do *not* constitute merely a sequence of directive acts, but also involve assertions such as the concluding sentences of our steps (a), (f) and (h).[11] In what follows, we will meet with other language acts inserted into texts for algorithms. Two additional remarks on our text will prove useful below.

The first remark bears on the designation of the operations. Using this text requires that the reader has some competence. When, as in step (f) for example, the text prescribes a multiplication, this calls for the use of another procedure to carry out this operation on 108 and 9. However, the text does not include elsewhere a description of how to perform this computation. Nor, more importantly for us, does it enter into that level of description here. On the one hand, this fact therefore indicates that the text assumes certain aptitudes in the reader—in this case, knowing how to multiply integers. More generally, the text of an algorithm reflects assumptions regarding the competences possessed by the user, and the kinds of competences required vary according to how it is formulated. This dimension will be of interest for us in the context of this chapter.[12] On the other hand, the text being examined brings several similar operations into play, all of which are of the same level of integration (division, incorporation). It, thus, presents a certain kind of uniformity in the level of detail chosen for writing down the algorithm. We refer to this dimension of the text, by introducing the notion of "granularity" of the description. In these terms, the granularity of the text is somewhat uniform. In that respect, the text reflects a more general fact: the text of an algorithm involves higher-level operations, which encapsulate other procedures of a lower level. In fact, the way in which these operations are constituted and introduced depends on the social and historical context in which the text is produced.[13] The competences required of a reader or a user can, at

---

[11] Most of the articles published about algorithms in ancient China echo this assumption. Let me signal, without dwelling on this issue for the moment, that these assertions are not the only elements of the translated text that do not refer to an event on the surface. The division to be carried out is qualified as being "in return." This indication points to the motivation for which the division is used here. We come back to such phenomena later.

[12] J. Virbel, "Speech Act Theory and Instructional Texts," this volume, Sect. 2.3, raises this issue by indicating that the aid provided by an instructional text can take "multiple forms" and "be relative" to the user. In *Ibid.*, Sect. 2.3.3.2, he points out that "maximum efficiency," when the concern is "to give to H (KC, the user of the text) as much aid as possible," may not be to provide as many details as possible. When Wittgenstein emphasized the import of the surveyability (übersichtlichkeit) of a proof, he pointed out a concern of the same type. I. Hacking quoted and discussed these remarks made by Wittgenstein in his lecture, the text of which can be found at http://www.college-de-france.fr/media/ian-hacking/UPL384701617840960135_4___D__monstration.pdf, pp. 10–11, accessed on April 14, 2015.

[13] Let me illustrate this statement by referring to the example of an operation used in present-day mathematical practice, i.e., "reducing to the same denominator." This operation invites the reader to apply a procedure composed of several multiplications. The text of an algorithm can use either the expression "reduce to the same denominator," leaving to the reader the task of understanding the actions required, or only give the list of actions. In the latter case, the user possibly remains unaware that the operations fulfill this aim. Another important remark here is that, in the context of ancient Chinese sources, some texts of algorithms introduced operations of the same level as

least partly, be approached through the list of higher level operations required to be known.[14] Again, we come back to these issues below.

Our second remark deals with how the text is formulated more globally: it mentions explicitly the values to which the computations are to be applied (21, 5, 3, …), and it occasionally also indicates the "meaning" of the values: "numerator," and "denominator." However, the text is to be interpreted as referring *not only* to this particular case (multiplying 21 and 3/5 by 9), but also to all similar cases, in a tradition of the expression of general procedures that goes back at least as early as *The Nine Chapters*.[15] This remark reveals another competence required of the reader in order to use this text: s/he must know how to interpret it in such a way as to determine the set of cases for which the procedure is valid. It can be assumed that the degree of generality of a procedure was determined on the basis of its text. In addition, the reader also probably relied on the position given to a particular problem and procedure among the set of problems and procedures presented for carrying out a given task. In our case, the limits of the validity of the procedure were thus also highlighted by the problems that followed and the procedures attached to them.

9.1.2   Ways in Which the Text for a Procedure Refers to the Layout on the Instrument

The previous analysis led us to note that the terms on which the operations were to be applied were designated either by their value in the paradigm, or by their "meaning" with respect to the situation being dealt with. When examining the problems

---

"reducing to the same denominator" to achieve the same result. However, the way in which they group actions to be carried out differs from what "reducing to the same denominator" does. What for us corresponds to "reducing to the same denominator" is prescribed there by means of two higher-level operations, one referring to the action on denominators, the other to the actions on numerators. We thus see how the constitution of such operations may vary according to the context.

[14] Here, the competences involve knowing how to lay out the computations on the surface and how to execute the operations. Such competences were necessary for using the written text. Either such knowledge was a prerequisite to reading the text or it was meant to be provided through oral explanations accompanying the text. In any case, the written text did not record any further indications. These questions were outlined in Chemla (2009a). Since we know nothing about the use for which the written text was composed, we cannot go any further here. The question on which we chose to focus, i.e., describing the relationship between the text of the algorithms and the corresponding flows of computation, allows us to bypass this range of consideration in a first phase. However, the conclusions reached could cast light on the issue of the function and use of the writings.

[15] We can prove that this is how the commentator Liu Hui, the earliest reader whose reading of a problem can be observed, interpreted a similar text of algorithm in *The Nine Chapters*, see Chemla (2003). In the case of this passage from the *Mathematical Classic by Zhang Qiujian*, this interpretation is confirmed by Li Chunfeng's commentary to be examined below. Note that the reader must also know, in such cases, how to transfer the procedure to a different problem. On this issue compare Volkov (1992), (2008), Chemla (2010b).

and procedures that follow immediately afterwards in the *Mathematical Classic by Zhang Qiujian,* we will see that there are two additional main options. Moreover, and more importantly here, these subsequent problems and procedures will constitute a basis necessary when considering, below, Li Chunfeng's commentary on this opening section of the book. Problem (1.2) reads as follows:

(1.2) "以二十一、七分之三乘三十七、九分之五。問得幾何。

答曰：八百四、二十一分之十六

草曰：置二十一，以分母七乘之，內子三，得一百五十。又置三十七，以分母九乘之，內子五，得三百三十八。二位相乘得五萬七百為實。以二分母七、九相乘得六十三而一，得八百四，餘六十三分之四十八。各以三約之，得二十一分之一十六。合前問。

One multiplies 37 and 5/9 by 21 and 3/7. One asks how much it yields.

Answer: 804 and 16/21

Detailed procedure (*cao*):

(a) One puts 21, one multiplies it by the denominator 7, and incorporates (the result) into the numerator 3, hence yielding 150.

(b) Again, one puts 37,

(e) one multiplies this by the denominator 9, and incorporates (the result) into the numerator 5, hence yielding 338.

(f) The two positions being multiplied by one another yield 50,700, which makes the dividend.

(g) The two denominators 7 and 9 being multiplied by one another yield 63,

(h) by which one divides, hence yielding 804, remainder 48/63. Simplifying each of them by 3 yields 16/21. This conforms to the previous questioning."

The problem now requires multiplying two quantities, each combining an integer and a fraction. The computations prescribed can be represented as follows:

$$\left(21+\frac{3}{7}\right)\left(37+\frac{5}{9}\right)=\frac{7.21+3}{7}\cdot\frac{9.37+5}{9}=\frac{150.338}{7.9}=\frac{50700}{63}$$

$$=804+\frac{48}{63}=804+\frac{16}{21}$$

In this case, as in the previous one, the sequence of actions to be executed to solve the problem is clear, since each verb of the text corresponds either to the execution of an operation or to the statement of a result. On the surface, we can suggest the way of restoring the corresponding flow of computations represented in Table 9.2:[16]

---

[16] The *Mathematical Classic by Master Sun,* one of the ten mathematical classics brought together by Li Chunfeng in the seventh century, provides key information for restoring the arrangement of the values here. In particular, it explains that a multiplicand and multiplier were respectively put in the upper and lower positions of the surface, whereas the dividend and divisor were placed in the middle and lower positions, respectively. These pieces of information guide my way of restoring the computations on the surface here. Compare Chemla (1996). Below, we will see that this is coherent with the information contained in the texts of all the procedures.

Table 9.2 The flow of computations for Problem (1.2)

Upper	Upper	21							
	Middle	3	150	150	150				
	Lower	7	7	7	7	7			
Middle								804	804
	Dividend					50,700	50,700	48	16
	Divisor						63	63	21
Lower				37					
				5	338	338	338	338	338
				9	9	9	9	9	9

The sequence of actions carried out here differs from the one prescribed for the previous problem. We will discover below why this point is interesting. In fact a comparison with the previous procedure shows that the formulation of the corresponding text presents two new features, related to the designation of the terms of the operations, which are worth evoking.

To begin with, in step (f), the text designates one of the intermediary results as "dividend." Strictly speaking, this clause ("which makes the dividend") does not correspond to an event on the surface, but combines a declarative and an assertive act, in a combination typical of this kind of text.[17] This act allocates a name to the value just obtained ("dividend.") Furthermore, the name given asserts the function played by this value in one of the subsequent operations to be executed: in step (h), the value "by which one divides" is that obtained immediately beforehand, whereas the value to be divided is that designated upstream as "dividend." Such a designation of a value by the functional role that it plays within the algorithm is one of the four main techniques used to refer to a value in the text of an algorithm. Here the term "dividend" probably also indicates a position on the surface—that is, the middle position. If this position is really to be inferred from this sentence, it is evidence of another competence required of the reader.

This leads us to our second remark. Also in step (f), instead of referring, as above, to the operands of the operations by their value (150, 338), or their "meaning" (for instance, "numerators"), or even their function (like "dividend"), the text designates them by the "positions," in which they can be found on the surface. This latter mode constitutes the fourth main option used to designate the terms of an operation. It is important to analyze here how this type of reference works. As can be seen in Table 9.2, the content of positions is changed by the flow of computations. The values thus designated are precisely those to be found in these positions at the very moment that the positions are mentioned in the text. Translating the text efficiently into action thus requires that the practitioner relies on the flow of computations that his or her own actions create on the surface, while using the text of

---

[17] Compare J. Virbel, "Speech Act Theory and Instructional Texts," this volume, Sect. 2.3.4.2.

the algorithm. Otherwise, the action deriving from the text at that point would not be correct. Moreover, turning the text into action also requires that the practitioner knows how to lay out the values on the surface, how to use and thereby transform them and which positions are meant by the text.[18] Interestingly, here the text does not provide any information on this aspect of the computations, thereby disclosing the expectation that the user's competences allow him or her to complement what the text keeps silent. Otherwise, the reference to values by means of indicating their position would not work.

On that count, the text for the next procedure is completely different. Its verbs do not only designate arithmetical operations, or assert some of the transitory results obtained or grant names to values, as we saw for the previous problems. In addition to mentioning that values are placed on the instrument, some verbs also indicate *how* to place them on the surface.[19] The latter clauses, on positioning, thus make explicit that part of the actions that the practitioner carries out when s/he produces a flow of computations on the surface. Immediately afterwards, the text makes use of the positions introduced to designate the values to be found in them. In correlation with these features, the text for this procedure can *only* be used in relation to the counting instrument employed in ancient China. By contrast, the first text of procedure examined above can correspond to actions carried out on any instrument.[20] Let us now analyze in greater detail how positioning is described and how the names of positions are used, in the text of this procedure. The text in question, which illustrates another way of writing down a procedure, is placed in the *Mathematical Classic by Zhang Qiujian* after problem (1.3), the last in a sequence related to the task of multiplying various values by each other. This problem introduces yet another variation of the situation under consideration, which extends the range of situations referred to. It reads as follows:

---

[18] In Table 9.2, I suggest interpreting these "positions" as being "upper" and "lower," respectively. More precisely, when the values placed in a position are integers followed by fractions, these positions become zones in which the quantity is displayed in three rows. Moreover, in my view, computations are carried out in these zones, according to principles, for the layout, that are the same as those described by the *Mathematical Classic by Master Sun* for the higher level. The term "positions" in the text of the algorithm will designate the value in the central row of each of the zones.

[19] Note that what is described in terms of positioning is in affinity with what has been restored above. In particular, these indications confirm that the information found in the *Mathematical Classic by Master Sun* remains valid in our context.

[20] These remarks illustrate one feature of the possible localization of instructional texts—one of the types of difference between such texts introduced by J. Virbel. Compare J. Virbel, "Speech Act Theory and Instructional Texts," this volume, Sect. 2.3.1, point 10.

(1.3) "以三十七、三分之二乘四十九、五分之三、七分之四。問得幾何。
答曰：一千八百八十九、一百五分之八十三
草曰：置三十七，以分母三乘之，內子二，得一百一十三。又置四十
九於下[21]。別置五分於下右，之三在左。又於五分之下別置七分，三
分[22]之下置四。維乘之，以右上五乘左下四得二十，以右下七乘左上
三得二十一，併之得四十一。以分母相乘得三十五。以三十五除四十
一，得一，餘六。以一加上四十九得五十。又以分母三十五乘之，內子
六，得一千七百五十六。以乘上位一百一十三，得一十九萬八千四百
二十八為實。又以三母[23]相乘得一百五為法。除實得一千八百八十
九，餘一百五分之八十三。合所問。

One multiplies 49 and 3/5 and 4/7 by 37 and 2/3. One asks how much this yields.
Answer: 1889 and 83/105.

Detailed procedure (*cao*):

(a) One puts 37, one multiplies this by the denominator, 3, and incorporates (the result) into the numerator, 2, hence yielding 113.

(b) Again, one puts 49 *below*[24] (that is, in the lower position, translator's note);

(c) moreover, one puts 5 parts *below* on the right hand side and 3 parts to the left hand side; again, under the 5 parts, one further puts 7 parts, and under the 3 parts, one puts the 4. One *cross-multiplies*: multiplying the 4 below on the left hand side by the 5 above on the right hand side yields 20; and multiplying the 3 above on the left hand side by the 7 below on the right hand side yields 21; summing these (the results) yields 41; the denominators being multiplied by one another yield 35. Dividing 41 by 35 yields 1 and remainder 6.

(d) Adding 1 to 49 *above* yields 50.

(e) Again, one multiplies it by the denominator, 35, and incorporates (the result) into the numerator 6, hence yielding 1756.

(f) One multiplies by this the upper position, 113, hence yielding 198,428, which makes the dividend.

---

[21] Like Guo Shuchun 郭書春 and Liu Dun 劉鈍 (2001), p. 343, footnote 4, I consider that the text contained in our sources can be understood and I keep it in the edition of this passage (Chemla 1996). Qian Baocong 錢寶琮 (1963, p. 333), footnote 1, considered it was corrupted and replaced here 下 "below" par 上 "above."

[22] This is the text as given by all documents. Qian Baocong 錢寶琮 (1963, p. 333), footnote 2, considered it was corrupted and replaced it by "之三". Guo Shuchun 郭書春 and Liu Dun 劉鈍 (2001, p. 343), footnote 5, consider that "三分" is a corruption of the original "三," hence suppressing 分, which they consider to be an interpolation. If we relied on the evidence provided, for instance, in Chap. 12 of Li Ye 李冶's *Ceyuan haijing* 測圓海鏡, we could consider that Qian Baocong was right. However, the text can be understood as it stands. Moreover, the same kind of formulation can be found in the commentary on problem 4.24 in *The Nine Chapters* (Chemla and Guo Shuchun 2004, p. 378, fn. 375, 534).

[23] Our sources contain here "分母三母". Qian Baocong 錢寶琮 (1963, p. 333), footnote 3 considered that the original text contained here "三分母". I follow the editorial suggestion made by Guo Shuchun 郭書春 and Liu Dun 劉鈍 (2001, pp. 343–344), footnote 6. However, there are reasons to be discussed below for considering that this point should be revised.

[24] Here, and below, my emphasis.

Table 9.3 The flow of computations for Problem (1.3). The cells in grey correspond to the sub-procedure embedded

	Upper	37					(*)	(**)	
Upper	Middle	2	113	113	113	113	113	113	
.	Lower	3	3	3	3	3	3	3	
Middle	Dividend								198428
	Divisor								105
Lower				49	49	49	49	50	
							1		
				3 5	21 5	41	6	6	1756
				4 7	20 7	35	35	35	35

(g) Again, the three denominators being multiplied by one another yield 105, which makes the divisor.

(h) Dividing (by this) the dividend yields 1889, and remainder 83/105. This conforms to the previous questioning."

The situation dealt with in problem (1.3) differs from that in problem (1.2) in one key feature: one of the two quantities that is multiplied contains not merely one, but in fact two fractions after the integer. The procedure solving it is composed of the procedure following problem (1.2), that is, steps (a), (b), (e), (f), (g), and (h). In addition, between steps (b) and (e), a procedure amounting to steps (c) and (d) was embedded.

The procedure constituted by steps (c) and (d) computes the sum of two fractions, yielding the result in the form, traditional for results greater than 1, of an integer followed by a fraction (see Table 9.3 and in particular the grey cells, which correspond to these two steps). This procedure corresponds precisely to the segment of the whole text containing greater detail on the actions linked to the lay out of the computations. In other words, there is a correlation between the fact that the text of one procedure was embedded in another, and the new textual feature on which we are focusing. Note that this indicates that overall, the text of the procedure is not uniform in the way it relates to the actions carried out: details on positioning are given in only one of its segments. Before we analyze this segment, we should therefore first examine how the embedding is carried out.

One problem—adding together two fractions—was grafted onto another problem—multiplying quantities. Infixing the text of one procedure within another is what allowed the reduction of the situation to the one solved by the procedure given after problem (1.2). Accordingly, after steps (c) and (d), the second part of the procedure for problem (1.2) is used to complete the solution of problem (1.3). Problems and procedures composed in this way appear quite often in some extant Chinese mathematical sources from the time of the *Mathematical Classic by Zhang Qiujian* onwards, and less so in others. The procedure solving this kind of problem embeds the solution of the former into the solution of the latter, in such a way that the former is left basically unchanged.

Indeed, if we examine how the embedding was implemented, we see that the procedure resulting from the combination of the two procedures could have been simplified. The computations corresponding to the columns marked by (*) and (**) in Table 9.3 could have been skipped: they cancel each other out.[25] The indirect way of proceeding that characterizes the text of the procedure given after problem (1.3) has two related properties. The meaning of the procedure can be interpreted through its individual steps (Chemla 2010a). In addition, the text shows a way of combining different procedures, without blurring the limit marking the end of one and the start of the other.[26] As a result, the actions taken in relation to step (c) form a coherent whole in a specific zone of the surface.

This remark brings us back to the issue of how this segment of the text of the algorithm refers to the surface on which the corresponding actions are taken. On the one hand, the segment gives greater detail on the actions that put numerical values onto the instrument. On the other hand, subsequently, the terms of the operations are designated by their value in the paradigm that problem (1.3) constitutes *as well as* by their position on the surface. In other words, the interpretation of the text in terms of actions requires that the practitioner looks at what happens on the surface throughout the process of computation and more precisely how the numbers placed in the positions change.[27] If we observe more closely how the text makes use of positions to designate values, we discover that it does so according to two different modalities.

In some cases, positions are designated in an absolute way: see, for instance, the case where, in step (b), one is asked to put the integer 49 "below". This refers to placing 49 in the lower space on the surface. In fact, it will become clear later that this refers more specifically to the upper position, for the integer, within the lower zone: the user of the text again requires competences to be able to interpret it.

In other cases, positions are referred to in a "relative" way: this is the case, for instance, when in step (d), the same 49 is designated as the position "above." 49 was indeed placed above the fractions linked to it. The reader of the text must therefore

---

[25] More precisely, once the sum of the numerators, 41, and the product of the denominators, 35, were obtained, the procedure could have directly used step (e): multiplying 49 by 35 and incorporating the result to 41. Instead, the procedure given divides by 35, which yields the result of the sum of the fractions. It is followed by an operation canceling its effect: a multiplication by 35. Such procedures are typical of those that the commentators presented to account for why a procedure given in *The Nine Chapters* was correct. In a first step, they presented the unnecessary operations that brought to light the meaning of the procedure, before, in a second step, deleting them through valid transformations that led to yielding the procedure as found in *The Nine Chapters* (Chemla 2010a).

[26] Table 9.3 shows clearly that the column marked (*) is grey and belongs to the inserted procedure, whereas the column marked (**) is not grey, and belongs to the procedure linked to Problem (1.2).

[27] Note that the numbers by which one multiplies (the multipliers) remain on the surface, whereas the values multiplied (the multiplicands) are replaced by the results (see Table 9.3, fifth column). By contrast, the operations formulated in a symmetrical way with respect to their terms correspond to an execution that deletes the initial values and leaves only the result on the surface (probably the result of the sum replaces 21, and the result of the product is placed in the row between the values multiplied (5 and 7)).

have the competence to determine whether a term refers to a position in an absolute or relative way.[28]

Moreover, parallel to the fact that a procedure was embedded in another procedure, the layout of the former was grafted onto the layout of the latter. Within the lines for the numerator and the denominator in the lower zone, the layout for adding two fractions is deployed, reproducing the usual layout on a smaller scale. The text makes the positioning explicit precisely for this aspect of the computations. The consequences of the grafting in regards to the layout can be observed in the sequence of transformations shown in Table 9.3, through which I suggest how the computations developed on the surface could be restored. (I do not indicate how the computation ends, since it is similar to what was shown above).

The text we have just examined, like the texts analyzed previously, has, for the most part, a straightforward relationship to the actions that execute the algorithm on the surface. Its verbs designate arithmetical operations or prescribe placing values, sometimes giving greater detail on their positioning. The latter feature illustrates how the text for an algorithm can provide lesser or greater detail on the actions to be carried out. In addition, this text, like the previous ones, contains assertions on the values of provisional results and carries out declarative-assertive acts, naming values by a function they have in a subsequent computation (steps (f) and (h)). Regarding the designation of the terms of the operations, the text illustrates the four modes already brought to light: using the value in the paradigm, its "meaning," its function or its position on the surface—in ways that were described above.

However, two sections of the text do not fit with this description. They introduce the first phenomena that disturb the representation that we may have spontaneously—and anachronistically—formed about the relationship between the text of an algorithm and the corresponding actions. Let us now consider them.

9.1.3 First Difficulties in the Relationship Between the Text of a Procedure and its Related Actions

The first segment on which I invite the reader to concentrate is the following sentence, in step (g): "the three denominators being multiplied by one another yield 105." Indeed, the value 105, i.e., the product of the three denominators, is needed to conclude the computation in the same way as in the previous procedure, after problem (1.2). In step (g) of the latter, one could read: "The two denominators 7 and 9 being multiplied by one another yield 63." However, if, for the same reasons, in the context of problem (1.3), 105 is definitely to be computed and corresponds to the product of the three denominators occurring in the numerical values, is it actually computed by the operation of multiplying the three denominators by one another? In fact, let us observe the values present on the surface, at the step represented by the column marked (**), in Table 9.3. This is precisely the step in which the computation of 105 is to be carried out. In the lowest line, we find 35, obtained

---

[28] When Qian Baocong considered here the text as corrupted (see footnote 21), he assumed that positions were designated only in an absolute way.

previously by multiplying two of the denominators, 5 and 7—note that in step (e), the text refers to this value of 35 as "the denominator," in relation to the fact that it is considered as the denominator of the quantity that has appeared in the lower zone of the surface.[29] In the upper zone, we see 3, the third denominator. One is thus tempted to believe that 105 is computed by multiplying 35 and 3, rather than by multiplying the three denominators by one another, as the text has it. The conclusion is all the more tempting as at the time that the column marked (**) is on the surface, the denominators 5 and 7 are probably no longer there, to be multiplied by 3. These arguments rely on the physical layout of the computations. There is another argument in favor of the same conclusion, which invokes the logic of embedding of one procedure in another, as described in the previous section. From this perspective, it would seem reasonable that, following the procedure for (1.2), 35 and 3 are multiplied together here. If we are convinced by this group of arguments and consider that 105 is yielded by multiplying 35 and 3, there appears an interesting consequence with respect to how the text relates to the action carried out: there must be a dissociation between the way in which the text prescribes the operation and the way in which the action is taken.

So what could be the benefit of writing down the operation in a way different from the way in which it is carried out? Two hypotheses can be offered to explain this disparity between the actual way of computing and the way in which the text formulates it. We can observe that the way step (g) is formulated marks a difference between the corresponding steps in this procedure and the procedure placed after problem (1.2). Stressing the differences resulting from changes to the problem may have been a benefit in which the author of the text could have been interested. However, this hypothesis seems to me less probable than the second one: such a formulation of step (g) indicates the "meaning" of the value 105 with respect to the initial problem—we see below that various texts of algorithms regularly reveal an interest in this property of the formulation.

Yet, before drawing a conclusion too hastily on this question, let us take the opportunity of this example to remind ourselves of a fundamental fact that has an impact on the entire discussion: the text of step (g), in our translation, is an artifact of the critical edition (see footnote 23 of this chapter). Our whole discussion thus depends on philological considerations. Had the corrupted text been edited in

---

[29] This detail is quite revealing. It indicates that the text is written down in such a way as to indicate by intermittences the meaning of the operations prescribed: the fact that the value computed, 35, is referred to as a "denominator" implies that the result of the computation yielding it is interpreted with respect to the situation, whether this is done by sheer interpretation or by reading the position in which the value is placed on the surface. Moreover, not only is the interpretation achieved, but the outcome of the interpretation of the result is also used in the next step of the text, to refer to the value obtained. Therefore, the user of the text must formulate for himself or herself the meaning of the sequence of results computed in order to be able to use the text. The interpretation is certainly a condition for embedding a procedure into another procedure: the main procedure needs a value for the denominator of the quantity placed in the lower position. Chemla (2010b) reaches the same conclusion from a completely different perspective.

another way, our analysis of the relationship between the text and the computations on the surface would have been different. If we had suggested that the original text of step (g) was, for instance, "以分母相乘 The denominators being multiplied by one another," or "以二母相乘 The two denominators being multiplied by one another," the relationship between the text and the computations would have been straightforward. This general philological problem cannot be underestimated in the kind of analysis developed here. Conversely, developing such an analysis can provide tools for philology.

Let us turn to the second segment of our text that disturbs the representation of a straightforward relationship between the text of the algorithm and the actions on the surface, that is, the introduction, in step (c), of the "cross-multiplying *(weicheng)*" operation. The key point here is that in the text of the procedure given after problem (1.3), the name of the "cross-multiplying *(weicheng)*" operation is inserted, but, in fact, there *it does not correspond to an event on the surface*. In other words, we have a verbal expression in the text which does not correspond to a concrete operation. There is no one-to-one correspondence between actions to be taken on the surface and items in the list of operations enumerated by the text. What then is the function of this expression in the text of the algorithm? If we observe the operations that follow the insertion of the term, "multiplying the 4 below on the left hand side by the 5 above on the right hand side yields 20; and multiplying the 3 above on the left hand side by the 7 below on the right hand side yields 21," we see that on the surface they involve two diagonal multiplications. These multiplications thus trace a "cross." The meaning of the expression "cross-multiplying" thus correlates to what happens on the surface in the computations that follow. The expression introduces the multiplications in question and highlights a feature of their material execution, thereby associating the operations one with the other. It is used in exactly the same way in the "detailed procedure" following problem (3.22).[30] However, in its two other occurrences in the same book, the same expression is employed in a different way, which nevertheless confirms what was just said: in these other occurrences, "cross-multiplying" is used to *prescribe* a subprocedure involving such a pair of multiplications. In conclusion, this set of documents thus shows how difficult it appears, at first sight, to determine what the user of a text did when s/he encountered the expression "cross-multiply:" did s/he observe the shape of the following computations on the surface, or did s/he bring a subprocedure into play? The way in which the text relates to the actions changes, depending on which text we observe.

"Cross-multiplying" is associated with another phenomenon that is essential for our purpose, since it reveals another way in which the relationship between the text of the procedure and the actions which correspond to it can be disrupted. In fact, in the last two contexts mentioned above, the sets of actions to which the expression "cross-multiply" corresponds differ. In other words, again within the same book, the "meaning" of the expression "cross-multiply"—if we formulate the "meaning"

---

[30] See Qian Baocong 錢寶琮 (1963, p. 392).

in terms of the related actions carried out—is not always the same and differs according to the context. Let us describe the phenomenon in greater detail. It can be shown that, in its second occurrence, in the text of procedure which follows problem (3.22), "cross-multiplying" refers *only* to such a single pair of multiplications.[31] By contrast, in the "detailed procedure" following problem (1.6)—translated below, in Sect. 9.2.1—, the operation of "cross-multiplying (*weicheng*)" can only be interpreted by understanding that the corresponding actions go beyond the two multiplications and include what, in the context of problem (1.3), covers the whole following subprocedure: "multiplying the 4 below on the left hand side by the 5 above on the right hand side yields 20; and multiplying the 3 above on the left hand side by the 7 below on the right hand side yields 21; summing these (the results) yields 41; the denominators being multiplied by one another yield 35." Here too, one may legitimately wonder how the practitioner uses the text and how s/he derives actions on the basis of the verb prescribing operations.

These facts call for several remarks.

First, let us concentrate on the second type of use of the term "cross-multiply," when it corresponds to the prescription of a subprocedure. With it, for the first time, we meet an example where the text of an algorithm makes use of a higher level operation. By this, I mean that the name of the "cross-multiplying (*weicheng*)" operation in the text corresponds, as far as actions are concerned, to bringing into play an autonomous procedure composed of several ordinary arithmetical operations. In this respect, it is similar to the case of "reducing fractions to the same denominator," evoked in footnote 13. Indeed, such a situation could be compared to the use of terms such as "multiplying", since there too, the term for the operation corresponds to a subprocedure. It is typical of the text of an algorithm to make use of such terms for higher level operations. However, the "granularity" of the procedures understood from the terms varies. This is illustrated here precisely by the difference between "multiplying" and "cross-multiplying." We have here a first example where we can evidence *two distinct ways of writing down a text for what corresponds to the same flow of computations* (one more detailed and the other more condensed). We see more examples below, and of a much more remarkable kind. As a consequence of the choice of formulation, the relationship between the text and the set of actions taken on the surface varies. Moreover, this example raises a key question: in which ways do the meanings of two texts corresponding to the same set of actions differ in relation to the differences in the formulations used? Here, the answer clearly depends on the meaning associated with the term "cross-multiplying." Regardless of the meaning of the list of operations, the specific expression of "cross-multiply" that was selected to refer to the subprocedure correlates to a material feature of the

---

[31] This conforms to the use of the operation "cross-multiplying" in *The Nine Chapters*, in exactly the same context, see Chemla and Guo (2004, p. 562). In that text too, there is a correlation between the fact that the text makes the positions explicit and the fact that it employs such expressions as "cross-multiplying."

execution of the list of operations—the physics of the computation, if you will.[32] The modalities in which such names are formed and the meanings they convey will prove below to vary in quite an interesting way. With this first set of remarks, we start inquiring into the way in which the directives, among the speech acts composing the text of a procedure, are performed: *what* the acts refer to, in terms of actions, and *how* they refer to it.

My second remark is inspired by the phenomenon, indicated above, that the actions to which "cross-multiplying" correspond differ according to the context. The corresponding subprocedures share the fact of involving a pair of multiplications, to which the name refers in a certain way. However, either the subprocedure contains only these multiplications (problem 3.22), or it goes further (problem 1.6). In the latter case, the name thus falls under the rubric of the synecdoche: the way in which part of the subprocedure develops materially on the surface provides the name for the whole subprocedure. In what follows, we encounter other modes for naming operations of a higher level, which condense subprocedures. In addition, we will see that the same phenomenon as that just observed recurs: depending on the context, the same verb in a text will correspond to either a whole subprocedure or part of it. This phenomenon brings us back to the issue of the competences that the reader, or the user, of the text must possess to make sense of "cross-multiplying." In fact, the competence appears to be two-fold. On the one hand, it includes knowing what "cross-multiplying" refers to. On the other hand, it also involves being able to choose, among the different subprocedures that the name embraces, which one is to be applied in a given context.

My final remark addresses the issue of how such competences were acquired. This issue invites us to return to the question we raised about the function of the first kind of use of "cross-multiplying." Let us recall that in the first cases mentioned, the expression "cross-multiplying" was used in texts, where it did not correspond to an action, but where it was followed by a list of operations, which it refers to precisely elsewhere. One may assume that one of the functions of such texts may have been to introduce the procedural meaning(s) of the term. The insertion of the name would then be a way of introducing the subprocedure that it condenses. If this is the case, we could interpret the sentence "one cross-multiplies" as carrying out a declarative as well as an assertive act: it would both name the procedure that follows and assert a property of the sequence of actions carried out. The language act would constitute the mode of introducing a higher-level operation. Such phenomena would account for *how* and *why* the relationship between the text and the corresponding flow of computations is no longer straightforward.

For the reader of the text to understand that "cross-multiplying" was *not* a prescription, among other prescriptions, but was rather *explained*—and was meant to be explained—by the list of operations following the term, it was probably essential

---

[32] In addition, when the text makes use of the expression "cross-multiplying", the term discloses a structure in the list of computations, in that two successive multiplications are shown to be linked and parallel to each other. See below the discussion on the property asserted by the introduction of the term "cross-multiplying."

that s/he be aware of the genre to which the text being read belonged—another kind of competence necessary for using the text.

This remark leads us now to widening our scope and saying a few words about the sections of the book from which I have extracted the pieces of text read so far. In the *Mathematical Classic by Zhang Qiujian*, all the texts for algorithms examined above were introduced by a term specifying their status: "detailed procedure (*cao*)." We have already indicated that these sections were written in the sixth century by Liu Xiaosun. Usually, in the *Mathematical Classic by Zhang Qiujian*, they come after another type of text dating from the second half of the fifth century and associated with the same flow of computations: a "procedure (*shu* 術)." In this book, in general, we therefore have two texts of algorithms corresponding systematically to the same flow of computations, a situation which provides us with quite a rich source material for our topic. It is interesting to note, for instance, that, with respect to problem 3.22, the use of "cross-multiplying" as the prescription of a subprocedure occurs within the "procedure", whereas the use of "cross-multiplying" not corresponding to an event on the surface, but followed by the list of operations meant, occurs in the "detailed procedure." The confrontation between the two texts in the list of operations following the latter occurrence of "cross-multiplying" allows the extent that it covers to be clearly determined. More generally, one may expect that secondary texts such as "detailed procedures" or "commentaries" would deliver explanations of that sort. [33] However, things are not so simple, as we now see.

In fact, for reasons that remain unexplained, the first six problems of the *Mathematical Classic by Zhang Qiujian*—so far, we have read the first three—are not followed by a "procedure", but only by "detailed procedures." After the first three problems, and again after the next three, the commentator, Li Chunfeng, immediately draws attention to the lacuna, and formulates a "procedure *shu*" for them. Note that, by contrast to the *Mathematical Classic by Zhang Qiujian*, in this context Li Chunfeng gives a single procedure for each group of three problems. Examining it will thus give us precious source material for the topic of this chapter. On the one hand, we will be in a position to observe how a text referred to as a "procedure" differs from a "detailed procedure"—that is, another way of composing a text for an algorithm—, when both deal with the same subject. On the other hand, in this case, Li Chunfeng reworks the three algorithms of the *Mathematical Classic by Zhang Qiujian* in relation to the intention of producing a single "procedure" to solve the general problem. It is therefore interesting for us to use this unique piece of evidence in order to observe how he operates this transformation and how the resulting text relates to the actions carried out on the surface in the various cases.

---

[33] Note that, more fundamentally, a knowledge of the textual operations involved in writing down texts of algorithms, in the culture in which the texts were composed, also appears to have been a competence required for using them.

9.1.4 Producing a Text for a Procedure that Covers all Possible Cases

In the *Mathematical Classic by Zhang Qiujian*, the topic of multiplying quantities having an integer, a fraction or a combination of both is dealt with through three problems, representing three cases, and the related "detailed procedures," which provide means for solving each problem, respectively. The original book probably also included three procedures, but they were not handed down. If this was the case, it may be assumed that, like everywhere else in the book, they corresponded to the kind of problem with which they were associated.

The texts of the "detailed procedures" analyzed above share certain common features, for instance in the way in which operations are prescribed in general, and their terms designated. Moreover, they also share a common structure. The main purpose of all these "detailed procedures" is to obtain a multiplicand and a multiplier (steps (a) and (e)). On this basis, a dividend and a divisor are computed (steps (f) and (g)), and the whole process is concluded by a division (step (h)).

By contrast, when Li Chunfeng reacts to the absence of a "procedure (*shu*)" attached to these problems, he describes a *single* procedure for the general problem of multiplying such quantities. In other words, he composes a single text that covers all possible cases. This, in fact, means a single text for a procedure solving problems (1.1) and (1.2), since he does not address the combined problems that constitute 1.3.[34] Let us translate the text, before analyzing its features:

"臣淳風等謹按：以前三條，雖有設問而無成術可憑。宜云，分母乘全內子，令相乘為實。分母相乘為法。若兩有分，母各乘其全內子，令相乘為實。分母相乘為法。實如法而得一。

Your subject (Li) Chunfeng et al. respectfully comment: Although with the three preceding clauses, there are problems displayed, there is no complete procedure on which one can rely. One should say: *the denominator multiplies the integers* and (the results) are incorporated into the numerators. One makes them *multiply by one another* to make the dividend. The denominators are *multiplied by one another* to make the divisor. If *both have parts*, the denominators, respectively, multiply the integer corresponding to them and (the result) is incorporated into the (corresponding) numerator. One makes them (the results) multiply by one another to make the dividend. The denominators are multiplied by one another to make the divisor. Dividing the dividend by the divisor yields the result." (My emphasis, Qian Baocong 錢寶琮 1963, p. 332)

The translation and interpretation we give require some explanation. Let us start with the second half of the text, which begins by the condition: "If *both have parts*..." "Parts" is the technical term designating fractions, which, in ancient China, were conceived of as composed of parts (2/3 being "two of three parts.") The condition implies that, as in problem (1.2), both quantities to be multiplied contain not

---

[34] The same holds true for division below.

only an integer, but also a fraction—whether the size of the parts may be the same or not is not clear, we come back to this question below.

As is explicitly indicated by the term "respectively," the first sentence following the condition (that is, "the denominators, respectively, multiply the integer corresponding to them and (the result) is incorporated into the (corresponding) numerator") corresponds to two actions, carried out successively but in parallel, on the two quantities to be multiplied. In other words, what, in the "detailed procedure" following problem (1.2), was two sublists of operations (steps (a) and (e)) has now become a single sentence. Here we encounter one of the means used, in this specific type of text, to make a single sentence correspond to different actions. We observe below other such techniques achieving the same goal.

The fact that (at least) two parallel series of actions correspond to this single sentence is confirmed by the next sentence, which prescribes taking the results of these operations to "multiply them by one another": This detail indicates that the computations corresponding to the previous sentence yielded more than one result. Such a use of "respectively" is quite common in the texts of procedures in *The Nine Chapters*. Its occurrence generally correlates with the fact that a sentence corresponds to at least two parallel computations.[35]

In this segment of the procedure, the relationship between the text and the actions taken differs from what we have seen above for the "detailed procedure," where each clause corresponded to one action or a single sequence of actions. Perhaps in correlation with this difference, Li Chunfeng's text no longer mentions numerical values, but only magnitudes.[36] Furthermore, as our text does not give any detail about how to place the numbers it may be assumed that the practitioner must have known the principles. We need to bear in mind that the procedure we are reading does not occur in a separate book, but in a commentary on the text we have already read. The issue of competences must be discussed in this new context. Let us thus illustrate this part of the "procedure" by the sequence of operations as I restore it on the surface, using letters instead of numerical values and placing operations corresponding to a single sentence in the same column (Table 9.4).

We can see that, although the text differs in its formulation and organization from the "detailed procedure" given after (1.2), the corresponding flows of computations on the surface are the same.

By contrast to the case dealt with in the second half of Li Chunfeng's text, it seems that the first part—that is, the three sentences following "the denominator

---

[35] Note that in the texts analyzed above, we indicated how the technical term "cross-multiplying" was used to refer to two related multiplications. Here, the association of computations, which are parallel and related by the fact that they are motivated by similar reasons, is carried out in a different way. In other texts, one can find rather long lists of operations corresponding to such distinct sequences of parallel computations.

[36] This last feature characterizes more generally another way of writing down a procedure that can be used for a class of problems. We stressed above the fact that the use of particular values in its text did not in and of itself impair the generality of a procedure. In the case examined here, such a formulation is essential to allow a sentence to correspond to two different and parallel computations.

Table 9.4 The computations for the general case, according to the second half of Li Chunfeng's procedure

Upper	Upper	a			
	Middle	b	$ac+b$		
	Lower	c	c	c	
Middle	Dividend			$(ac+b)(a'c'+b')$	$(ac+b)(a'c'+b')$
	Divisor				cc'
Lower		a'			
		b'	$a'c'+b'$	$a'c'+b'$	$a'c'+b'$
		c'	c'	c'	c'

multiplies the integers"—deals with the case when an integer (a') is multiplied by a quantity having both an integer and a fraction ($a+b/c$). This would correspond to problem (1.1), a hypothesis that seems to be confirmed by the text of the procedure. However, if we compare this segment of the text to the "detailed procedure" following problem 1.1, we see that the sequence of actions given to deal with such cases has been modified. We need to understand *how* and *why* this transformation was carried out. Let us examine this point in greater detail.

The second sentence of this first part of the text reads: "One makes them *multiply by one another* to make the dividend." Again, the "by one another" indicates that, as in the second part of the text examined above, the first sentence of the first part yields at least two results and thus corresponds to more than one computation. This assumption is confirmed by the third sentence of the first part of the text, which reads: "The denominators are *multiplied by one another* to make the divisor." How can we reconcile the fact that, in this case, there is only one denominator with these two features in the first part of the text? In my view, the only way to account for this point is to assume that the integer (a') is considered as being followed by a fraction, whose numerator is equal to 0 and whose denominator has any value (see footnote 56). For reasons linked to the grammar of classical Chinese (see footnote 37), I believe that the value chosen here is the same as the denominator of the other quantity (c). The computations to which the first part of the text corresponds, as I restore them, can therefore be illustrated as shown in Table 9.5.

According to this interpretation, the first sentence of the text does not correspond only to one, but in fact two multiplications by the single denominator c: this explains why I translated it as "*the denominator multiplies the integers* and (the results) are incorporated into the numerators".[37] Now, this interpretation allows us to make sense of the opposition that Li Chunfeng's text makes between the two cases:

---

[37] According to the grammar of classical Chinese, the terms "integer," "numerator" as well as the verb "incorporate" can be interpreted in the singular or in the plural. Note that such a choice between the two alternatives, and more generally the interpretation of the text, requires a competence from the user of the text. Again here, such competences are needed for a correct use of the "procedure." Theoretically, one could also understand the first sentence as follows: "*the denominators multiply the integers* and (the results) are incorporated into the numerators." However, if the denominators were different, I believe the sentence would instead be the following: "*the denominators respectively multiply the integers corresponding to them* and (the results) are incorporated into the numera-

Table 9.5 The actions, on a surface, corresponding to the first half of Li Chunfeng's procedure

Upper	Upper	a			
	Middle	b	$ac+b$		
	Lower	c	c	c	
Middle	Dividend			$(ac+b)(a'c)$	$(ac+b)(a'c)$
	Divisor				cc
Lower		a'			
		0	$a'c$	$a'c$	$a'c$
		c	c	c	c

the condition "If *both have parts*...," which introduces the second part of his text, probably refers to the case where no numerator vanishes and where, therefore, the denominators of both fractions are explicit.[38]

If, for cases like problem (1.1), we compare the "detailed procedure" and the first half of Li Chunfeng's text, we observe two layers of difference: Li Chunfeng has modified not only the *kind* of text used for the procedure, but also the *sequence* of actions to be executed to yield the solution. As a result, for these cases, the algorithm is now clearly much more complex than the one corresponding to the "detailed procedure," and it introduces computations that are not necessary. What is the benefit of this transformation? One consequence of this reformulation is clear: the solution for both types of problem now follows exactly the same sequence of actions. In contrast to the "detailed procedure" following (1.1), which exploited the particularity of the type of case to offer a simplified algorithm, the steps corresponding to the "procedure" described by Li Chunfeng are the same for (1.1) and (1.2). The "procedure" suggested is truly general, in the sense that the same uniform sequence of operations can be used for all cases.[39]

---

tors." Also note that the effect of the superfluous multiplication by c is cancelled by a superfluous division by c at the end of the sequence of computations.

[38] The reader has understood that I rely on all possible details of the text to restore the corresponding actions. This is the only possible method, if we are to interpret the text in the narrowest context possible. It must be assumed that the practitioner relied on another stock of knowledge to derive actions from the text s/he used to apply for any case encountered.

[39] In *The Nine Chapters,* the procedure given to solve this category of problems is called "Extending generality 大廣." It differs from the one formulated here by Li Chunfeng in two ways. The formulation is slightly different, in a fashion that is not essential for us here. Moreover, and more importantly, for all the problems, it contains *only* what corresponds to the second part of Li Chunfeng's "procedure" here. Li Chunfeng, who also commented on *The Nine Chapters*, offers an interpretation on the name of the operation. It turns out that this interpretation stresses the unicity of the procedure for all cases, and it is exactly along the same lines as what we suggest here, for the procedure he formulates, that we can understand his own comments on the generality of the procedure "Extending generality" (Chemla and Guo Shuchun 2004, pp. 172–173). More precisely, Li Chunfeng interprets the formulation of the procedure in *The Nine Chapters* as integrating, in a single text, procedures for solving all possible cases. As a result, if we follow Li Chunfeng's interpretation of *The Nine Chapters*, in the way in which the procedure for "Extending generality"

If we summarize what we have found thus far, we have seen two fundamentally distinct ways of covering all the different cases in which a general problem—multiplying quantities—can vary. Specific types of text correspond to each of these distinct ways. In the *Mathematical Classic by Zhang Qiujian*, there was a problem for each case, and each one was followed by a specific procedure, valid for the whole class of situations of the same category, despite the fact that the text was written with reference to a particular case. By contrast, Li Chunfeng described a general procedure, for which a single text covered all possible cases. In the *Mathematical Classic by Zhang Qiujian*, these distinct modes of writing down a text for an algorithm were distinguished through the contrasting pair of terms: "procedure *shu*" and "detailed procedure *cao*." This pair of terms is not used in the oldest known mathematical sources. In correlation with this fact, the only term used to designate a procedure, *shu*, refers to texts of the two types described. Analyzing the various texts for "procedures" in the most ancient sources from the perspective just outlined would certainly prove profitable. But the task exceeds the scope of this chapter.

To go back to the *Mathematical Classic by Zhang Qiujian* and its commentary, we have also seen that Li Chunfeng reworked the algorithm associated with one case, with the result that the list of operations solving all cases achieved a greater *uniformity*. The ensuing integration of all cases into a single procedure was made possible by making the solution of the first problem more complex. In other words, there is a correlation between describing a single procedure integrating all cases and solving cases like (1.1) in a more convoluted way.

So far, we have analyzed each case and its corresponding section in the text of Li Chunfeng's procedure separately. If we now consider the text of the procedure as a whole, we discover that its structure is unexpected and calls for our attention. Instead of giving a single list of operations covering all cases, as in the corresponding text for the same operation in *The Nine Chapters*, Li Chunfeng's text starts by dealing with a particular case (exemplified by problem (1.1)), for which it offers a list of operations; then introduces a condition ("if both have parts"), followed by a second list of operations. The text thus has the structure of an enumeration, with two items. In fact, the nature of the first item is not made clear, until it can be deduced by means of a contrast with the second item. It is by comparison with the second item, and the condition introducing it, that the first item appears to refer to the simpler case. The text of the second item begins by making clear which case the corresponding procedure applies to.

---

is formulated, the first part of the text discussed here is understood as being subsumed under the second part. In other words, the first part is contained in the very formulation of the second part: this is another way of achieving uniformity. This remark hence confirms our interpretation of the procedure Li Chunfeng formulates in his commentary on the *Mathematical Classic by Zhang Qiujian*: in this procedure, the first part is reshaped in such a way that now a uniform procedure solves all cases. In this commentary, Li Chunfeng has a reason to keep the first and second parts separate, which still has to be made clear. Note also that this type of generality is to be distinguished from the fact that, in contrast to the "detailed procedures," Li Chunfeng's "procedure" is described in abstract terms. The interest for uniformity in the treatment of the various cases in ancient China can also be captured in the way in which the commentator Liu Hui deals with problem 6.18 in *The Nine Chapters*, see Chemla (2003).

How can we infer how the practitioner handled such a text, when addressing a given problem? It seems necessary for the reader to start reading the text in the middle, where the condition was. There, s/he would encounter two options: either the problem being dealt with conformed to the condition, in which case the operations to be used were those that followed the condition; or the problem did not conform, and the actions to be executed were those corresponding to the first half of the text, taken from the beginning. Accordingly, to account for how the text was used, we need to introduce the idea that the reader *circulated* in the text according to an acquired convention. This is not surprising, since present-day readers also carry out a specific circulation technique for using texts designed for algorithms. However, the types of circulation are not always the same, which reveals that such features depend on the scholarly culture in the context of which these texts were produced and used. These ways of translating the text into actions belong to what may be called the uptake of the instructional text.[40] Moreover, the circulation in such a text sharply differed from what has been described so far: the sequence of operations appearing in Li Chunfeng's text no longer corresponded to the sequence of actions to be carried out, in a straightforward relation like the one analyzed above. Within a given scholarly culture, various types of texts for algorithms may have existed, some presenting unexpected features. We find ourselves taken further and further away from the common assumptions regarding texts of algorithms. In addition, as has already been emphasized, in each section of Li Chunfeng's text, some clauses correspond to more than one sequence of operations executed in parallel. We see that for at least two reasons, the way of translating the text into actions required a specific treatment of the clauses read, one that differed from what has been described so far.

Such texts may look strange, especially when we know that, in *The Nine Chapters*, a simpler text was available for writing down the same general list of operations. Yet, as we see below, texts for algorithms with a similar structure will prove to be frequent in our sources.

This said, the text by Li Chunfeng examined here nevertheless remains specific, because the first and simpler case is dealt with through the same procedure as the later case. Note, furthermore, that the final operation common to both cases, "One divides the dividend by the divisor," is placed at the end of the text, and concludes the flows of computation of the two types. Even though the cases were somehow dissolved through the shaping of a single list of actions for all,[41] the text still distinguishes cases. All these features would remain difficult to account for, if we could not suggest the hypothesis that the composition of the text was modeled on a pattern of enumeration that was common for algorithms encompassing different cases. The remaining part of this chapter establishes this fact beyond doubt and shows that such a way of writing down algorithms can be evidenced in sources composed

---

[40] For the notion of uptake in relation to texts as well as the pragmatics of instructions, see J. Virbel, "Speech Act Theory and Instructional Texts," this volume, Sects. 2.2 and 2.3.

[41] The rewriting or transformation of procedures with the intention of dissolving cases can be evidenced elsewhere in the earliest extant sources from China, see Chemla (2003).

on a time span covering more that nine centuries. To start with, we show that, in fact, after the three problems dealing with division in the *Mathematical Classic by Zhang Qiujian*, Li Chunfeng introduces a single "procedure" of a similar type, which shares the same pattern, but not the same mode of integrating the cases. Let us now turn to examining the contrast between these two types of general algorithms. This will give us opportunities to evidence other modes of grouping distinct computations into operations of a higher level.

9.2 Dividing: A Contrast Between Two Texts for a General Procedure

The three problems and "detailed procedures" through which the *Mathematical Classic by Zhang Qiujian* deals with the division of integers and fractions by quantities of the same type, are similar, in most of their features, to what has been described above. Our main objectives in Sect. 9.2 are to confront the "detailed procedures" of the *Mathematical Classic by Zhang Qiujian* and the "procedure *shu*" described by Li Chunfeng and, also, to confront this second procedure and Li Chunfeng's procedure as analyzed above.

9.2.1 The Procedures to be Later Subsumed in a Unique Text

To achieve this objective, let us first translate the problems and detailed procedures and illustrate the moves on the surface that can be restored as corresponding to the text:

(1.4) "以十二除二百五十六、九分之八。問得幾何。

答曰:二十一、二十七分之十一。

草曰:置二百五十六, 以分母九乘之, 内子八, 得二千三百一十二為實。又置除數十二, 以九乘之, 得一百八為法。除實得二十一。法與餘俱半之, 得二十七分之十一。合所問。

One divides 256 and 8/9 by 12. One asks how much this yields.

Answer: 21 and 11/27

Detailed procedure:

(a) One puts 256,

(c) one multiplies this by the denominator, 9, and incorporates (the result) into the numerator, 8, hence yielding 2312, which makes the dividend.

(d') Again, one puts the quantity that divides, 12, one multiplies by 9, thus yielding 108, which makes the divisor.

(e) Dividing (by it) the dividend yields 21. The divisor and the remainder are both halved, thus yielding 11/27, which conforms to what was asked. (Table 9.6)"[42]

---

[42] Note that the halving is carried out twice to yield the result stated.

Table 9.6 Dividing a quantity by an integer. The computations on the surface, as restored for Problem (1.4)

Upper					21	21
Middle dividend	256					
	8	2312	2312	2312	44	11
	9	9	9			
Lower divisor			12	108	108	27

Except for some minor details, this sequence of computations corresponds to the one derived from the texts of the algorithm the *Book of Mathematical Procedures* gives for carrying out the same operation—one of them being entitled "Directly sharing 徑分."[43] The formulation of the text of procedure placed after problem (1.4), however, differs from "Directly sharing 徑分": here again, different texts correspond to the same sequence of actions. Such documents set us the task of determining criteria that can allow us to capture the system of differences between them. As far as the formulation is concerned, the "detailed procedure" following (1.4) is more closely related to the procedure proposed by Liu Hui in his commentary on the algorithm that *The Nine Chapters* contains under the same name of "Directly sharing." In fact, after having explained the text of the procedure "Directly sharing"—an explanation which we come back to below—, Liu Hui offers another procedure as an alternative to that in *The Nine Chapters*. On the one hand, the "detailed procedure" placed after problem (1.4) in the *Mathematical Classic by Zhang Qiujian* strongly evokes the formulation adopted by Liu Hui. On the other hand, it differs from Liu Hui's procedure because its sequence of operations has been simplified to deal with the particular set of cases represented by problem (1.4).

By contrast, the following "detailed procedure" given in the *Mathematical Classic by Zhang Qiujian* for the category of problems represented by (1.5), reproduces Liu Hui's procedure more faithfully, even though, in contrast to Liu Hui's text, it is formulated with respect to a paradigm.[44] It reads as follows:

(1.5) "以二十七、五分之三除一千七百六十八、七分之四。問得幾何。
答曰：六十四、四百八十三分之三十八。
草曰：置一千七百六十八，以分母七乘之，內子四，得一萬二千三百八十。又以除分母五乘之，得六萬一千九百為實。又置除數二十七，以分母五乘之，內子三，得一百三十八。又以分母七乘之，得九百六十六為法。除之，得六十四。法與餘各折半，得四百八十三分之三十八。得合所問。

---

[43] More precisely, in the *Book of Mathematical Procedures* the first segment of the procedure—the transformation of the dividend composed of an integer and fraction(s)—is referred to as a change of unit, from which multiplications derive, see Peng Hao 彭浩 (2001, pp. 45–46, 48).

[44] This opposition between Liu Hui's "procedure" and this "detailed procedure" is the same as the one sketched in Sect. 9.1.4 between the formulation in the "detailed procedure" and in Li Chunfeng's "procedure". On Liu Hui's procedure, see Chemla and Guo (2004, pp. 168–169). Its text also has the structure of an enumeration comparable to the one described in Sect. 9.1.4.

One divides 1768 and 4/7 by 27 and 3/5. One asks how much this yields. Answer: 64 and 38/483.

Detailed procedure:

(**a**) One puts 1768,

(**c**) one multiplies this by the denominator, 7, and incorporates into the numerator, 4, hence yielding 12,380.

(**c'**) Again, one multiplies this by the denominator of what divides, 5, thus yielding 61,900, which makes the dividend.

(**d**) Again, one puts the quantity that divides, 27, one multiplies this by the denominator 5, and incorporates (the result) into the numerator, 3, hence yielding 138.

(**d'**) Again, one multiplies this by the denominator 7, yielding 966, which makes the divisor.

(**e**) Dividing by this yields 64. The divisor and the remainder are each halved, hence yielding 38/483. What is yielded conforms to what was asked."

In fact, this "detailed procedure" conforms to the second—and most general— clause in the algorithm for division suggested by Liu Hui in his commentary following problem 1.18 in *The Nine Chapters*. This clause is given for the same class of cases as problem (1.5) and can be represented by the following sequence of transformations: [45]

$$\frac{a + \dfrac{b}{c}}{a' + \dfrac{b'}{c'}} = \frac{\dfrac{ac + b}{c}}{a' + \dfrac{b'}{c'}} = \frac{\dfrac{(ac + b).c'}{c}}{a'c' + b'} = \frac{(ac + b).c'}{(a'c' + b').c}$$

The computations on the surface corresponding to the "detailed procedure" following problem (1.5), translated above, can be restored as in Table 9.7.

Finally, the last problem of the opening section of the *Mathematical Classic by Zhang Qiujian*—problem (1.6)—illustrates the case when the dividend does not contain only one, but in fact two fractions. The "detailed procedure" provided after the statement of the problem embeds the procedure for adding fractions, as a subprocedure, within the previous procedure—the one given after problem (1.5)—, exactly as was done for the "detailed procedure" following problem (1.3). The same analysis holds, as can be checked in the translation of the following text:

(**1.6**) "以五十八、二分之一除六千五百八十七、三分之二、四分之三。問得幾何。

答曰：一百一十二、七百二分之四百三十七。

草曰：置六千五百八十七於上。又別置三分於下右，之二於左。又置四分於三下，之三於左。維乘之，分母得十二，子得一十七。以分母除

---

[45] The first clause of Liu Hui's procedure deals with the case when a quantity of the type *a/b* is to be divided by a quantity of the same kind, *c/d*. It presents common features with another procedure in the *Book of Mathematical Procedures* for a similar kind of division: "Detaching the length 啟從". See Peng Hao 彭浩 (2001, pp. 113–115). I will return in a future publication to these various algorithms. Let us stress only that different authors distinguish cases in different ways and thereby shape different general procedures. Moreover, in correlation to the fact that, in his second clause, Liu Hui groups parallel operations in a single sentence, the order in which the computations are to be executed is modified.

Table 9.7 Dividing integers and fractions by one another: the problem (1.5)

Upper							64	64
Middle dividend	1768							
	4	12,380	61,900	61,900	61,900		76	38
	7	7	7	7				
Lower divisor			27					
			3	138	966		966	483
			5	5				

子得一，餘五。加一上位得六千五百八十八。以分母十二乘之，內子
五，得七萬九千六十一。又以除數分母二因之，得一十五萬八千一百
二十二。又置除數五十八於下。以二因之，內子一，得一百一十七。
又以乘數分母十二乘之，得一千四百四為法。以除實得一百一十二。
法與餘俱半之，得七百二分之四百三十七。

One divides 6587 and 2/3 and 3/4 by 58 and 1/2. One asks how much this yields.

Answer: 112 and 437/702.

Detailed procedure:

(**a**) One puts 6587 above.[46]

(**b**) Again, furthermore, one puts the 3 parts on the right hand side below, the 2 of them on the left hand side. Again, one puts the 4 parts below the 3, and the 3 of them to the left hand side. Cross-multiplying them yields 12 for the denominator, and 17 for the numerator.[47] Dividing the numerator by the denominator yields 1 remainder 5.[48] One adds 1 to the position above, thus yielding 6588.

(**c**) One multiplies this by the denominator 12, and incorporates (the result) into the numerator, thus yielding 79,061.

(**c'**) Again, one multiplies this by the denominator of the number that divides, 2, thus yielding 158,122.

(**d**) Again, one puts the number that divides 58 below. One multiplies it by 2 and incorporates into the numerator 1, thus yielding 117.

(**d'**) Again, one multiplies this by the denominator which is the quantity (yielded by) the multiplication,[49] 12, hence yielding 1404, which makes the divisor.

---

[46] Probably, this means the upper line in the middle zone of the surface, where the dividend usually lies. In terms of positioning, we follow the same principle as for multiplication, which we derived from the *Mathematical Classic by Master Sun*.

[47] The term "cross-multiplying" in the text corresponds to the performance of a sequence of arithmetical operations on the surface. This occurrence of the term makes clear what the subprocedure includes here. The assertion of the result appears to provide a way of determining the operations covered by the subprocedure.

[48] The way of embedding the algorithm for the addition within the algorithm for division is exactly the same here as that described above for the "detailed procedure" following problem (1.3).

[49] Or "which is the quantity that multiplies." One may think that the text is corrupted here and originally contained, instead of "乘數," "除數," the whole expression hence meaning the denominator "of the quantity that divides." However, one may also remark that this editorial problem occurs

(e) Dividing the dividend by this yields 112. The divisor and remainder are both halved, thus yielding 437/702."

The algorithm for dividing, onto which an algorithm for adding fractions is grafted, is the same as the one described after problem (1.5). If we except the use of the operation of "cross-multiplying," the texts of the three "detailed procedures" just examined are in most aspects similar to those dealing with multiplication. The reader can check, on the translation, that the use of the term "cross-multiplying" fits with what has been said above in the analysis developed in Sect. 9.1.3.

9.2.2 Comparing Texts for Procedures that Cover all Possible Cases

Here again, as after problem (1.3), Li Chunfeng notices that no "procedure *shu*" is provided in the *Mathematical Classic by Zhang Qiujian* for the three problems dealing with division. Consequently, instead of providing a "procedure" for each of them, he also sets out to formulate a single one, through which he reshapes the flows of computation solving the various cases at the same time as he composes another type of text. This text is particularly rich for the insights it yields on the issues dealt with in this chapter. We will examine it in some detail, pursuing in our analysis of his text the four following related objectives.

First, we want to analyze the textual techniques used to compose a text for an algorithm, from which distinct sequences of actions can be derived, depending on the case to which it is applied. Secondly, we need to examine how this procedure relates to the "detailed procedures" given in the *Mathematical Classic by Zhang Qiujian* and for which ends Li Chunfeng reshaped them, when he did. In particular, how do the actions on the surface to which the procedure refers relate to those corresponding to the "detailed procedures"? Thirdly, we must compare the text formulated to the similar one Li Chunfeng composed for multiplication—analyzed in Sect. 9.1.4 above. What is the structure of the "procedure" and how does it compare with that of the previous general procedure formulated by Li Chunfeng? Lastly, as we will see, this text exemplifies a new way of carrying out directives when prescribing operations. We will describe this type of act, which will prove essential to analyzing similar ways of prescribing operations in other texts for algorithms composed in ancient China. Before we set out to develop our analysis, let us read the commentary in which Li Chunfeng presents this "procedure *shu*":

"臣淳風等謹按：此術以前三條亦有問而無術。宜云置所有之數，通分内子為實。置所除之數，以分母[50]乘之為法。實如法得一。若法實俱有分及重有分者，同而通之。

---

precisely at the same place, for the text of division as the one discussed above, in footnote 23, for multiplication. The two problems may therefore have to receive a joint solution.

[50] The ancient sources have here "三分". I follow Qian Baocong, when he suggests that the text is corrupted and should be restored as "分母."

Your subject (Li) Chunfeng et al. respectfully comment: As for the procedures here, with the three preceding clauses, similarly there are problems, but there is no procedure. One should say: one puts the quantity of *what one has, one makes the parts communicate* and incorporates (the result) into the numerator, which makes the dividend. One puts the quantity which one eliminates (i.e., by which one divides), one multiplies this by the denominator, which makes the divisor. Dividing the dividend by the divisor yields the result. If the divisor and the dividend both have parts and there are repeated parts,[51] one *equalizes them and makes them communicate.*" (My emphasis)

---

[51] It seems to me that the Chinese sentence can be understood here in two ways. Either it lists two types of conditions for which the procedure that follows is to be used: the first condition would correspond to the case when the dividend and the divisor both contain a fraction ($a+b/c$ and $a'+b'/c'$, respectively), the second condition would correspond to the case where there is a sequence of different types of parts attached to either of the two terms of the division ($a+b/c+b'/c'$). To express this option, one could emphasize the translation of the conjunction as follows "If the divisor and the dividend both have parts and also in case there are repeated parts." Or—second option—the text defines here the cases for which the second procedure should be applied by the combination of two conditions: the fact that the dividend and the divisor both have fractions and, furthermore, these fractions have different denominators. It is then translated as follows: "If the divisor and the dividend both have parts and there are repeated parts (i.e., with different denominators)." I opted for the first interpretation in Chemla (1992). The comparison with the first procedure by Li Chunfeng now incites me to opt for the second option. The benefit would be twofold. First, in contrast to what I suggested in my previous article, we would not be compelled to introduce another interpretation for the meaning of the technical expression *chong you fen*, which would here, as in *The Nine Chapters*, consistently be understood as "there are different types of parts." Secondly, the impact on the interpretation of the first part of the procedure would be interesting. We suppose that, as above (Sect. 9.1.4), the cases dealt with in the first part of the text of the procedure are defined by contrast with the condition that marks the beginning of the second part of the text. If we choose the second option, the cases covered by the first part of the procedure would be either of the following: either only one of the dividend and the divisor has a fraction, or they both have fractions, but with the same denominator. Contrary to what the first option would entail, as we will see below, the interpretation based on the second option would make the procedure described here consistent: the second case would in any event be reduced to the first case. Moreover, the computations would be more straightforward. The only problem raised by this interpretation is that it would make the procedure different from that given in *The Nine Chapters* after a sequence of problems similar to the three problems in the *Mathematical Classic by Zhang Qiujian*. Let us explain why this constitutes a difficulty, by reminding the reader of the situation of our sources (see Appendix): *The Nine Chapters* is a key classical text, which Li Chunfeng et al. put together with other classics, including the *Mathematical Classic by Zhang Qiujian*, when they assembled and annotated the collection of the *Ten Mathematical Classics*. Why then would Li Chunfeng give a procedure in his commentary on the *Mathematical Classic by Zhang Qiujian* comparable to, but different from, the one given in *The Nine Chapters*? The difference could be the following: in the procedure in *The Nine Chapters*, according to my interpretation (see below), the corresponding step (last case) deals with either the case when the dividend and the divisor have different fractions or one of them has two fractions with different denominators. This would correspond to the first option in the interpretation of Li Chunfeng's procedure, which, for the reasons explained above, I would rather discard in favor of the second option. Since the first procedure given by Li Chunfeng after problem (1.3) and the formulation of this procedure both suggest that the commentator relies on *The Nine Chapters*, this difference would require explanation. One possible solution for this puzzle would be to interpret that neither for multiplication, nor for division, does Li Chunfeng's "procedure" take into account cases when two fractions with different denominators follow each

Like the "procedure" for multiplication, discussed in Sect. 9.1.4, Li Chunfeng's "procedure" for division calls for several remarks. We concentrate on those that add to what has been said in the previous section.

9.2.2.1 Carrying out Directives

Let us start with a range of remarks bearing on the formulation of the text for the algorithm and, particularly, the way in which directives are carried out, by contrast to the "detailed procedures." For this, we concentrate on the step by which the text begins: "one puts the quantity of *what one has, one makes the parts communicate* and incorporates (the result) into the numerator, which makes the dividend." Regarding the terms of the operation, we note the locution "what one has," which refers, in an abstract way not evidenced before, to the quantity to be divided. The same term is also found in the text of some procedures in *The Nine Chapters*.

What is more fundamental for our purpose here is the *way in which* the speech act is carried out in the directive: "one makes the parts communicate." The action to be taken—the illocutionary point, in the terms introduced by J. Virbel ("Speech Act Theory and Instructional Texts," this volume, Sect. 2.2)— is the same as what, in the "detailed procedure" placed after problem (1.4), was indicated by the multiplication in the following passage: "One puts 256, one multiplies this by the denominator 9...." In other words, what, in the latter statement, is prescribed through the name of an arithmetical operation is, in the text of the "procedure," formulated in an indirect way. The main question for us here is to understand how this is done precisely, that is, with the terms introduced by J. Virbel, "how the illocutionary point is achieved."

The term "make communicate" occurs on several occasions in *The Nine Chapters*, and we can rely on the commentaries to capture what is at issue in its use.[52] Note here a point of method that will prove essential in what follows: For some early Chinese sources for which ancient commentaries have been handed down, we can rely on the commentators to determine how a reader in the past (H, in J. Virbel's notations) understood a given language act that s/he encountered. In the case of "making communicate," in relation to fractions, if we follow Liu Hui's explanation, we understand that when two fractions have the same denominator, they communicate. Such a state implies that, when viewing fractions as "parts"—what I call their material meaning—, these fractions are expressed with "parts" of the same size and can hence enter together in arithmetical operations. The commentator Liu Hui introduced the term, in the context of his proof that the procedure given in *The Nine Chapters* for adding fractions was correct. There, Liu Hui showed that "mak-

---

other. In this sense, this also holds true for the alternative procedure offered by Liu Hui in his commentary on *The Nine Chapters*. We come back to the comparison between the procedure in *The Nine Chapters* and Li Chunfeng's below.

[52] On this term, the reader is referred to my glossary of terms occurring in the oldest Chinese mathematical texts in Chemla and Guo (2004, pp. 994–998).

ing communicate" expressed the aim that the algorithm had to pursue, i.e., enabling the fractions which were to be added and which were expressed with parts having different sizes, to communicate, by refining each part in such a way that all the parts share the same size and the fractions thereby "communicate." In a following step of this argument, Liu Hui showed how the algorithm actually fulfilled this task. By designating this goal with the term "making communicate," Liu Hui did not capture it in a way that made sense only within the context of adding fractions. In fact, in other parts of his commentary, he also showed how some other procedures had the aim of "making" some other types of entities "communicate." As a consequence, using the term in the context of fractions does not only express the "intention of the algorithm" on a material level, but it also signals that the main operations at play in the procedure for adding fractions fall under a much more general rubric of transformation carried out by algorithms, through the action of which entities enter into communication.[53] In other terms, by the mere use of the term "making communicate," Liu Hui's commentary highlighted that, seen in a formal way, the transformation could be compared to other transformations occurring within other mathematical contexts, and in fact beyond. We go back to this issue below. We only need to stress here that the term captures the effect of the operations on parts, both in the specific context of fractions and more generally, in a formal way.

In *The Nine Chapters* and its commentaries, the term "making communicate" is also used in relation to another operation within the context of fractions, one that bears on an integer followed by a fraction. This latter use is of great importance for our discussion here. In his commentary in which he sketches a proof of the correctness of the procedure "Small width" of *The Nine Chapters*,[54] Li Chunfeng makes the "meaning" of a step explicit by using this term. He writes about the operation that, in a quantity of the type $a + b/c$, transforms a into ac: "*The reason why* one multiplies the integral (numbers) of *bu* by the denominators[55] is that *one makes* the corresponding parts *communicate*. 以分母乘全步者, 通其分也" (My emphasis). We can interpret the commentary along the same lines as we did above. As in the previous context, the term "making communicate" captures the intention of the operation prescribed (computing ac), or, in other terms, interprets what it achieves in this context. The term indicates the transformation carried out by the operation, when one considers the transformation of the integer from a material point of view. What happens when the units of an integer a are split in such a way, is that the resulting

---

[53] In each context in which the term is used, "making communicate" has a meaning specific to the context per se—here, making parts share the same size—as well as a formal meaning—"making" entities "communicate." The term chosen to refer to the operation in our specific context thus indicates the formal meaning. More generally, in the same context, other terms function in the same way. See Chemla (1997b). We will have to come back to this issue from two different perspectives below, in Sect. 9.3. In particular, the second, formal meaning will be best grasped when compared to another way of referring to the same operation.

[54] See Chemla and Guo (2004, pp. 342–343).

[55] Here again, in fact, in the context within which this sentence occurs, the term can be understood as a singular or a plural, see footnote 4 of the translation into French, p. 798. Since it is not a central issue for us here, I also refer the reader to it for a discussion of the other possible version of the text.

pieces share the size of the parts expressing the fraction *b/c*: their parts are made to communicate. As a result of this transformation, now seen from a numerical point of view, the integer, expressed in this way, can be added to the numerator. As above, interpreting the transformation as "making communicate" refers both to a specific goal—the "material intention," in relation to units that are disaggregated—, and to a more general goal that is formally at stake in so doing—the "formal intention."[56]

This part of Li Chunfeng's commentary is essential for our purpose. This quotation gives us evidence to analyze more precisely what is at stake when Li Chunfeng writes in the "procedure" inserted in the commentary on the *Mathematical Classic by Zhang Qiujian*: "one makes the parts communicate," in place of the operation that, in the "detailed procedure," is formulated as: "One multiplies this (that is, 256) by the denominator 9...." We see that Li Chunfeng indicates the operation to be carried out—the illocutionary point of the directive—by means of what, in another context, that of proving the correctness of a procedure, was used to formulate an interpretation of the effect of the intended multiplication. In the terms used by J. Virbel, Li Chunfeng carries out the directive, by "asserting ...the state of affairs in the world that S (here the commentator, KC) wants H (here, the user of the text of the procedure, KC) to ... achieve" (J. Virbel, "Speech Act Theory and Instructional Texts," this volume, Sect. 2.3.2.3). As we see in the following section, in *The Nine Chapters* the term "make communicate" is also used in the same way.

What we now need to understand better is the way in which this kind of directive is performed here, since we show below that this can be done in various ways. The operation of multiplication (*ac*) is designated by a verb that captures its aim within the context in which it is used—"making the parts" of the integer "communicate" with those of the fraction. What is essential in this case, as will appear later, is that this aim is grasped in a formal way. As a consequence, the way of achieving the directive indicates the formal intention for using the multiplication and therefore the reasons why, from a mathematical point of view, the operation is correct.

To summarize what we have observed so far on this aspect, whether we compare this speech act in Li Chunfeng's "procedure," to the corresponding one in the "detailed procedure," or to any of those in Li Chunfeng's first general "procedure," we see that we have identified two different ways of carrying out a directive: some verbs refer to an arithmetical operation by means of a term designating the operation in question, whereas others refer to the operation by asserting the world situation to be achieved and thereby indicating the reasons for the correctness of the procedure. As we will soon see, the latter phenomenon is in fact quite general in our sources. Indirect ways of prescribing operations make use of various ways of

---

[56] We could be tempted to see no difference between adding an integer and a fraction, on the one hand, and adding two fractions, on the other. To understand the difference better, one should recall that in ancient China, an integer was most probably not understood as the numerator of a fraction, whose denominator is 1. Instead, the general type of rational number had the form of $a+b/c$ and, within this context, an integer was approached as $a+0/c$. Seen from this perspective, interpreting the operation on the integer by the same term of "making communicate" as the one used, in the previous context, to interpret the operations on the numerators of two fractions establishes a relationship between the two. The use of "making communicate" stresses that, in the motivations, if not in the computations, the two transformations have something formal in common.

formulating the effect that the operations have in the context in which they are used and hence the reasons why they are relevant. It is clear that a competence is required to understand such speech acts. Interestingly enough, this competence is linked to knowing standard ways of interpreting the meaning of the operations.[57]

In fact, we met with a similar phenomenon earlier, but so far, for want of the relevant tools, I left it without commenting on it (see footnote 11). On the basis of the analysis just developed, we can now go back to this phenomenon. In the "detailed procedure" following problem (1.1), in step (h) (p. 321), one could read: "If, *in return*, one divides by the denominator, 5,..." (the emphasis is mine). Here the operation of division is indicated by an expression meaning "to divide." However, a particle, "in return," qualifies it, in a way that is essential for us. Like "making communicate," the particle indicates the motivation why, in this context, the division must be carried out: it cancels the effect of the inverse multiplication, which was prescribed upstream. However, even though the two passages have in common the fact that they refer to the motivations to carry out the operations, they also present differences in this respect. "In return" evokes the motivation in a different way, since it does not assert the world situation to achieve. Rather, it relates the operation to another one, executed earlier, making the point that the *reason* to use the division is to cancel the effects of that operation. Accordingly, the character "in return" does not *refer* to the operation, but rather *qualifies* the term designating the operation. Both ways of prescribing operations with a reference to their intention are regularly attested in our sources.[58] Here, we focus on examples of the first type, illustrated by "making communicate," since our corpus evidences other ways in which language acts are carried out in a similar way.

We encounter our second example of the same kind, if we widen our scope and now read Li Chunfeng's second general procedure globally. As in the first "procedure," examined in Sect. 9.1.4, after a preliminary sequence of operations, Li Chunfeng inserts a condition in the middle of the text: "If the divisor and the dividend both have parts and also in case there are repeated parts." For the latter cases, the list of operations following the condition is to be used. This list reads: "one equalizes them and makes them communicate." The text of the general procedure hence also has the structure of an enumeration, with two items—we analyze its structure later. For now, let us keep our focus on the terms for operations that it involves. These terms call for several remarks.

---

[57] This is the second kind of indication we have that evidences that readers formulated an interpretation of the effect of the operations at the same time as they turned the sentences into computations—or, to formulate it from the point of view of the authors of procedures: this is the second hint we have that the authors interpreted the meaning of the operations as they wrote them down. If we recall, in footnote 29 above, we stressed that the fact that the text designates "35" as a "denominator" implied that the meaning of the value had already been interpreted. Similarly, here, the fact of designating the multiplication by "making communicate" implies that the intention of the operation achieved is present in the mind of the author of the procedure and must be present in the mind of the practitioner using it.

[58] See "to divide in return 報除," "to divide together 并 除," "to divide at a stroke 連 除," in my glossary, in *Les Neuf Chapitres*.

Here again, clearly a competence is required to interpret the operations meant. In fact, this list of operations quotes part of the procedure given for exactly the same general problem in *The Nine Chapters*. We examine its interpretation in greater detail in the next section, since in this other context, we can rely on Liu Hui's commentary to acquire the necessary competence. This will put us in a better position to examine the relationship between these two general procedures given for the same operation of division. However, without entering here into further detail, we see that the list of operations uses "making communicate" together with another operation, "equalizing." It turns out that the language act achieved here, by the use of this term, is of exactly the same type and can be approached in exactly the same terms as we developed above. The term "equalize" refers to operations to be executed by stating the world situation to be achieved, to keep using J. Virbel's terms mentioned above. This situation to achieve is captured both in its material dimension—equalizing denominators—and in its formal features—equalizing entities. Again, more on this later.

This second occurrence of "making communicate" bears witness to another range of phenomenon and thus requires further analysis. To state it bluntly, although the term "making communicate" recurs in this context, the operations to which the term refers in this second occurrence go beyond a simple multiplication, as was the case with the previous use of "making communicate" analyzed above. In this second occurrence, in fact, the term encapsulates a larger sequence of operations. If we represent the effect of the operations in modern terms, as will be shown in Sect. 9.3, starting from a situation that can be represented[59] as:

$$\frac{a + \dfrac{b}{c} + \dfrac{b'}{c'}}{a' + \dfrac{b''}{c''}}$$

"equalizing" leads to

$$\frac{a + \dfrac{bc'c''}{cc'c''} + \dfrac{b'cc''}{c'cc''}}{a' + \dfrac{b''cc'}{c''cc'}}$$

while "making communicate" encapsulates the operations that lead to the final division

$$\frac{acc'c'' + bc'c'' + b'cc''}{a'cc'c'' + b''cc'}$$

---

[59] Whether, as in *The Nine Chapters*, the algorithm covers the case when there can be a sum of two fractions or not (as we explained in footnote 51) is a matter of interpretation. We leave this question aside here. Deciding over this issue does not affect my description of *how* the term "making communicate" is used. One could develop the same argument on "equalizing."

As a consequence, we see that with the term "make communicate," we are again confronted with some of the phenomena described above with respect to "cross-multiplying." However, the two cases also present some differences. Let us compare these two cases.

First, like "cross-multiplying," depending on the context, the same term "making communicate" can refer to a shorter or a longer list of operations. In both cases in which "making communicate" occurs in our example, the list of operations meant includes the multiplication of an integer by a denominator that makes the integer share the same constitutive parts as the fraction involved. It is this core operation that, strictly speaking, carries out making entities communicate. However, in one occurrence, the expression "making the parts communicate" only refers to this single multiplication. In the other, "making them communicate" includes the transformation of several integers as well as the addition of the integers transformed into the corresponding fractions. In the latter case, the expression referring to the operations also falls under the rubric of the synecdoche. It also illustrates a case in which we do not have a one-to-one correspondence between the events in the world—the surface—and the verbs listed in the enumeration that the text carries out. More generally, as will be seen in Sect. 9.3, the operations meant by "making communicate" vary. Whatever the case, the operations brought together under the verb of "make communicate" are associated by the fact that their combination achieves the world situation asserted by the name of the subprocedure.

This leads me to my second set of remarks on this issue. In contrast to "cross-multiplying," the name "making communicate" does not derive from how the computations are executed on the surface, but from an interpretation of their effects on the entities, to the components of which the transformations are applied. Also in contrast to "cross-multiplying," "making communicate" sometimes rewrites a single operation, "multiplying," and sometimes encapsulates a list of transformations. In the latter case, like "cross-multiplying," it constitutes an operation of a level higher than any ordinary arithmetical operation. These remarks bring to light a key fact: the reader, or the user, of the text must here again possess a specific competence to determine the extension of the procedure covered by the name. Conversely, whether the author of a text of a procedure makes use of the term "making communicate" to prescribe a sequence of operations, or rather embeds the list of arithmetical operations to be executed, as in the "detailed procedure" described, this choice reveals different ways of conceptualizing the flow of events occurring on the surface. Such a way of achieving the directives characterizes the difference between the second procedure and the related "detailed procedure," in contrast to the first procedure by Li Chunfeng. Here again, we have different texts referring to the same list of computations. The meanings the two kinds of text convey differ in relation to how the operations are prescribed, that is, in relation to the different assertions made by the directives on the world situation to be achieved.

9.2.2.2   Contrasting the "Procedure" and the "Detailed Procedure"

Let us now turn to a second range of questions: how does Li Chunfeng's "procedure" compare with the "detailed procedures," regarding both the text of the algorithms and the corresponding sequences of actions taken on the surface? In fact, Li Chunfeng composes a text for solving the general problem in a highly integrated way in at least two senses. On the one hand, he uses higher-level operations. On the other hand, a single text corresponds to different procedures when applied to all possible cases. What was described as different computations is now captured by the same "procedure." In correlation with this reformulation, Li Chunfeng changes the way in which the procedures tackle the different cases and thus departs from the "detailed procedures." Let us observe these differences.

Our preliminary observation will be hypothetical: it depends on the way in which the condition "If the divisor and the dividend both have parts and also in case there are repeated parts" is interpreted (see footnote 51). If the condition encompasses both cases represented by problems (1.5) and (1.6), Li Chunfeng's procedure deals with the two of them in one single list of operations: "one equalizes them and makes them communicate." Therefore, what corresponded to two different "detailed procedures" is now captured by a single list of higher-level operations. While we cannot be certain that this is the case here, it does appear, however, that the same list of operations in *The Nine Chapters* is interpreted in this way by the commentator Liu Hui. We will thus come back to this issue in our Sect. 3 and concentrate for now on how Li Chunfeng's "procedure" relates to the "detailed procedures" formulated, respectively, after problems (1.4) and (1.5).

The contrast between the first part of Li Chunfeng's "procedure" and the first "detailed procedure" is one of sheer formulation, which was analyzed above. On the one hand, the "procedure" is abstract, making no reference to the paradigm. On the other hand, where the "detailed procedure" has the prescription "multiplying by the denominator," the "procedure" has "making the parts communicate". However, these different formulations correspond to the same computations on the surface. Locally, we see here a clear illustration of the phenomenon we announced in the introduction: the same set of events can correspond to different texts for the algorithm. On the basis of our analysis of the term "making communicate," the importance of the difference appears clearly. There may yet be an additional subtlety here, which we will come back to, after having discussed the two algorithms given for such problems as (1.5).

The same does not hold true for the second part of the "procedure": the computations to which it leads on the surface differ from what corresponds to the "detailed procedure." If we suppose that we deal with the case:

$$\frac{a+\dfrac{b}{c}}{a'+\dfrac{b'}{c'}}$$

on the basis of an argument to be developed in Sect. 9.3, we suggest that "equalizing" leads to

$$\frac{a+\dfrac{bc'}{cc'}}{a'+\dfrac{b'c}{cc'}}$$

while "making communicate" corresponds to the operations

$$\frac{acc'+bc'}{a'cc'+b'c}$$

By contrast, as was suggested above, the "detailed procedure" can be formally represented by the sequence of equalities:

$$\frac{a+\dfrac{b}{c}}{a'+\dfrac{b'}{c'}}=\frac{\dfrac{ac+b}{c}}{a'+\dfrac{b'}{c'}}=\frac{\dfrac{(ac+b).c'}{c}}{a'c'+b'}=\frac{(ac+b).c'}{(a'c'+b').c}$$

We are thus clearly confronted with a case for which, in correlation to the formulation of a text for an integrated procedure, Li Chunfeng reworked the procedure for this range of problems. Note that the reformulation makes the quantity cc' appear explicitly—this is the one "equalized." Moreover, the "procedure" brings to light the fact that the two integers a and a' are transformed in exactly the same way. By contrast with Li Chunfeng's "procedure" for multiplication analyzed above, the reworking of the computations does not introduce a procedure that involves useless operations that have no bearing on the result. Yet, the computations corresponding to the "detailed procedure" are less complex than those reworked in Li Chunfeng's procedure. What is the benefit of this increase in the complexity of the operations?

My assumption is that, as in the case of multiplication, this increase in complexity correlates with achieving the integration of sequences of operations needed for the various cases into a single "procedure." However, the integration is done differently and it has a different effect. Let us examine the process of integration in greater detail.

A first aspect relates to the way in which the two main cases are brought into relation to each other. Indeed, if we observe the result of the first operation— $\dfrac{a+\dfrac{bc'}{cc'}}{a'+\dfrac{b'c}{cc'}}$ —,

we may suggest that it leads to a situation that falls within the range of the first part of the procedure.

To understand this remark, we need to turn to the question of determining this range. This question brings us back to the analysis of the structure of Li Chunfeng's procedure. Let us, to make things simpler, opt for the interpretation of the condition that introduces the second part of the "procedure" as "If the divisor and the dividend both have parts and there are repeated parts," that is, as a conjunction of two conditions.[60] Then, the range covered by the first part of the "procedure" should be deduced by *complementarity* to this condition. We were led to this hypothesis on the structure of such texts for algorithms (two clauses complementing each other, the second one being introduced by a condition) by our analysis of Li Chunfeng's first "procedure." Clearly, similar enumerations recur in our sources for the texts of algorithms and this hypothesis holds for all of them. If we adopt this interpretation for similar textual structures in general, this amounts to suggesting there was a way in which these specific modes of enumerating cases to formulate general algorithms were to be read. In what follows below, we analyze two further examples of enumerations of this kind before coming back to this hypothesis in the conclusion.

In the case of division, among the problems requiring division between quantities in which fractions occur, by complementarity the first clause in Li Chunfeng's procedure should hence be defined by a disjunction and cover two sets of cases: either only one of the dividend and the divisor has a fraction (problem (1.4)), or, if both have fractions, the denominators are the same. This latter point is crucial here, for my argument about the integration of algorithms carried out by the "procedure."

It is clear that, if I am correct, the case $\dfrac{a + \dfrac{bc'}{cc'}}{a' + \dfrac{b'c}{cc'}}$ falls under this latter rubric.

One essential feature to be stressed in this situation is that, in relation to the production of the procedure integrating all "detailed procedures," the range of cases covered by the first part of the "procedure" was modified in comparison to the "detailed procedure" following (1.4). The first part of the "procedure" does not *only* include problems such as (1.4)—as we thought until now—, but also other problems for which the same computations can be used. This means that in addition to reformulating the prescription of the main operation to be carried out for problems like (1.4), as was analyzed above, Li Chunfeng also extended the range of cases for which it was given. This is the additional subtlety to which I alluded above, which adds to our list of transformations that can be evidenced when a text relating to computations is rewritten.

If we recapitulate the kinds of transformations we have seen so far, we find the following: first, the text for an algorithm may be reformulated. Secondly, the computations corresponding to the new text may be reshaped. Thirdly, the range of cases covered may be modified. In the case of Li Chunfeng's procedure that we have analyzed, these three kinds of transformation occur in relation to a fourth more global transformation that we will analyze right now: the formulation of a procedure integrating previous "detailed procedures."

---

[60] This was the second option discussed in footnote 51, that is, that the condition refers to the case when the dividend and the divisor both have fractions and these fractions have, in addition, distinct denominators.

Now, let us go back to the issue of the relation between the transformations observed on the two "detailed procedures": the extension of the range of cases covered by the first part of the "procedure" correlates with a reshaping of the computations to which the second part of the "procedure" (for problems such as (1.5)) corresponds. This remark puts us in a position to grasp a benefit that derived from this reshaping, when compared to the "detailed procedure" following problem (1.5). Once "equalizing" has reduced the cases treated by the second part of the text to a case that the first part can deal with, the computations to be applied next to complete the task are the same, whether we consider them in the first part of the procedure or in the final segment of the second part. Again, a certain *uniformity* has been achieved in that, not only are the cases reduced one to the other, but also the procedure for the former is concluded by that for the latter. The lists of operations for the various cases were interlocked into another. A further consequence would be that the second occurrence of "making communicate" in Li Chunfeng's text would in fact simply encapsulate the subprocedure corresponding to the initial segment of the first part of the "procedure," the conclusion of the whole procedure being provided by a division.

In conclusion, as was said above, an integration of the various algorithms for the different cases was achieved, as for multiplication. This integration was done in a specific way. It involved using higher-level operations. It went along with an increase in uniformity within the set of procedures solving the various cases. As a result, the list of computations that were prescribed for some cases could become the list of operations concluding the solution of the other cases: procedures were interlocked into each other in such a way as to make them fit within the kind of specific enumeration described above. There appears to be a correlation between the mode of integration and the kind of text through which the integrated procedure is formulated.

Seen from this angle, the almost invisible modification carried out on the first part of the procedure and the reshaping in the second part mesh with each other, and their combination appears to correlate with the intention of shaping a "procedure" integrating the sequences of operations needed for any possible case, as Li Chunfeng has done here. The ensuing structure of Li Chunfeng's second "procedure," for division, can now be fully described, and this leads us to our third range of remarks, bearing on how the "procedure" compares with the previous one enunciated by Li Chunfeng.

9.2.2.3 Comparing the Two "Procedures" Formulated by Li Chunfeng

The text of Li Chunfeng's second procedure can be described as follows: A first list of operations—the first part of the text of the "procedure"—is given for a set of cases, which are determined by contrast with the condition explicitly formulated at the beginning of the second list of operations. However, as with multiplication, the first part of the text does not begin by the formulation of a condition. The second part of the procedure, which starts with the enunciation of a condition, in fact embeds the first part in that the sequence of operations it prescribes can be decomposed into two segments: the first segment transforms the cases dealt with into cases covered by the previous part of the "procedure" and the second segment repeats the initial segment of the first part of the procedure in a condensed way. The conclusion of the procedures dealing with the cases introduced by the condition corresponds to the final section of the first part of the procedure, that is, the statement of the division. It is in this way that, as a result, the "procedure" covers all cases.

Seen from a textual point of view, its text has the same enumeration structure as the first general "procedure," and the enumeration of the two items is carried out in the same way. Its interpretation seems to require that the reader have a specific competence to deal with such texts, in which a first procedure solving a specific but not specified case is followed by another where the cases are specified. Such a hypothesis is supported by the fact that the two procedures formulated by Li Chunfeng in the commentary on his *Zhang Qiujian suanjing* share similar modalities in stating a general "procedure." Below, we will see other instances in *The Nine Chapters* and the *Book of Mathematical Procedures*. This new source material confirms that we are confronted with a type of text typical of the description of general algorithms in ancient China.[61]

Another point common to the two "procedures" formulated by Li Chunfeng which was brought to light is that, here as for multiplication, the integration of two distinct "detailed procedures" was carried out through a reshaping that yielded higher uniformity. However, the uniformity achieved for division is of a type different from that for multiplication. It is no longer the same list of computations that allows all cases to be dealt with in the same way, no matter how complex it may be for some cases.[62] Rather, a first set of cases serves as a basis, to which all the other cases are reduced.[63] As a result, the list of computations solving cases in this first set will be extended by operations prefixed to it in order to yield the procedure for the other cases. Accordingly, the procedures for all cases share a common conclusion.[64]

---

[61] In addition to the instances discussed below, bringing to light this type of organization for algorithms clarifies the interpretation of several other texts of algorithms in *The Nine Chapters*, since they share a similar structure. One may think of the algorithms for root extraction (square roots as well as cube roots, in Chap. 4) and the algorithms linked to what was later called in the West 'rules of false double position' (Chap. 7). Even the structure of a chapter like *fangcheng* (Chap. 8) can be better understood from this perspective. The first algorithm it contains solves a specific and fundamental case. Later on in Chap. 8, conditions will be added and the algorithm will be made more general. We cannot discuss these examples further within the scope of this chapter.

[62] Let us recall that in case only one term in a multiplication had a denominator, the procedure for different denominators was used, by artificially considering that the integer to be multiplied was followed by a fraction with a zero numerator and the same denominator.

[63] It is remarkable that this echoes with the series of problems through which Liu Hui comments on the procedure "Multiplying parts." Each problem offers a question that an always larger final section of the procedure solves, thereby gradually shaping an interpretation of the final result (Chemla 2009b). I'll inquire into this parallel in another article.

[64] In correlation with this opposition, the text for multiplication presents two parallel and uniform subprocedures before concluding with a common conclusion. In division, however, a complete procedure is formulated, followed by how it is to be modified for the second set of cases. As a

In contrast to multiplication, we thus see that the integration of all the procedures is carried out by resorting to another technique. Yet, the interest in uniformity is clearly the same. It correlates to the production of texts integrating various procedures, the structure and interpretation of which is similar.

This comparison leads us to an unexpected conclusion. An essential property common to the two texts emerges, beyond the differences between Li Chunfeng's two procedures. The list of operations promoted at the beginning of the text, forming the first item of the enumeration and introduced without a condition, is in all cases common to all procedures solving any of the particular problems dealt with. Moreover, this list allows a set of particular cases to be solved, which for this reason are fundamental for the general problem under consideration. The texts introduce conditions after this initial statement, common to all procedures. The cases corresponding to these conditions are the only ones for which one may need to add some other operations to a common trunk. Such a conclusion accounts for the fact that the first item is never introduced by a condition. It also confirms, and explains, what we said above regarding the circulation that the practitioner has to carry out within the text in order to produce, based on this, the relevant list of operations for a given case. The practitioner starts using the text at the point where the conditions are placed, and not from the beginning. In case the problem to be solved fits with the conditions, the practitioner first makes use of the list of operations following the conditions. Interestingly, this opposition between the core list of operations and its extension echoes what we said above regarding the expression "making them communicate." The operations corresponding to it may vary. However, they always include a core operation that gives the whole list its identity.

Such textual practice may seem strange, simply because it differs from the textual practices we are used to. However, if we considered the texts used today to write down algorithms from this point of view, we would discover that they manifest similar features, even though they are different in the detail. Moreover, if we do not attempt to restore the practices of the kind that were in use in the contexts in which our sources were produced, we may fail to interpret them correctly. If this were the case, it would cause many problems, since, as we will now see, such texts for procedures were used for a period of time that spanned several centuries.

In conclusion to this section of the chapter, let me stress a first main result that we have established: we have seen that our sources contained at least two markedly distinct types of texts for algorithms. It is important to stress that this distinction corresponded to categories used by some of our actors. The *Mathematical Classic by Zhang Qiujian* and its commentaries, with some excerpts of which we illustrated these differences, use two different terms to refer to such texts: "procedure *shu* 術" as opposed to "detailed procedure *cao* 草."

---

consequence, the relationship between the procedures for the first and second sets of cases recalls the relationship between the "detailed procedure" following problems (1.5) and (1.6). In the latter case, a procedure, that for adding fractions, was grafted onto another procedure, reducing the more complex case to a simpler one. The text of the algorithm for division is built according to the same logic.

With respect to the texts of "procedures," if we follow Li Chunfeng's use of the term, we have also highlighted that they do not conform to our expectations as present-day readers. We brought to light that these texts had a peculiar structure. Accordingly, we have shown that their users had to bring a specific kind of reading into play in order to derive from them the correct list of computations for any given case encountered. In particular, we revealed the unexpected circulation required between the various segments of the text. Moreover, we observed that there was variation in the way in which the texts prescribed arithmetical operations or grouped a sequence of operations under a single verb. In particular, we brought to light that some of our sources sometimes carried out directives by asserting in a specific way the situation that was to be achieved. As a result, the text of the algorithm expresses in and of itself the reasons why the algorithm was correct.

In what follows, we will concentrate on these two main lines of inquiry. By turning, first, to a text for an algorithm for division included in *The Nine Chapters* and to its commentary by Liu Hui, we will discover another instance of a text similar to the "procedures" described above and integrating the treatment of several cases. In addition to confirming that we are analyzing a kind of text for algorithms typical of mathematical sources of ancient China, this case will be essential for our purpose. Indeed, it gives us the opportunity to observe how a commentator read a text of that kind and thereby to reveal evidence confirming our conclusions. Further, the commentary will also bear witness to how a reader of the past interpreted directives of the kind we have already met. This will allow us to analyze such language acts in greater detail.

9.3 Focusing on Two Texts for a General Procedure, one Handed Down and one from a Manuscript

The purpose of this section is mainly to provide evidence that the phenomena brought to light in the previous two sections are not unique. By showing how they recur, we will place ourselves in a better position to argue that such analyses can be useful for history of science. This specific goal explains why, from now on, we will not analyze the texts examined with the same amount of detail as we did previously, limiting ourselves to focus on the two main lines of inquiry presented above.

9.3.1 The Text for a General "Procedure" as Interpreted by a Commentator

In Chap. 1 of *The Nine Chapters*, the issue of dividing between quantities having integers and fractions is dealt with by means of a general procedure comparable to the one formulated by Li Chunfeng that was analyzed above.[65] The operation carried

---

[65] The text is translated and analyzed in Chemla and Guo (2004, pp. 166–169). I refer the reader to this other publication for greater detail. See also Chemla (2012).

out by this procedure is called "Directly sharing 經分"—note that the first character differs slightly from the one used in the *Book of Mathematical Procedures* to refer to the same operation, which we mentioned above. To begin with, let us observe the structure of the text of the procedure.

9.3.1.1   The Text of a Procedure as Enumeration and its Interpretation by a Commentator

In *The Nine Chapters*, the general procedure is given after the statement of two problems:

(1.17)

"Suppose one has 7 people sharing 8 units of cash and 1/3 of a unit of cash. One asks how much a person gets.

Answer: a person gets 1 unit of cash and 4/21 of a unit of cash.

今有七人, 分八錢三分錢之一。問人得幾何。

荅曰:人得一錢二十一分錢之四。

(1. 18)

Suppose again one has 3 people and 1/3 of a person sharing 6 units of cash, 1/3 and 3/4 of a unit of cash. One asks how much a person gets.

Answer: a person gets 2 units of cash and 1/8 of unit of cash.

又有三人三分人之一, 分六錢三分錢之一、四分錢之三。問人得幾何。

荅曰:人得二錢八分錢之一。"

Despite slight differences, on which I do not dwell, problems (1.17) and (1.18) present cases comparable to those of, respectively, problems (1.4) and (1.6) in the *Mathematical Classic by Zhang Qiujian*. The procedure included in *The Nine Chapters* and placed after the two problems allowed practitioners to solve them both and even solve other problems. In other words, it is placed with respect to the problems in a position, and plays a part, both comparable to that of Li Chunfeng's procedure in the *Mathematical Classic by Zhang Qiujian*. The text of the procedure for "Directly sharing" reads as follows:

"經分

術曰:以人數爲法, 錢數爲實, 實如法而一。有分者通之; 重有分者同而通之。

Directly sharing

Procedure: One takes the quantity of people as the divisor, the quantity of cash as the dividend, and one divides the dividend by the divisor. If there is one type of part, one *makes* them *communicate*. If there are several types of parts, one *equalizes* them and *makes* them *communicate*." (My emphasis)

Several elements brought to light in the previous analysis give us clues for tackling some of the key features of the present text.

To begin with, it is clear that its structure is of the same type as was described above. After a subprocedure, formulated without an introductory condition, a condition is stated, which I interpret as: "If there is one type of part 有分者." Quite interestingly here, after providing a list of operations following this condition, the text goes on, with the formulation of a second condition, which I interpret as: "If there are several types of parts 重有分者." My interpretation of the conditions relies on two factors, that is, on the one hand, on their formulations in the text, but also, on the other hand, on the fact that they must constitute a system of conditions in the algorithm. This is why the expression "if there are parts 有分者" is interpreted in this context as "If there is one type of part 有分者." We will see that this is also how the commentator Liu Hui interprets the condition.

Altogether, we have a text for the general procedure that is of the same type as above: an enumeration of three items, the first of which is not prefixed by a condition. The text is thus simply longer than those described above, in the sense that it develops a sequence of two conditions, each followed by a list of operations.

Secondly, the structure of the text allows the author to integrate the treatment of three types of case in the same way as above.

The first segment, placed before the statement of a condition, allows one to solve the case where one has to divide two integers by one another. In fact, it yields the text of a subprocedure common to all the procedures derived from "Directly sharing" for dividing between quantities composed of any combination of integers and fractions. In other words, if we consider the first segment of "Directly sharing" from the point of view of the structure of the text, it still corresponds to a fundamental case, in the sense that, like the texts for "procedures" analyzed above, it includes the sequence of operations concluding the solution of all the cases that can be encountered, when solving the general problem according to "Directly sharing." So from a syntactic point of view, the first segment of the enumeration plays the same part as was described previously.

However, if we now examine the *content* of the initial segment of the text, it is no longer the same as in the description by Li Chunfeng analyzed above: another fundamental procedure was placed at the beginning of the text, corresponding to a new fundamental case, that of the division of two integers one by the other. Note that it does not correspond to any problem formulated in the Classic.[66] In short, the "meaning" of the initial item of the enumeration is the same as above—it still yields the procedure for the fundamental case to which any case will be reduced—, but its content differs—the fundamental case was changed. This supports the hypothesis that we are describing a type of text that was regularly used in the mathematical tradition from ancient China examined and that it is specific to it.

The comparison with Li Chunfeng's text for division, as we interpreted it, suggests that the first condition formulated after the introductory subprocedure in *The Nine Chapters* corresponds to the item that, in Li Chunfeng's general procedure, was placed at the beginning, prior to any conditions. By contrast, the second condition in the procedure for "Directly sharing" appears to be less limited than the one

---

[66] When Li Chunfeng comments on the procedure for multiplying quantities which is called "Extending generality," as evoked above (See footnote 39), he interprets the text as incorporating three cases (multiplying two integers, multiplying two fractions, multiplying two integers followed by fractions). However, there too, there is no problem in the immediate context which raises the case of multiplying two integers.

formulated by Li Chunfeng: it does not require that "the dividend and the divisor both have parts." In addition to this latter case, it may encompass the case when an integer is followed by two fractions as well, as in problem (1.18) in *The Nine Chapters*. Again, we find the phenomenon where the same list of operations may correspond to a larger set of cases.

If we now turn to the operations in the second and third items of the text for "Directly sharing", the operation in the second item reads "one *makes* them *communicate*," whereas, in the third item, we read: "one *equalizes* them and *makes* them *communicate*." Before we turn to this modality for carrying out the directives, several remarks can be made. Clearly the text of the operations in the third item includes that of the second item. Moreover, a comparison of these two items with Li Chunfeng's "general procedure" shows that, although the formulations partly differ, the same computations are referred to. In particular, both items refer to computations that reduce the various cases to the fundamental one, placed at the beginning.[67]

In other words, for each of the second and third items of "Directly sharing," the procedure prescribed for the cases covered by the condition amounts to prefixing the operations that, in the text, follow the condition to the first part of the general procedure. The further one proceeds in the enumeration, the greater number of operations one prefixes to the fundamental procedure. Moreover, in the enumeration, the cases are arranged in such an order that the procedures solving them are embedded into one another in an order that follows the presentation in the text. In terms of the sequence of actions that a practitioner derived from "Directly sharing," the list of operations solving the cases covered by item 3 contains the procedure solving the cases covered by item 2, which in turn contains the procedure described in item 1. We see that here too the procedure achieves a uniformity in the treatment of the various cases. Moreover, this uniformity corresponds to the fact that the author used the format of an enumeration comparable to those described previously for the text of the procedure. We thus have the same correlation between this format and the valuing of uniformity. However, in terms of text, the operations in items 2 and 3 are to be followed, not by the operations in the previous item, but by those in the initial item.

As a result of this brief analysis, we see that the text of the procedure for division included in *The Nine Chapters* is of the same type as the one identified in the two preceding procedures by Li Chunfeng. In addition, the text makes use of modes of carrying out directives comparable to some of those examined in Sect. 9.2. Again here, we are interested in how readers in ancient China interpreted both the terms for the operations and the text of such an enumeration. In order to determine what the meaning of such an enumeration may have been for them, we will follow Liu Hui's commentary on this procedure.

---

[67] Some slight differences can be noted in this respect. In Li Chunfeng's text, the operations prescribed in the second item overlapped with the beginning of the operations listed in the first item. However, the procedure for the cases covered by the second item still had to be completed with the last operation prescribed in the first item. In the text for "Directly sharing" included in *The Nine Chapters*, the operation for the second item is part of those given for the third, whereas the two types of procedures are concluded by the operations placed at the beginning of the text.

With this aim in mind, we will in particular examine from this perspective the directives "one *equalizes* them" and "one *makes* them *communicate*" used in the text of "Directly sharing." In the previous section, we approached the question of their interpretation by confronting various ways of prescribing the same computations. At one point, though, we had to make use of a commentary by Li Chunfeng on a procedure in *The Nine Chapters* to cast light on how he used the term "make communicate." Here, with Liu Hui's commentary on the procedure for "Directly sharing," we can approach the directives much more systematically.

At first sight, his commentary deals with a different question, since it accounts for the correctness of the procedure. The connection of this concern to the issues addressed in this chapter will prove to be essential. To develop his proof of the correctness of the procedure, on the one hand, the exegete makes the actions highlighted by the directives explicit in terms of ordinary arithmetical operations. Liu Hui thus provides the interpretation of the directive as made by an actor of the third century. On the other hand, he brings to light the "meaning" of the operations with respect to the entities to which they are applied. Making such "meanings" explicit is an essential task in proving correctness. The key point for us is that the two facets of the commentary allow Liu Hui to reveal a correlation between the way in which the speech acts are carried out and how he apprehends the effects of the operations—the change of the situation of the world operated by the actions prescribed. Let us illustrate this claim while reading the commentary step by step.

Interestingly enough, Liu Hui treats the operations in an order that is the reverse to the one in which they were introduced in the general procedure: he deals first with "equalizing," then with "making communicate," and, finally, with the entities "dividend" and divisor." He thus comments on the fullest procedure—the one embedding all the others—, starting from the last case—the one considered in item 3. Within this item, Liu Hui treats the operations in the order in which they present themselves. This appears to confirm our conclusion regarding the circulation in the text: a reader started approaching such an enumeration by the conditions, that is, in this case, by the last condition. With respect to the proof conducted, if such can be the case, this means that the correctness of the procedure for the general cases corresponding to the first two items in the text can be treated as part of the proof of the procedure for the third item. In other words, the interlocking of the procedures one into the other corresponds to the embedding of the reasons for the correctness.

Let us now read his commentary and focus on what it tells us on the directives.[68]

---

[68] I will not enter into all details, for this the reader is referred to the critical edition, the French translation, and especially to the footnotes and the glossary through which I comment on the procedure and the commentaries (see Chemla and Guo Shuchun 2004, respectively, pp. 166–169, 766–767, and 993–998).

9.3.1.2 How Liu Hui Reads the Directives

The first two sentences of Liu Hui's commentary on "Directly sharing" relate to the directive in "Directly sharing": "one equalizes them." The commentator's formulation refers the reader to the part of the commentary where, in an earlier section, Liu Hui had introduced the operations "equalizing" and "homogenizing." They read:

"If "the denominators multiply the numerators that do not correspond to them", it is to homogenize the corresponding numerators. If the "denominators are multiplied by each other", it is to equalize the corresponding denominators. 母互乘子者[69],齊其子。母相乘者,同其母。"

Liu Hui thus quotes the operations as formulated by *The Nine Chapters* in the procedure for the addition of fractions (Chemla and Guo 2004, pp. 156–161): "the denominators multiply the numerators that do not correspond to them," the "denominators are multiplied by each other." With respect to a pair of fractions a/b and a'/b', the former clause corresponds to the parallel computations of, respectively, ab' and $a'b$, whereas the latter corresponds to the computation of bb'. In his proof of the correctness of the procedure given by *The Nine Chapters* to add fractions, which we alluded to previously, Liu Hui had brought to light the "meaning" of these operations. In his terms, the latter "equalizes": it computes a denominator "equal" for all the fractions involved, that is, the size of a part with respect to which all can be expressed. Moreover, the former operations "homogenize" the numerators, that is, transform them, in correlation to the change of denominator, so as not to damage the fractions involved.

It is interesting that, in the first section of his commentary on "Directly sharing," as an echo to the use by *The Nine Chapters* of the directive "one equalizes them," Liu Hui quotes this previous section of his commentary. He thereby relates the formulation of the directive in *The Nine Chapters* to an operation he had introduced earlier for the sake of proving the correctness of an algorithm. In terms of interpretation, Liu Hui thus connects the way in which an operation is prescribed—"one equalizes," instead of "the denominators are multiplied by each other"—and the way in which he previously accounted for the meaning of that operation. This is exactly the same phenomenon as we described in Sect. 9.2, with respect to how Li Chunfeng uses "making communicate." Moreover, and to be more precise, in terms of the operations meant, Liu Hui seems to understand that the two operations which, in this other context of the addition of fractions, were accounted for as "homogenizing" and "equalizing" are now prescribed by the single directive "one equalizes," as used here by the procedure "Directly sharing" in the Classic.

To recapitulate, in "Directly sharing" Liu Hui interprets "equalizing" as standing for "homogenizing" and "equalizing."[70] Moreover, he understands that the compu-

---

[69] I suggest here departing from what was suggested in footnote 18, p. 158, in our critical edition and, rather, follow Dai Zhen (after 1776, see Guo Shuchun in Chemla and Guo 2004, pp. 158, fn. 18), Qian Baocong (1963) and Li Jimin (1993, p. 162, fn. 26), in considering that "知" that is the reading in all ancient sources is a corruption of "者."

[70] In my glossary, I accounted for this phenomenon by suggesting that the terminology may have changed between the time of *The Nine Chapters* and that of Liu Hui. However, the previous analysis seems to reveal that it may have been common to use a single term to refer to different pro-

tations are indicated by a term that, again in J. Virbel's words, asserts the world situation that they are to achieve: "equalizing" denominators. Indeed, if such is the state that the computations achieve for the fractions, this indicates why they are relevant for the procedure and contribute to its correctness. The term "equalizing" brings to light the effect that the set of computations will have on the entities to which they are applied.[71] We thus meet again a phenomenon described in Sect. 9.2 of this chapter, in relation to "making communicate." The parallel is even more complete than explained so far. Let us explain why.

In other contexts in his commentary, Liu Hui uses the pair of terms "equalizing"/"homogenizing" to capture the "meaning," that is, the "intention," of other operations. However, if "equalizing" always refers to the fact that some "quantities" are equalized, the actual, material meaning of this "equalization" varies according to the context. In terms similar to those we used when, above, we analyzed "making communicate," "equalizing" captures the aim in two ways: in a material way—equalizing denominators—and in a formal way—equalizing entities. The same holds true for "homogenizing." As a consequence, the directives are formulated in both cases with terms that refer to the reasons for the correctness in a formal way. In Sect. 9.3.2, we will see that this is not always the case in the commentary and nor was it always the case in ancient China.

Let us now turn to the subsequent sentences in Liu Hui's commentary, devoted to the second directive of the procedure: "making communicate." On the basis of the previous segment of the commentary, we can now consider that, the denominators having been equalized, we are dealing with the case in which there is only one kind of denominator, that is, that we are within the framework of the first condition formulated by the text. We thus see how the interlocking of procedures corresponds to the reduction of cases, from the general case captured by the last condition to the fundamental case, passing through the case corresponding to the first condition.

We will show that Liu Hui's account for the term "making communicate" is more complex than that which we deduced above from Li Chunfeng' various uses of the term. To bring this point to light more clearly, we will divide the following part of Liu Hui's commentary into two segments. The first reads as follows:

"以母通之者, 分母乘全內子。乘, 散全則爲積分, 積分則與分子相通, 故可令相從。

"With the help of the denominator "making them communicate" is multiplying the integers (or: integral parts) by the denominator and incorporating these (the results) into the numerators. By multiplying, one disaggregates the integers, thus making the parts of the product (*jifen*). The parts of the product and the numerators thus communicate with each other, this is why one can make them join each other."

---

cedures, provided that they all shared the key operation referred to by the term. Therefore, we may also have here another occurrence of the same general phenomenon.

[71] This echo seems to indicate that there is a correlation between how Liu Hui develops proofs for the correctness of procedures and what he reads in the Classic. Moreover, he seems to read *The Nine Chapters* as possibly referring to reasons of the correctness by the way in which directives are carried out. We come back to this key issue below. It is developed in greater detail in Chemla (1992, 1997b, 2010a).

Liu Hui begins his commentary on the directive, by reformulating "making them communicate" in a way that carries out the speech act differently: although the same illocutionary aim is meant, the exegete indicates the computations to which the text refers, by means of terms designating ordinary operations. He thereby provides the readers, or the users, of the text with elements that they may be missing, in case they do not have the competence required for interpreting the directive. Commentaries appear as a kind of text, through which the competences linked to the interpretation of complex speech acts are transmitted. In this case, we see that Liu Hui's interpretation of "making communicate" corresponds to what we suggested in Sect. 9.2.2.1.

In the quotation above, the commentator continues with a *material* explanation of the effect of the operations, seen from the point of view of what happens to the "parts" involved. The effect of the "multiplication" on the integers is a "disaggregation" of their parts (units), in such a way that the resulting portions, collectively designated by the technical term "parts of the product," share the same size as those of the fraction present in the situation—remember there is only one type of part, since there is only one kind of fraction. Liu Hui then reinterprets this effect in a *formal* way: "they communicate with each other"—this recalls the first use of "making communicate" by Li Chunfeng analyzed above. It is striking that this is given as a reason accounting for the fact that the values, now having an adequate internal structure, can be used together in an arithmetical operation and thus be added. It is therefore clear *how* prescribing these computations with the term "one makes them communicate" is a way of prescribing them by asserting the situation of the world to be achieved and making clear how the subprocedure contributes to the correctness of the procedure.

However, this segment of the commentary does not exhaust what Liu Hui has to say about "making communicate." The second segment on the term reveals the additional meaning he sees in the fact that *The Nine Chapters* uses the term here—this is the meaning that our analysis of Li Chunfeng's usage of the term in Sect. 9.2.2.1 of this chapter could not disclose. Let us summarize the main point, before we go on reading the commentary.

In the previous paragraph we saw that Liu Hui focused on the computations that transform the terms of the division in the procedure "Directly sharing." These terms are quantities of the type $a + b/c$. The procedure requires transforming integers such as a into ac so that they can be added to b. However, in the case in which only one type of fraction appears in the values to be divided, another operation is required: suppose one divides $a + b/c$ by a' (or $a' + b'/c$), a' must be transformed into $a'c$ (or $a' + b'/c$ into $a'c + b'$). One may understand that this operation is also meant in the previous prescription of "making communicate." In other words, the same operation would simultaneously transform $a + b/c$ into $ac + b$ *and* $a' + b'/c$ into $a'c + b'$—we have already met with similar prescriptions for parallel computations in a single sentence. This is in fact how my translation of Liu Hui's commentary above interprets "making communicate" (see footnote 37). Yet the correctness of this parallel transformation *as such* has not yet been addressed.[72] Why is it that the two quantities provided by the statement need to be transformed simultaneously in this way? How does one account for the correctness of these parallel operations when taken as a whole? What is essential is that Liu Hui also addresses this issue now, and *also* that he does so in terms of "making communicate." This seems to indicate that he also reads this additional meaning in the prescription of the whole subprocedure by means of this term. Much more importantly, he tackles this dimension in the most general way, by introducing the key concept that is crucial to the whole idea of "communication," that of *lü*.[73] In the context we are examining, the dividend and the divisor exemplify the *lü*. As a consequence, any multiplication or division of one is to be carried out on the other, which it "communicates" with, for the pair to keep its meaning. In Liu Hui's commentary here, the concept of *lü* appears to be much more general. Let us read the following part of the commentary with this idea in mind:

"凡數相與者謂之率。率者[74],自相與通。有分則可散,分重疊則約也。等除法實,相與率也。故散分者,必令兩分母相乘法實也

"Whenever quantities are given/put in relation with each other, one calls them *lü*. *Lü*'s, being by nature in relation to each other, *communicate*. If there are parts, one can disaggregate; if parts are reiterated superpositions, one simplifies. A divisor and a dividend, divided by the equal number (g.c.d.), are *lü*'s put in relation with each other. Therefore, if one disaggregates the parts, one necessarily makes both the two denominators multiply the divisor and the dividend." (My emphasis)

It is not necessary to enter here into greater detail on the interpretation of this text. My sole aim is to point out the fact that the commentator reacts to the way in which the directive "making communicate" is carried out in the Classic, by developing several layers of meaning associated with it, all being related to the proof of the correctness of the algorithm. At this point of his reasoning, a key part is played by the fact that one is dealing here with a "dividend" and a "divisor." These identities of the elements involved are stated in the first item of the procedure for "Directly sharing." One can thus see how the proof develops progressively along the line provided by the procedure for the last case it dealt with.

These remarks conclude the points I wanted to make on the basis of a sketchy analysis of "Directly sharing."

---

[72] In the first segment of Li Chunfeng's general procedure for division analyzed in the previous section, this operation was not prescribed in relation to "making" entities "communicate."

[73] The term of *lü* qualifies quantities that represent the values of magnitudes only relatively to each other. These quantities can hence be multiplied or divided by the same amount, without damaging the property of the quantities obtained to also represent the relationship between the magnitudes. On this term, see my glossary in Chemla and Guo (2004, pp. 956–959). I refer the reader to Chemla (2010a, p. 274) for a more detailed analysis of the second layer of meaning that Liu Hui reads here in "making communicate."

[74] As in footnote 69, I follow Dai Zhen's idea, after 1776, in thinking that "知," which the ancient sources contain here, is a corruption of "者."

With respect to the text of an enumeration, the previous analysis suggests an answer to the question we raised regarding the meaning of what now appears to have been a standard way of writing down texts for algorithms in ancient China. Enumerations of the type that our various general "procedures" embody may have been organized in such a way so as to have the reason of the correctness unfold, as one tackles cases that require longer and longer procedures.

With respect to directives, the last example of "making communicate" is the most complex one. What is encapsulated, in terms of operations and in terms of meaning, in the use of the synecdoche "one makes them communicate" is now displayed more fully. By contrast to the use of "making communicate" where the term refers merely to a multiplication, this last usage is read by Liu Hui as designating a whole set of elementary operations, by reference to the program they fulfill and the facts from which their correctness derives. The operations are grouped on the grounds that their set as a whole has a meaning, which is indicated by the term with which they are prescribed. This meaning is captured in a formal way. This conclusion also reveals, more generally, how the same word can be used with varying levels of complexity. The use of the same term does not always represent the same kind of speech act.

The previous analyses highlight one of the ways in which the formulation of the directives in the text of an algorithm can indicate the reasons why the algorithm is correct. By turning now to a similar text of procedure found in the *Book of Mathematical Procedures*, we will observe another modality for this, as well as another, much earlier, example of the kind of integrated text on which we concentrate here. Again, we will limit our analysis of this complex text to these two features only.

9.3.2 Another Modality of Relationship Between Directives and Proof—Another Modality of Integration in a much Earlier Text Taken from the Book of Mathematical Procedures. Similarities & Differences

The composition of the *Book of Mathematical Procedures*, recently excavated from a tomb sealed in *ca.* 186 B.C.E., is much less standardized than *The Nine Chapters* or the *Mathematical Classic by Zhang Qiujian*. The procedure on which we will focus is recorded on bamboo strips 74–75 and, on these strips, it is given outside the context of a problem. In correlation with this, as we will soon see, the formulation of the procedure has a type of generality comparable to the general procedures examined above. Let us provide, to begin with, some information about the procedure, in order to be in a position to focus, later, on the two main points raised above.

The title on bamboo strip 74, "Determining the standard price (*lü*)[75] on the basis of the *shi*," is the name of the task that the procedure executes. It is the same task

---

[75] The same term *lü* is used here in a verbal sense. It also occurs with the same meaning and function in *The Nine Chapters*, in a similar set of problems, which is placed in the second half of

that the terms of a problem, placed by the editor Peng Hao immediately after that procedure (bamboo strips 76–77), asks to carry out.[76] The problem statement and the procedure following it read as follows:

Strips 76–77

賈鹽 今有鹽一石四斗五升少半升, 買取錢百五十, 欲石衛 (率) 之, 為錢幾何」。曰:百三錢四百卅 (三十) 口分錢九十五。术 (術) /76/曰:三鹽之數以為法, 亦三一石之升數, 以錢乘之為實。/77/

Trading salt Suppose one has 1 *shi* 4 *dou* 5 *sheng* 1/3 *sheng* salt[77] and that when trading it, one obtains 150 cash. If one wants to "*lü*" them (to determine the standard price for it?) on the basis of the *shi*, how much cash does this make? One says: 103 cash 9[2]/43[6] cash. Procedure: One triples the quantity of salt, which is taken as the divisor. One also triples the quantity of *sheng* of 1 *shi*, and with the cash, one multiplies it to make the dividend."

Let us for the moment leave aside the description of the procedure, we come back to it later. The task suggested by the problem can be described as follows: one trades a certain amount of salt (expressed with respect to several units of capacity and fractions) for a given amount of cash. The question is to determine how much cash corresponds to a given unit of capacity—here the *shi*.

The idea applied by all the procedures of this kind is the same. It contains two key elements.

On the one hand, one applies a kind of rule of three, which can be captured by the following formula:

$$\frac{cash \ multiplied \ by \ 1 \ unit \ (shi)}{quantity \ bought}$$

Chap. 2. I have already dealt with this extremely interesting procedure from different angles (its connection to procedures in *The Nine Chapters,* the connection between the formulation of procedures and the proofs of their correctness, the prescription by analogy) in, respectively, Chemla (2006, 2010a, 2010b). I refer the reader to these other publications for a more detailed treatment which I do not repeat here. In particular, I do not comment on the editorial problems raised by the passages considered, nor do I account for how I interpret it by contrast to how other historians interpret it. Since my 2006 article was published, two new books on the *Book of Mathematical Procedures* have appeared: Horng Wann-sheng 洪萬生, Lin Cangyi 林倉億, Su Huiyu 蘇惠玉 and Su Junhong 蘇俊鴻 2006, 張家山漢簡『算數書』研究会編 Chôka zan kankan Sansûsho kenkyûkai. Research group on the Han bamboo strips from Zhangjiashan *Book of Mathematical Procedures* 2006.

[76] See Peng Hao 彭浩 (2001, pp. 73–74). The reader should keep in mind the fact that the book was found as separate bamboo strips, while they were originally bound together in a roll. Each edition hence suggests different orders for arranging the strips. For instance, in the Japanese edition mentioned in the previous footnote, the sections we are interested in are placed as the 47th and the 48th. I systematically refer the reader to the numbers attached to the bamboo strips as provided by Peng Hao 彭浩 (2001) and I follow, for the time being, the organization of the book as suggested in this publication.

[77] These are three units of capacity, and the relationships between them are: 1 *shi* 石=10 *dou* 斗=100 *Sheng* 升.

This can explain why the procedures that correspond to this are placed in the second chapter of *The Nine Chapters*, devoted to the rule of three.

On the other hand, one transforms the units in the dividend and the divisor simultaneously before carrying out the operations of multiplication and division displayed in the formula above. Again this transformation of units can be represented by the following formula:

$$\frac{\textit{cash multiplied by } 1 \, \textit{unit} \, (\textit{shi}) \, \textit{expressed in the same unit as the divisor}}{\textit{quantity bought expressed with respect to a unit into which the quantity becomes an integer}}$$

The sequence of operations that lead from the first formula to the last one can be represented as follows:

$$\frac{\textit{cash multiplied by } 1 \, \textit{unit} \, (\textit{shi})}{\textit{quantity bought}} = \frac{\textit{cash multiplied by } 1 \, \textit{unit} \, u_1}{q_1 u_1 + q_2 u_2 + \dfrac{m}{n} u_2}$$

$$= \frac{\textit{cash multiplied by } n.1 \textit{unit} \, u_1}{n q_1 u_1 + n q_2 u_2 + m u_2}$$

and if $u_1 = k_1 u_2$

$$= \frac{\textit{cash multiplied by } n.k_1 u_2}{n q_1 k_1 u_2 + n q_2 u_2 + m u_2}$$

With these ideas in mind, the procedure provided after the problem quoted above is clear. "Tripling the quantity of salt" corresponds to expressing what will constitute the divisor with respect to units that are thirds of *sheng*. However, this interpretation is not indicated in the text of the procedure. The same transformation is prescribed for the "1*sheng*" that occurs in the dividend ("One also triples the quantity of *sheng* of 1 *shi*.") The speech acts are directives that are expressed by means of ordinary arithmetical operations.[78] They indicate the computations by making specific reference to the particular problem after which the text of the procedure is placed. It is interesting that, by the use of an "also", the text of the procedure states that the operations just evoked are linked to each other. The word has no use in carrying out the directives. Its only purpose is to indicate the correlation between two operations in the procedure. This link refers in fact to the reasons why the multiplications are

---

[78] Here, to be more precise, the sentence "One also triples the quantity of *sheng* of 1 *shi* 亦三一石之升數" prescribes two operations. Transforming "1 *shi*" into *sheng* is prescribed by the statement of the result ("the quantity of *sheng* of 1 *shi* 一石之升數"). Such prescriptions are frequent in the texts from ancient China. This first one is combined with the prescription of tripling.

carried out: they correspond to the same transformation.[79] This "also" reveals a meta-mathematical view of the procedure (Chemla 2010b.) Beyond the structure thereby indicated in the procedure, nothing more specific indicates the reason why the operations should be carried out.

The procedure provided by bamboo strips 74–75—"Procedure for determining the standard price (*lü*) on the basis of the *shi*"—, which is more general than this one, is quite different in this respect, in a way that is precisely why it is interesting for us here. Moreover, its text has the same type of structure as the procedures analyzed in the previous sections of this chapter. It has a first segment stated without a condition. This first segment is followed by two segments, each starting with a condition. Strikingly enough, it has a relationship to the specific procedure following it which can be compared to the relationship between the "detailed procedures" in the *Mathematical Classic by Zhang Qiujian* and the general "procedure" formulated later by Li Chunfeng. Again, we have two texts corresponding to the same sequence of actions for solving the "trading salt problem." Our aim here will be to capture the differences between them in terms of their formulation as well as the differences between the meanings conveyed.

The corresponding procedures in *The Nine Chapters* are associated with another name for the operation: "Directly *lü*-ing (determining the standard price?) 經率". It is interesting that this name is designed on the same pattern as "Directly sharing." This is all the more interesting as "Directly sharing" corresponds to a text for a general procedure that has properties similar to the scheme on which we are concentrating in this chapter. We are hence in fact exploring a range of related procedures, sharing similar kinds of text.[80]

As mentioned earlier, the following analysis of the text of the "Procedure for determining the standard price (*lü*) on the basis of the *shi*" will here again only focus on the same two themes: the way of expressing the directives and the modality of carrying out the integration of distinct sequences of operations required to address the various cases through a single text. As a whole, the text of the procedure reads as follows:

Strips 74–75

石衛(率)　石衛(率)之術曰:以所買_(買(賣))為法,以得錢乘一石數以為實。其下有半者倍之,少半者三之,有斗、升、斤、兩、朱(銖)者亦皆//破其上,令下從之為法。錢所乘亦破如此。

"Determining the standard price on the basis of the *shi*　Procedure for determining the standard price on the basis of the *shi*: One takes what is exchanged as divisor. One multiplies, by the cash obtained, the quantity of 1 *shi*, which is taken as dividend. Those for which, in their lower (rows), there is a half, one doubles them; those for which there is a third, one triples them. Those for which there are *dou* and *sheng, jin, liang* and *zhu*, one *also breaks up all* their upper (rows), one makes the

---

[79] I have dealt with the history of the use of the term "also" in the texts of procedures known from ancient China in Chemla (2010b).

[80] There are fundamental reasons for this, which I examined in Chemla (2006). The formulation that the text for "Directly *lü*-ing" receives in *The Nine Chapters* is quite interesting too. However, its analysis exceeds the scope of this chapter.

(rows) below join them, (yielding a result) which is taken as divisor. What the cash multiplies is *also broken up like this.*" (The emphasis is mine)

In contrast to the procedure analyzed above, which was given under the title "Trading salt," the text is less specific, in a way that we will now examine more closely. As a whole, it thus seems to require to be read as general and representing the general case. However, for the sake of clarity, let us offer our interpretation of this text by reference to the problem just quoted. This problem is all the more convenient for us in that it presents all the specificities required to understand all the conditions in the general procedure.

Let us repeat the text of the general procedure section by section—we will number the sections to refer to them more easily—and analyze the actions meant as well as the way of referring to them, while focusing more specifically on the two main topics of our chapter.

To begin with, the interpretation of the first section of the text requires understanding the task to be performed by the procedure, that is, determining the price in cash of a unit of a given commodity. It may be that the task is stated through the title of the section or through the problem statements linked to the procedure, like the one quoted above. The first section reads as follows:

1. 以所買_(買(賣))為法,以得錢乘一石數以為實。

"One takes what is exchanged as divisor. One multiplies the quantity of 1 *shi* by the cash obtained, which is taken as dividend."

The formulation does not mention any specific commodity or value, except for the "1 *shi*." This detail echoes the title of the pair of bamboo strips examined. Given the formulation of the procedure, it is clear that this "1 *shi*" has a paradigmatic value. However, how is it to be understood? Here we encounter the general question of how the meaning of paradigms was determined in ancient China (Chemla 2003). Is this "1 *shi*" to be interpreted as standing for any unit of capacity, as in the "trading salt" problem? In fact, *shi* is the name of the highest unit in the system of measuring units for weights as well as that for capacities. As we see below, the text of the procedure makes it clear that it applies equally to weights and capacities, and thus paradigmatically to any system of measuring units. However, although the procedure may seem to be applicable to any kind of unit, it in fact applies to only the highest measuring unit in a given scale.[81] This "1 *shi*" is thus meant to refer to the highest measuring unit in any system of units.

Like all the texts of the same kind analyzed above, this one starts with a section that addresses the fundamental case among all possible cases. For "Determining the standard price on the basis of the *shi*," this is the case when the quantity in the divisor—the thing traded—is expressed in *shi*, that is, in the same unit as the "1 *shi*."[82] At least, this is how one is tempted to interpret it against the background of the other texts of the same kind analyzed in the first two parts of this chapter, and this ainterpretation will prove below to be most probably correct. The analysis developed above thus provides us with some tools for interpretation.

---

[81] I cannot develop this point here. I refer to the publication in which I will analyze the system of such procedures that the *Book of Mathematical Procedures* contains.

[82] This corresponds to the initial item of the procedure for "Directly sharing" in *The Nine Chapters*.

Table 9.8 Positioning the values

Quotient	Below, this line won't be indicated			any longer
Dividend	1 **shi**			
			multiplied by	
	150 cash			
Divisor	1		*shi*	
	4		*dou*	Upper
	5		*sheng*	Middle
	1	3	*sheng*	Lower

For this fundamental case, the dividend and the divisor can be computed directly, as the text prescribes, and division yields the value sought. In the text, this section (1) is, however, followed by other sections (2–5) that deal with other cases and prescribe computations to be carried out on the divisor and part of the dividend, *before* the operations mentioned in section 1 are carried out. Again, the fundamental case is the one for the solution of which the procedure is provided in the first section of the general procedure and constitutes the end-segment of the procedures for all cases.

To make this point clear, I will present my interpretation of the actions corresponding to each section of the general procedure, as I said above, by reference to the "trading salt problem." The first segment of the general procedure indicates for this case the function that the various quantities will play in the overall computation. By means of these functions, the text also probably indicates a positioning of the quantities on the instrument on which computations were carried out. The *Book of Mathematical Procedures* contains indications that computations were carried out on a surface similar to the one evoked above, and it is not clear to which extent the principles for using the surface that can be recovered from later sources hold for this earlier use of the instrument. It will prove useful to restore the related configuration on the surface for each section. The first section for our example leads to the configuration on the surface as shown in Table 9.8.[83]

As we will see below, the key point is that in the case of the example displayed, the operations prescribed in the first section of the text must only be executed at the end of the process of computations, and not at the beginning, where the prescription occurs in the text. We come back to this point later. This feature derives precisely from the fact that we are dealing with a text similar to those of the general "procedures" examined above. This section is followed by two items, each starting with a condition:

2. 其下有半者倍之，少半者三之 "Those (divisors) for which, in their lower (rows), there is a half, one doubles them; those for which there is a third, one triples them."

In the case of the problem we are following, this yields the configuration shown in Table 9.9.

---

[83] Here, the actual configuration of numbers on the surface is indicated in the middle column. To help the reader follow, I add words in the middle column, as well as in the left and right columns. However, there is no reason to believe that anything was written on the surface.

Table 9.9 Getting rid of fractions

Dividend	1 **shi**			
			multiplied by	
	150 cash			
Divisor	3		*shi*	
	12		*dou*	Upper
	15		*sheng*	Middle
	1	(3)[a]	*sheng*	Lower

[a] I do not know whether the multiplication by 3 makes the "3" disappear from the surface or not, which is why I indicate it between brackets

Table 9.10 Getting rid of higher units

Dividend	1 **shi**		
		multiplied by	
	150 cash		
Divisor	300	*sheng*	
	120	*sheng*	Upper
	15	*sheng*	Middle
	1	*sheng*	Lower

The condition asking whether "there are parts" in the divisor is itself expressed by an enumeration of two specific cases—a mode of expression of something general that will recur below. The parallelism between the two sentences shows the pattern of the operation to be applied in all cases: one should multiply the whole quantity in the divisor by the denominator.

The next item starts with another condition, bearing on the possible presence of various units in the divisor. Its formulation is quite important for one of our purposes. It reads:

3. 有斗、升、斤、兩、朱(銖)者亦皆//破其上 "Those for which there are *dou* and *sheng, jin, liang* and *zhu*, one *also breaks up all* their upper (rows)." (My emphasis) (Table 9.10)

This key step calls for several remarks.

First, as stated before, we see that the possibility that there are various units in the divisor is expressed by a list of units that has a given structure: *dou* and *sheng* are units of capacity, whereas *jin, liang* and *zhu* are units of weight. The numerical relationships between the former are decimal. This does not hold true for the system of measuring units for weights. Whatever the case, again, the possibility that there could be any sequence of units in the divisor is expressed by two specific distinct cases, given here each in the form of enumeration, one having two items, the other three. Note that this indicates clearly that, although the procedure is formulated with respect to the *shi*, it is meant to be used for any similar unit.[84]

The second point in the sentence just translated that requires emphasis is the "also." Which similarity does it point to? In fact, most interestingly for our purpose,

---

[84] By "similar," I mean any highest unit in a scale. In fact, in the *Book of Mathematical Procedures*, other similar procedures are given for measuring units having other positions in a given scale. This kind of problem plays a key role in the book.

it expresses an analogy between the operation formulated in the previous step ("multiply") and the one that follows: "break up." The "also" thus indicates that "break up" is carried out by multiplying. This designates a multiplication to be applied to all the quantities attached to the higher units, which are placed in the higher rows of the divisor. This multiplication converts them into quantities of the basic measuring unit, the one placed in the middle row of the lower zone of the surface, that of the divisor. Note that what is to be multiplied is made precise thanks to a specific use of the positions in the lower zone of the surface ("all their upper (rows).")

Quite importantly for us, this directive is expressed by stating the world situation to be achieved: the multiplication breaks up these higher measuring units into smaller pieces. Remember that this interpretation in terms of disaggregation was precisely the one used by commentators on *The Nine Chapters,* when they expressed the meaning of operations on parts. We thus have here a conceptual continuity between the *Book of Mathematical Procedures* and the commentaries. However, when it comes to how an operation is prescribed by stating the world situation to be achieved, a difference appears. Although we meet here another instance in which an operation is prescribed—or a directive expressed—by means of the reason for carrying out the operation, the nature of the term used to do so is quite different from what we analyzed above. Here, "break up" refers to the *material* meaning of the operation, whereas, above, "make communicate" refers to the same operation by means of its formal meaning. We hence see that there appear here a set of similarities and differences between various expressions of directives in an indirect way.

But, let us go back to this "also". It does not only indicate that "break up" is to be interpreted as a multiplication. Conversely, it expresses the fact that the multiplication in section 2 had the meaning of "breaking up" units. In section 2, the directive, the illocutionary point of which is a multiplication, is formulated, though indirectly, by means of an elementary arithmetical operation. In section 3, the illocutionary aim is expressed by stating the world situation to be achieved for the entities in the context in which the operation is to be applied. The meaning of the operation, that is its "intention," is retrospectively assigned to the former multiplication. As a consequence, the text reveals the fact that the operations on fractions and various units are interpreted in the same terms.

As a result of the operations prescribed in sections 2 and 3, all the components in the quantity placed in the divisor zone are expressed with respect to the same basic unit. The next step seems to begin a section of the procedure common to all cases covered by sections 2 and 3—it is to be applied when condition one holds, or when condition two holds, or when both conditions hold. It reads:

4. 令下從之為法。"One makes the (rows) below join them, (yielding a result) which is taken as divisor. (Table 9.11)"

Note that the same term "divisor" now designates another value. The *Book of Mathematical Procedures* also makes use of the assignment of variables in the formulation of texts for algorithms.

The next section of the procedure, also common to cases for which at least one condition holds, is also formulated in quite an interesting way. It reads:

5. 錢所乘亦破如此 "What the cash multiplies is *also broken up like this.*" (The emphasis is mine.)

Table 9.11 Adding all values corresponding to the same unit

Dividend	1 **shi**	
		multiplied by
	150 cash	
Divisor	436	*sheng*

We have a third "also"—in fact, this time it corresponds to the "also" in the specific procedure given after the problem on bamboo strips 76–77. As above, it indicates that the prescribed operation that this "also" qualifies is motivated by a reason similar to an operation carried out earlier in the process of computation. In this case, the "also" expresses an analogy between the transformations applied to the divisor and that to be applied to "1 *shi*" (Chemla 2010b.) We will come back to its function. Before that, two new elements should be noted and they will allow us to conclude on the two key points we are focusing on in this chapter.

First, the value to which the operation "break up" is now to be applied is "1 *shi*." However, instead of being indicated through the meaning it has in the problem, this value of "1 *shi*" is designated by the part it played in the operation in which it was involved in section 1. This has an interesting consequence, which we have already indicated above, but which we can understand only now: for cases like the "Trading salt" problem, the operation prescribed in section 1 of the procedure can be executed *only after* the operations indicated in sections 2–5 have been carried out. The reason for this is simple: the multiplication in section 1 can be executed only when its terms have reached the values that need to be multiplied. We now see that one of its terms is modified in step 5. It is only when the operations needed by the cases meeting the conditions expressed in sections 2 and 3 have been performed that the multiplication indicated in section 1 is made possible. This conclusion is confirmed by the specific procedure formulated after the "Trading salt" problem. As a consequence, we see that, exactly in the same way as the "procedures" analyzed in the previous sections of this chapter, the "procedure for determining the standard price on the basis of the *shi*" requires a specific circulation in its text to yield the specific list of actions, for each case to which it can be applied. The circulation starts in the text with the conditions. The operations made necessary, depending on the conditions fulfilled, need to be carried out before going back to section 1 of the procedure and executing the multiplication and then the division. One should note that at the end of section 5, any case for the solution of which the procedure was applied would be reduced to a fundamental case, which could be solved by the operations prescribed in section 1. The "procedure for determining the standard price on the basis of the *shi*" is thus built according to the same principles as the other "procedures" examined above. It shows that such texts for procedures together with the reading they required existed as early as the beginning of the second century BCE.

Yet, the fact that section 1 is placed at the beginning of the text still seems to play a part. It seems to possibly indicate a positioning of the values on the surface, as we indicated in our succession of tables. The conditions then indicate the transformations that these values undergo before the operations of multiplication and division indicated in section 1 are executed. In other terms, part of what section 1 indicates is carried out, whereas another part, for instance the multiplication which is explicitly "prescribed," is not. This remark opens up a new perspective, showing that even the use of such a simple term as "multiplication" is by no means obvious in our sources. It may well be that, in the same way as a formula textualizes a set of operations in a condensed way, and that the operations are not executed before the formula has reached a certain form, the diagrammatic expression of algorithms on the surface that corresponds to the fundamental case, would serve as a basis for transforming its components via what the conditions indicate. The computation would be executed only when the configuration on the surface would have reached its final stage, corresponding to a fundamental case. Note, however, that it is precisely in this respect that the way in which the text for the specific procedure following the particular problem is formulated differs from that for the general procedure.

All these remarks confirm that our sources bear witness to the fact that we have faced two distinct types of procedures from as early as the time when the *Book of Mathematical Procedures* was composed—one linked to the prescription of specific computations and the other linked to the expression of reasons and integrating all specific procedures, the latter being regularly translated into the former. This concludes what we want to say about the structure of the text.

The second new element on which I will dwell appears when one considers the following question: what is the transformation of 1 *shi* that is prescribed by the final step? We recall that it is indicated by the expression "is also broken up like this." In the example followed, as we can see from the specific procedure, the transformation prescribed includes: multiplying by 3 (whose interpretation as "breaking all units" we discussed above) and multiplying the *shi* so as to break it up into *sheng*. The directive "is also broken up like this" combines two elements. On the one hand, the author S asserts the world situation that H's actions should achieve (break up). It does so in a way that was analyzed above. I will not repeat the analysis here. On the other hand, the way in which this "breaking up" should be achieved is indicated by analogy with the previous steps. It requires that step 2 be interpreted as "breaking up"—one of the uses of "also" in the text ensured that interpretation—and that step 3 is understood as asserted in the text for H to carry out the "breaking up" of 1 *shi* correctly.

As a result, in this second occurrence, the directive "break up" no longer rewrites a simple multiplication as was the case earlier. Here, its illocutionary point is the whole procedure that was applied to the divisor as a consequence of the way in which it met with the conditions listed in the procedure. We thus see that, as above for "making communicate," the procedure indicated by the directive "break up" varies. It can be longer or shorter, while it always keeps the meaning of "breaking up units." However, here, the determination of the procedure is highly contextual and essentially depends on the fact that the practitioner correctly interprets the meaning of the actions s/he has taken at each step of the procedure.

9.4 Conclusion

At the end of these analyses of various texts of procedures, what can we conclude?

Clearly, to begin with, the way in which operations were prescribed in ancient sources is a complex issue. Addressing this issue appears to be important for several reasons.

Firstly, in addition to showing how varied the ways of prescribing operations were, the previous analyses brought to light many complex phenomena in this respect. We saw that different texts, and even different kinds of text, had different ways of achieving the same illocutionary aim. Moreover, we saw that sometimes the same term ("*po* break up," "*tong* make communicate") corresponded to distinct procedures, depending on the context. We also saw that sometimes the simplest of all prescriptions ("*cheng* to multiply") did not correspond to an action at all, or, more precisely, that it corresponded to an action to be executed at a time yet to be determined (see Sects. 9.3.2). Describing the various ways of prescribing in general provides tools for interpreting the text of algorithms with greater certainty. The tasks, in particular, of grasping how to determine the actions meant by such terms as *cheng, po* or *tong* and of understanding better the distinct meanings conveyed by different ways of designating the same illocutionary aim prove essential if we are to interpret our sources more accurately.

Secondly, beyond the variety of modes for prescription we encountered, we could also identify, in the Chinese sources examined, recurring modes for prescribing and recurring phenomena in the ways of prescribing. It may be assumed that some of these modes were specific to the Chinese tradition that we are exploring. They may even constitute means for the historian to identify texts that were produced within the context of the same scholarly culture.

Thirdly, analyzing recurring ways of prescribing led us to put forward hypotheses regarding transformations in the way in which kinds of directives were carried out. I am thinking here of the recurring way that we encountered of prescribing by stating the world situation to be achieved. In the *Book of Mathematical Procedures*, in the second century BCE, the world situation was grasped in terms that referred to some of its material features, whereas in later texts the world situation was regularly approached in formal terms. This conclusion suggests that our analysis of speech acts could prove useful as a tool for historians, since it allows us to capture meaningful changes where previously they could not be perceived.

Fourthly, this last mode of prescribing constitutes a way of stating, through the very prescription itself, the reasons why the operations prescribed should be carried out. Our analysis thus gives further support to one of my earlier theses, namely, that the texts of algorithms can in and of themselves indicate the reasons for their correctness. However, seen from the angle from which we now see it, we have a better understanding of how this can be the case. In addition, the third point made above indicates that we could be able to grasp changes in the way of approaching the correctness of algorithms, by merely attending to the way in which algorithms were stated. Again, such tools are important for historians of science.

The previous set of issues represents only some of the problems raised by the analysis of texts of algorithms in general and in particular in the Chinese tradition to which all the sources examined are attached. We encountered other problems related to how the practitioner had to handle the texts of algorithms in order to derive from them the correct list of actions to be carried out for a given case. This issue was quite prominent for a particular set of texts that we identified and that, by contrast to other texts, had the common feature of including, within a single formulation, the lists of operations that should be applied to several distinct cases.

We have shown that these texts shared a common enumeration structure. Their initial section was introduced without a condition and described operations that were common to all the cases dealt with. It was followed by clauses introduced by conditions and followed by operations to be taken for the cases fulfilling the conditions. The practitioner had to know how to circulate, in a particular way that we could describe, between the various segments of the text in order to be able to know what to do for a given case.

It is interesting to note that in addition to these structural properties, these texts for algorithms share other properties. On the one hand, they are the very texts in which we find prescriptions by the statement of the world situation to be achieved. On the other, they handle the various cases in such a way that the way they are dealt with presents a kind of uniformity in the operations as well as in the reasons for using them.

The circulation required between the sections of the text, as well as the capacity to interpret the various prescriptions of operations alluded to above, bring to light the competences that a practitioner had to acquire or needed to possess to use these texts. Again, the description of these competences are a means for the historian not only to interpret these texts, but also to grasp features of the scholarly culture in the context in which they were written down.

We have shown that our sources contain these kinds of texts for algorithms from the second century BCE through to the seventh century CE. This remark indicates that part of the specificities of this scholarly culture in relation to the writing and handling of algorithms were handed down from generation to generation. On the other hand, it suggests that an analysis of the texts of algorithms can yield powerful tools to identify scholarly cultures and their transmission along traditions.

I read Jens Høyrup's analysis of cuneiform texts as a contribution to the same research program (Høyrup 1990, 2002). His results show how very different texts for algorithms can be between different traditions. These preliminary forays give hope that many new results can be extracted from our ancient sources. We may derive a better understanding of the various traditions to which our sources bear witness, as well as a better understanding of the texts of algorithms in general.

Appendix: The Sources Used in this Chapter

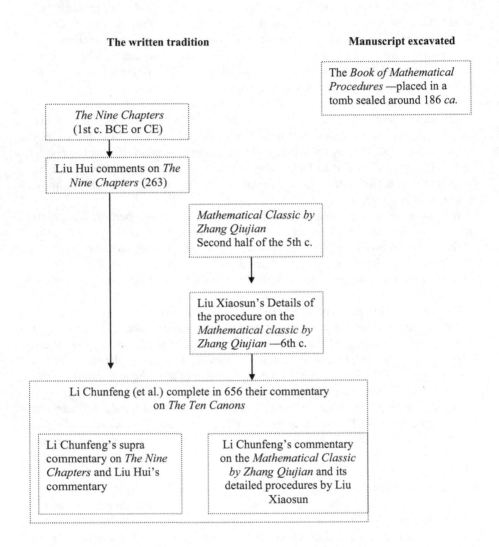

The written tradition   **Manuscript excavated**

The *Book of Mathematical Procedures* —placed in a tomb sealed around 186 *ca.*

The Nine Chapters (1st c. BCE or CE)

Liu Hui comments on *The Nine Chapters* (263)

Mathematical Classic by Zhang Qiujian Second half of the 5th c.

Liu Xiaosun's Details of the procedure on the *Mathematical classic by Zhang Qiujian* —6th c.

Li Chunfeng (et al.) complete in 656 their commentary on *The Ten Canons*

Li Chunfeng's supra commentary on *The Nine Chapters* and Liu Hui's commentary

Li Chunfeng's commentary on the *Mathematical Classic by Zhang Qiujian* and its detailed procedures by Liu Xiaosun

Bibliography

Chemla, K. 1992. Les fractions comme modèle formel en Chine ancienne. In *Histoire de fractions, fractions d'histoire*, eds. P. Benoit, K. Chemla and J. Ritter, 189–207, 405, 410. Basel: Birkhäuser.

Chemla, K. 1996. Positions et changements en mathématiques à partir de textes chinois des dynasties Han à Song- Yuan. Quelques remarques. In *Disposer pour dire, placer pour penser, situer pour agir. Pratiques de la position en Chine*, eds. K. Chemla and M. Lackner, 115–147, 190, 192. Saint-Denis: Presses Universitaires Vincennes.

Chemla, K. 1997a. Reflections on the world-wide history of the rule of false double position, or: How a loop was closed. *Centaurus. International Journal of the History of Mathematics Science and Technology* 39: 97–120.

Chemla, K. 1997b. What is at stake in mathematical proofs from third-century China? *Science in Context* 10: 227–251.

Chemla, K. 2003. Generality above abstraction: The general expressed in terms of the paradigmatic in mathematics in ancient China. *Science in Context* 16: 413–458.

Chemla, K. 2006. Documenting a process of abstraction in the mathematics of ancient China. In *Studies in Chinese language and culture—Festschrift in Honor of Christoph Harbsmeier on the Occasion of his 60th Birthday*, eds. C. Anderl and H. Eifring, 169–194. Oslo: Hermes Academic Publishing and Bookshop A/S. http://halshs.archives-ouvertes.fr/halshs-00133034, http://www.instphi.org/Festschrift.html.

Chemla, K. 2009a. Literacy and the history of science. Reflections based on Chinese and other sources. In *Cambridge handbook of literacy*, eds. D. R. Olson and N. Torrance, 253–270. Cambridge: Cambridge University Press.

Chemla, K. 2009b. On mathematical problems as historically determined artifacts. Reflections inspired by sources from ancient China. *Historia Mathematica* 36: 213–246.

Chemla, K. 2010a. Proof in the wording: Two modalities from ancient Chinese algorithms. In *Explanation and proof in mathematics: Philosophical and educational perspectives*, eds. G. Hanna, H. N. Jahnke and H. Pulte, 253–285. Dordrecht: Springer.

Chemla, K. 2010b. Usage of the terms "likewise" and "like" in texts for algorithms. Algorithmic analogies in ancient China. In *Analogien in Naturwissenschaft und Medizin*, ed. K. Hentschel, 329–357. Halle: Leopoldina (Acta Historica Leopoldina).

Chemla, K. 2012. Reading proofs in Chinese commentaries: Algebraic proofs in an algorithmic context. In *The history of mathematical proof in ancient traditions*, ed. K. Chemla. Cambridge: Cambridge University Press.

Chemla, K., and Guo Shuchun. 2004. *Les neuf chapitres. Le Classique mathématique de la Chine ancienne et ses commentaires*. Paris: Dunod.

Guo, Shuchun 郭書春, and Liu Dun 劉鈍. 2001. *Suanjing shishu* 算經十書. *Guo Shuchun, Liu Dun dianjiao* 郭書春,劉鈍 點校. Chonghe 中和 (Taipei county 台北縣): 九章出版社 Jiu zhang chubanshe.

Horng, Wann-sheng 洪萬生, Lin Cangyi 林倉億, Su Huiyu 蘇惠玉, and Su Junhong 蘇俊鴻. 2006. *Shu zhi qiyuan* 數之起源 *(The origin of mathematics)*. Taipei 台北: Taiwan Shangwu yinshuguan 臺灣商務印書館.

Høyrup, J. 1990. Algebra and naive geometry: An investigation of some basic aspects of Old Babylonian mathematical thought. *Altorientalische Forschungen* 17: 27–69, 262–324.

Høyrup, J. 2002. *Lengths, widths, surfaces: A portrait of old Babylonian algebra and its kin*. Sources and studies in the history of mathematics and physical sciences. New York: Springer.

Li, Jimin 李繼閡. 1993. *Jiuzhang suanshu jiaozheng* 九章算術校證 (*Critical edition of The Nine Chapters on Mathematical Procedures*). Xi'an: Shaanxi kexue jishu chubanshe.

Martzloff, J.-C. 1987. Histoire des mathématiques chinoises. Trans. S. S. Wilson. Paris: Masson.

Peng, Hao 彭浩. 2001. *Zhangjiashan hanjian «Suanshushu» zhushi* 張家山漢簡《算數書》注釋 (*Commentary on the Book of mathematical procedures, a writing on bamboo strips dating from the Han and discovered at Zhangjiashan*). Beijing: Science Press (Kexue chubanshe).

Qian, Baocong 錢寶琮. 1963. *Suanjing shishu* 算經十書 *(Qian Baocong jiaodian* 錢寶琮校點*)* (*Critical punctuated edition of* The Ten Classics of Mathematics*)*. 2 vols. Beijing 北京: Zhonghua shuju 中華書局.

Volkov, A. 1992. Analogical reasoning in ancient China: Some examples. In *Regards obliques sur l'argumentation en Chine. Extrême-Orient, Extrême-Occident. 14*, ed. K. Chemla, 15–48. Saint-Denis: Presses Universitaires de Vincennes.

Volkov, A. 2008. Raisonnement par analogie dans les mathématiques chinoises du premier millénaire de notre ère. In *L'analogie dans la démarche scientifique. Perspective historique*, ed. M.-J. Durand-Richard, 61–95. Paris: L'Harmattan.

張家山漢簡『算數書』研究会 Chôka zan kankan Sansûsho kenkyûkai. Research group on the Han bamboo strips from Zhangjiashan *Book of Mathematical Procedures*. 2006. 漢簡『算數書』 *Kankan Sansûsho. The Han bamboo strips from Zhangjiashan Book of Mathematical Procedures*. 京都 Kyoto: 朋友書店 Hôyû shoten.

Chapter 10
A Work on the Degree of Generality Revealed in the Organization of Enumerations: Poincaré's Classification of Singular Points of Differential Equations

Anne Robadey

Abstract This chapter attempts to bring to light and study an approach that emerges in Poincaré's earliest writings on differential equations. In the course of this research, Poincaré chooses not to consider cases that he calls "more special," by arguing that "they do not produce themselves if the polynomials [defining the differential equation being studied] are the most general of their degree." I wish to show that one can read in this a sign of a reflection on the degree of generality of the studied phenomena. This reflection is linked with a deliberate choice by Poincaré to study what happens in most cases, rather than in all cases. It can be shown that this direction of his work toward the essential continues to manifest itself in his later works in celestial mechanics.

10.1 Introduction

This chapter attempts to bring to light and study an approach that emerges in Poincaré's earliest writings on differential equations. In the course of this research, Poincaré chooses not to consider cases that he calls "more special," by arguing that "they do not produce themselves if the polynomials [defining the differential equation being studied] are the most general of their degree." I wish to show that one can read

Introductory Note: Chapter 10 was prepared by Anne Robadey on the basis of her Ph.D. Dissertation Robadey 2006, before she entered the CistercianAbbey Notre Dame d'Igny, in which she now lives. She could not revise the translation carried out by Rebekah Arana and prepared for publication by Pierre Chaigneau, Jonathan Regier and Myriam Smadja, whom we are glad to be able to thank here. We refer the reader to her dissertation for a French version and further developments of the ideas presented in this chapter (http://tel.archives-ouvertes.fr/tel-00011380/).

A. Robadey (✉)
Laboratoire SPHERE UMR 7219 (ex-REHSEIS),
Université Paris 7—CNRS, 75205 Paris Cedex 13, France
e-mail: chemla@univ-paris-diderot.fr

© Springer International Publishing Switzerland 2015
K. Chemla, J. Virbel (eds.), *Texts, Textual Acts and the History of Science*,
Archimedes 42, DOI 10.1007/978-3-319-16444-1_10

in this a sign of a reflection on the degree of generality of the studied phenomena. This reflection is linked with a deliberate choice by Poincaré to study what happens in *most* cases, rather than in all cases. Robadey (2006) shows that this direction of his work toward the essential continues to manifest itself in his later works in celestial mechanics.

However, in an essential *mémoire* that he published in two parts, (Poincaré 1881, 1882), Poincaré's view on generality is not yet thematized. It does not lead to the *definition* of a degree of generality, nor to the explicit introduction of a *criterion* which allows for a characterization of what is more general. Rather, Poincaré proceeds by distinguishing different cases and exploring them in order of decreasing generality. The only resource at our disposal for understanding how Poincaré assesses the generality of the cases for ordering them as he does is the text in which he enumerates them. One might despair of arriving at our goal with so little means to discern the reasons that might have led Poincaré to assert that one case is "more special" than another. However, the passage that we will study is structured very strongly in the form of an enumeration, which gives us access to precious information about his approach.

At first sight, an enumeration seems to be a simple object: a list of items, possibly numbered. In fact, linguists who are interested in enumerations have recognized the diversity that one may encounter in the study of this kind of text.[1] They have noted many attributes by which enumerations may be differentiated, and which therefore form important characteristics of a given enumeration. Some of these will prove useful in studying Poincaré's text.

Thus, the *classifier* used by Poincaré reveals its importance. By this term, linguists designate the nature of the enumerated objects: in the enumeration we are interested in, the classifier is the case, since Poincaré considers a first case, a second case, etc. I will show that listing "cases" rather than "kinds"—or genres, types, classes, etc.—of singular points is not an insignificant act: the classifier used by Poincaré emphasizes the approach that rules over the enumeration.

We will see besides that this enumeration is organized as a nested enumeration. Examination of its branching structure shed light on an initial form of hierarchy among the different cases.

More broadly, linguists have observed that very often, the different items in an enumeration are not—or not all—treated in identical or *parallel* fashion: either from the point of view of their syntax, or from the point of view of their visual structure, or even from the point of view of their content. We will see that the study of parallelism and the flaws of parallelism among cases at the same level in the tree structure highlights certain elements by which the cases differ, which correlate with the indications of generality given by Poincaré. Thus, the study of the structure of the enumeration gives the precious clues needed to bring to light the criteria used by Poincaré in determining the degree of generality of the different cases.

---

[1] See for example (Luc et al. 2000) and (Luc 2001). See also Chap. 5 in the present book.

These textual elements, considered in connection with the way that Poincaréanalyzes each case, will give us access to the mathematical work which ruled over the construction of the enumeration.

10.2   Texts and Context

Poincaré's writings on differential equations continue the work of Cauchy (1842a, b, c, d, 1843), as well as those of Briot and Bouquet (1856b).

Cauchy focused on demonstrating the existence of analytic solutions for differential equations—or systems of differential equations—of the form[2] $D_t x = X(x,t)$, where X is an analytic function in x and t (Cauchy 1842c, d). He shows that if the value of the solution corresponding to a fixed value t_0 of the variable t is given, this solution is defined by a convergent series in a neighborhood of t_0. He then applies the same method to show the existence of solutions of partial differential equations when the initial conditions are fixed (Cauchy 1842a, b, 1843).

In 1856, Briot and Bouquet continued with Cauchy's research. I am primarily interested in the second of three *mémoires* which they published together.

The first, "Étude des fonctions d'une variable imaginaire",[3] serves as an introduction to the other two. Briot and Bouquet intend to present in it the fundamental elements of the theory of functions in one complex variable. In particular, they specify the conditions under which a function of a complex variable can be expanded as a power series. In the third, "Mémoire sur l'intégration des équations différentielles au moyen des fonctions elliptiques"[4], they show that one can, under precise hypotheses, express the solution of differential equations of a certain class with the help of known functions, rational fractions, and simply or doubly periodic monodromic functions.[5]

In the *mémoire* which interests us, the second one (Briot and Bouquet 1856b), the authors study differential equations of the form $\frac{du}{dz} = f(u,z)$. They begin by giving a new, simpler proof of Cauchy's theorem: "if the differential coefficient $f(u,z)$ is a finite, continuous, monodromic, and monogenic[6] function for all values of u and z in a neighborhood of u_0 and z_0, the integral function u [which takes the value u_0 when $z = z_0$] is itself finite, continuous, monodromic, and monogenic for values of z in a neighborhood of z_0."

---

[2] $D_{t,x}$ denotes the derivative of x with respect to t. The two variables x and t are imaginary.

[3] Study of the functions of an imaginary variable.

[4] *Mémoire* on the integration of differential equations by means of elliptic functions.

[5] "Functions" as considered at that time were able to have several function values associated with the same value of the variable, as for example $f(z) = \sqrt[n]{z}$. A function of a complex variable was said by Cauchy, and after him by Briot and Bouquet, to be *monodromic* in a certain portion of the complex plane if it always takes the same value at the same point, regardless of which path is followed to arrive there, so long as one does not leave the portion of the plane under consideration.

[6] A function is said to be *monogenic* if it is differentiable on $\mathbb{C}$. Today, we say that it is *holomorphic* (AR).

They next concern themselves with what happens to the solution of a differential equation in the neighborhood of a point where the function $f(u,z)$ becomes infinite, is of the form $\frac{0}{0}$, or ceases to be monodromic. In other words, they study the solutions of differential equations in neighborhoods of points that one could call "singular points," according to the terminology that will be used by Poincaré.

Many of Poincaré's texts will be useful for my analysis. His very first article, (Poincaré 1878), intends to clarify the results obtained by Briot and Bouquet in the case where the differential coefficient is of the form $\frac{0}{0}$. He provides series expansions of certain non-holomorphic solutions which Briot and Bouquet had only shown to exist.

The following year, Poincaré defended his thesis before a jury composed of Bouquet, Bonnet, and Darboux, under the title: "Sur les propriétés des fonctions définies par les équations aux dérivées partielles"[7] (Poincaré 1879). He presented his work as the counterpart, for partial differential equations, of the study carried out by Briot and Bouquet for ordinary differential equations. He is, in this new framework, considering the cases where Cauchy's theorem no longer applies.

In 1880, Poincaré submitted a *mémoire* "Sur les courbes définies par une équation différentielle"[8] to the Académie des Sciences. This *mémoire* was announced by a *note* in the CRAS[9] (Poincaré 1880). It was withdrawn from the Academy before the designated committee made its report, and published in two parts in the *Journal de Liouville* (Poincaré 1881, 1882). Two supplementary sections were shortly later published in the same journal (Poincaré 1885, 1886). Results of these two final parts were announced in notes published in the CRAS in 1881, 1882, and 1884 (see Gilain 1977, p. 37, 53).

I will focus primarily on the first part of the *mémoire*, composed of chapters I to IV, which appeared in 1881. Poincaré starts from the theorem of Cauchy and from the results on the behavior of the solutions in a neighborhood of singular points obtained by Briot and Bouquet and himself. All of these works used to study the functions defined by differential equations "in the neighborhood of one of the points of the plane" primarily (Poincaré 1881, p. 3). Now, Poincaré proposes studying them "in the entire plane." The change in perspective signified by "the entire plane" is two-fold. On the one hand, Poincaré is interested in the solutions of the differential equation on the whole of their interval of definition and no longer only in the neighborhood of a point. On the other hand, he considers the group of all solutions, rather than *one* solution defined by an initial condition. On another level, he is interested in *real* solutions, which could be represented by curves, and gives a qualitative study of them. From the study of solutions in a neighborhood of singular points, for which he uses the results of Briot and Bouquet on the one hand and his thesis on the other, he demonstrates a formula linking the number of singular points

---

[7] On the properties of functions defined by partial differential equations.

[8] On curves defined by a differential equation.

[9] Les Comptes Rendus de l'Académie des Sciences.

of different types. He then obtains results on the global geometric forms which can be taken by the solution curves.

This research undertaken by Cauchy, Briot and Bouquet, and finally by Poincaré can be studied from various angles.

The thesis of C. Gilain explores one aspect: the transformation, with Poincaré's *mémoire* "Sur les courbes", of an analytic and local approach into a global study that involves the use of geometry (Gilain 1977, p. 35, 115). The study of this change in the approach, which is critical for the study of the history of differential equations, leads C. Gilain to particularly emphasize the innovation of the work presented since 1880 in the *Note* and the *mémoire* "Sur les courbes". From this point of view, however, the article from 1878, together with the thesis of 1879, still concern the analytical and local study. Returning to these texts to examine how the authors worked with respect to generality, I was led to propose a different periodization. Several of the features of the work of Poincaré which distinguish it from that of his predecessors begin in 1878 rather than only after the *mémoire* of 1880 "Sur les courbes".

My work will focus on a second aspect of the research of Briot, Bouquet, and Poincaré: the treatment of special cases.

Before beginning the analysis of this aspect, I would like to highlight a characteristic common to the works of our four authors, although it appears in a different fashion in Poincaré's work. As a matter of fact, this remark will permit me to emphasize a feature of Poincaré's work in the *mémoire* "Sur les courbes", which will be important for my analysis.

All the work that we have introduced so far concerns what C. Gilain called the "general study of differential equations," to which he restricted his analysis—I will also do so. This approach consists of studying the properties of the solutions of the differential equation directly from the equation itself, rather than trying to express the solution with the help of known functions. It is introduced and motivated from the first lines of the *mémoire* of Briot and Bouquet.[10]

> The cases where one can integrate a differential equation are extremely rare and must be viewed as exceptions. But one can consider a differential equation as defining a function, and set about studying the properties of this function based on the differential equation itself. (Briot and Bouquet 1856b, p. 133)[11]

This approach follows that of Cauchy's research, which he himself did not claim was original (Cauchy 1882–1974, s. 1, t. 6, p. 462).

Cauchy attempts to close a gap in earlier research: one did not prove the existence of solutions, but was content to seek series expansions for solutions that were assumed to exist, without any guarantee of the convergence of these series. Cauchy demonstrates this convergence from the differential equation itself, which ensures, based on previous work, that the functions thus defined are solutions.

---

[10] My work focuses almost exclusively on the second *mémoire* of these authors, (Briot and Bouquet 1856b). It is therefore that *mémoire* which is referred to in the absence of further clarification.

[11] Translated from French. See all the original versions of all quotations in the Appendix 1.

Poincaré expresses the same interest in the general study of differential equations at the beginning of the introduction to his first *mémoire* "Sur les courbes". He gives the same motivation for this as Briot and Bouquet.

> Unfortunately, it is obvious that in most cases that present themselves, one cannot integrate [the differential equations] with the help of known functions, for example with the help of functions defined by quadratures. If one wished therefore to restrict oneself to the cases which one could study with definite or indefinite integrals, the field of our research would be singularly diminished, and the great majority of questions which arise in applications would remain insoluble.
>
> It is therefore necessary to study those functions defined by the differential equations in themselves, without trying to reduce them to simpler functions. (Poincaré 1881, p. 3)[12]

This issue in the general study of differential equations, as opposed to the search for conditions under which one can integrate these equations, is therefore motivated by a search for generality which is common to our authors. It seems to me to belong to a movement, which remains to be studied as such, in the history of analysis in the nineteenth century.[13] This movement, illustrated in various ways by Cauchy, Briot and Bouquet, and Poincaré, seems, for example, to share significant similarities, from the point of view of the research of generality, with a turning point in the study of functions that can be expanded as a Fourier series. Riemann adopts, in his *Habilitationschrift* of 1854, a new point of view that was sketched out at the end of an article by Dirichlet (1829). He presents it as follows:

> The work that we have mentioned regarding this issue had as its goal to prove the Fourier series for functions encountered in mathematical physics [...]. In our problem, the only condition that we impose on the functions is that of being able to be represented by a trigonometric series; we search therefore for the necessary and sufficient conditions for such a mode of expansion of the functions. While previous work established propositions of the following type: "if a function has such and such a property, it may be expanded in a Fourier series", we ask ourselves the inverse question: "if a function can be expanded as a Fourier series, what results are there regarding the path of this function, about the variation in its values, when the argument varies in a continuous manner?" (Riemann 1854, pp. 246–247)[14]

Thus, Riemann aims to study a function directly from the hypothesis that it can be expanded as a Fourier series, although this definition does not consist of a finite and explicit formula. In the same way, general study of differential equations proposes an examination of the functions defined by the differential equations directly from the equations, without trying to express them in another form. In both cases, the direct study of functions from the way they are defined derives from a search for generality. Riemann wishes to study *all* the functions which can be represented by a Fourier series, and not only the Fourier series expansion of certain classes of functions. In the same way, general study of differential equations aims to examine

---

[12] Translated from French.

[13] A first step in this direction has been made within the context of a seminar organized by the team REHSEIS on generality, with, in particular, a talk by R. Chorlay.

[14] Translated from French, cited by R. Chorlay.

the solutions of *all* differential equations, without limiting itself to those equations which have solutions that can be expressed in a particular form.

The case of differential equations allows us to identify two possible modalities for this search for generality. The desire to not limit themselves to the exceptional equations which can be explicitly integrated drives Briot and Bouquet to study differential equations in a very general form, without restrictive hypotheses. Cauchy likewise considers "all differential equations or partial differential equations". It is again in terms of a lack of restrictions that Riemann characterizes his approach in the quotation above.

Poincaré begins, in his *mémoire* "Sur les courbes", by recalling the same demand: not "to restrict himself to the cases which can be studied using definite or indefinite integrals." But the research program that he specifies rests rather on a reformulation of this requirement. Briot and Bouquet already give a statement close to this second formulation: "to study the functions defined by the differential equations 'in themselves.'" But they do not make it central to their approach. They were rather careful to write the equations in the most general form possible in order to avoid unnecessary restrictions. On his side, Poincaré did not pretend to study the most general equation, nor even a very general equation. He affirmed to the contrary: "I have restricted myself to a very special case, the one that naturally arises first of all, *i.e.* the study of differential equations of the first order and of the first degree." (Poincaré 1881, p. 5). In fact, this is already the framework in which Cauchy, Briot, and Bouquet worked; but they did not present it as a very special case. Poincaré restricts even further the group of equations that he considers: he is interested in the curves defined by an equation of the form $\frac{dx}{X} = \frac{dy}{Y}$ "where X and Y are two polynomials with integer coefficents in x and y" (Poincaré 1881, p. 5). Where Cauchy, Briot and Bouquet would make only an assumption of analyticity on the function defining the differential equation, Poincaré limits himself to the equations defined by two polynomials.

In the *mémoire* "Sur les courbes", this restriction is exploited in a central way by Poincaré. It allows him—we will come back to this soon—to work in the projective plane. But already in his thesis, he formulates a similar hypothesis, without it being necessitated by the chosen approach (Poincaré 1879, p. LXV).

Thus, from this moment, the manner of his work differed from that of his predecessors, and he did not care to work on equations that were as general as possible.

We will return to Poincaré's choice to work with polynomials, rather than with analytic functions, for example. This remark highlights two methods of presenting the general study of differential equations, or more generally the study of an object in its generality. One may work directly on an object as general as possible. But one can equally attempt to develop on a possibly very restricted class of objects a method of study—here the study of the solution *on the differential equation itself*—appropriate for the desired generality. One finds a comparable approach in Poincaré's *mémoire*, "Sur les lignes géodésiques", discussed in the article (Robadey 2004).

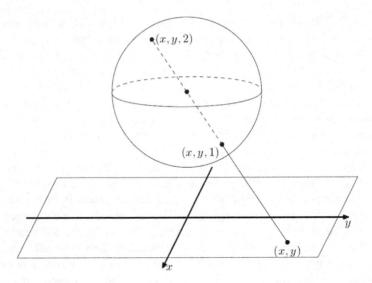

Fig. 10.1 Projection of the plane on the sphere

10.3   The Classification of Singular Points: A Work on The Degree of Generality of The Cases Considered

The *mémoire* "Sur les courbes" (Poincaré 1881) aims to study the curves defined by a differential equation of the form $\frac{dx}{X} = \frac{dy}{Y}$, where X and Y are polynomials in x and y. Poincaré calls these curves "*characteristic*"[15] Rather than studying them in the plane (x, y), where the study of infinite branches might introduce difficulties, Poincaré projects them onto a sphere (see Fig. 10.1): to each point (x, y) of the plane, he associates two points $(x, y, 1)$ and $(x, y, 2)$ on the sphere, where $(x, y, 1)$ and $(x, y, 2)$ are the points where the line passing through (x, y) and the center of the sphere intersects the sphere. The points on the equator of the sphere correspond to the points at infinity in the plane. In other words, Poincaré studies the considered equation in the projective plane, viewed as the sphere with opposite points identified. The fact that the equation is defined by polynomials permits, by a simple change of coordinates, the study of the *characteristics* in a neighborhood of the equatorial points: the equation in the new coordinates is of the same form as the initial equation.

Repeatedly, we will see Poincaré using some concepts and methods from projective geometry in this way. I will not attempt to analyze this phenomenon. It could undoubtedly be the subject of a separate work, on a broader body of work than only

---

[15] Poincaré uses the italics only at the moment where he introduces this terminology. As far as I am concerned, I will maintain the italics as necessary to permit the reader to avoid ambiguity between the mathematical sense given to the term by Poincaré, and the ordinary meaning that I will regularly use in the course of my analysis.

Poincaré's research, for studying the modalities of the introduction of projective geometry into the field of analysis.

The first chapter of the *mémoire* is devoted to the definitions and general properties of the *characteristics* and of curves drawn on the sphere.

It is to the details of the second chapter that we will now turn our attention. A table (Table 10.1) summarizing the contents of this chapter can be found in Appendix 2. Poincaré studies here the *characteristics* in the neighborhood of a point of the sphere, especially in the neighborhood of singular points. My objective is to show that the way in which Poincaré enumerates the types of singular points reflects a consideration of the degree of generality of the cases that he is led to distinguish. Furthermore, the study of the construction of this enumeration permits the identification of the guidelines of his approach and the means by which Poincaré evaluates the degree of generality of the different cases. As I have indicated in the introduction, it will be necessary for us to examine the different structuring elements of this enumeration with great precision. Having thus drawn Poincaré's working method which is visible in this text, we will be able to examine how he relies on, or distinguishes from, previous research.

10.3.1   The Division of the Text Into Sections

The chapter we focus on, entitled "study of the characteristics in the neighborhood of a point of the sphere", is composed of an introductory section followed by eight sections, each provided with a title in italics:

> *First case.*—[...]
>
> *Second case.*—[...]
>
> *First subordinate case.*—[...]
>
> *Second subordinate case.*—[...]
>
> *Third subordinate case.*—[...]
>
> *Fourth subordinate case.*—[...]
>
> *Special cases left aside.*—[...]
>
> *Points situated on the equator.*—[...]

This list of sections calls for several remarks.

From the point of view of the material formatting of the text, of the typography,[16] the eight sections are presented in the same way, at the same level. By contrast, the content of the titles shows a branching structure with at least two levels of depth: the *"subordinate cases"* are visibly subdivisions of the *"Second case."* The first six sections are clearly positioned in this way in the tree by the text of their title. This is

---

[16] The edition of Poincaré's *OE uvres* is faithful on this point to the typography of the original *mémoire*. The only difference is in the positioning of the page breaks.

not the case for the last two. It is necessary to take into consideration the contents of the sections for recognizing that the seventh, *Special cases left aside*, is situated—at least at first approximation, we will come back to this—at the same level as the four which precede it, *i.e.* under the *Second case*. The last section, as for it, corresponds to a hypothesis formulated even before the distinction between the *first* and *second cases* is made. Therefore we deal in reality with three levels of hierarchy.

[Points not on the equator]
>   *First case.*—[…]
>   *Second case.*—[…]
>>   *First subordinate case.*—[…]
>>   *Second subordinate case.*—[…]
>>   *Third subordinate case.*—[…]
>>   *Fourth subordinate case.*—[…]
>>   *Special cases left aside.*—[…]
> *Points situated on the equator.*—[…]

The outermost level, which corresponds in particular to the last section, interests us the least here. One can further note that the title of this final section is a little different than the others: it refers to "points" while the other sections mention "cases." Mathematically, Poincaré first shows how to study the *characteristics* in the neighborhood of a point not located on the equator. He is led to distinguish several cases, presented in the first seven sections. In the last section, he gives the change of variables needed to reduce the study in the neighborhood of a point on the equator into the former problem.[17] It is therefore not a different "case," but simply a preliminary manipulation which leads to the same list of cases. I will leave aside this level of the hierarchy, and in what follows, I will concern myself only with the first seven sections.

This elucidation of the hierarchy of the enumerations leads us directly to a second remark. Even before entering into the mathematical details of Poincaré's text, we see that the enumeration that he produces respect at first a perfect parallelism among the items, but that this parallelism disappears at the end of the enumeration. In particular, the first four subdivisions of the *Second case* all have titles formed in the same way: a cardinal number and the classifier "subordinate case." The title of the final subdivision is completely different. It still takes the classifier "case," but in the plural. So, the different items, the different subordinate cases, are not put on the same level: after four subordinate cases whose treatment is, at least if one believes the titles, similar, a group of other cases is collected and distiguished by the qualifier "special." This textual phenomenon, which appears from the simple observation of the section titles, is accompanied by many others, as we shall soon see. So Poincaré reserves a special treatment for the last cases which he considers. This raises two

---

[17] As I indicated at the beginning, this is possible because the equation is defined by two polynomials.

sets of questions which we seek to answer. The first concerns the interpretation of this textual phenomenon: how are these "special"? How are they different from the others? The second refers to Poincaré's work in producing this enumeration: how and why is Poincaré led to emphasize the specificity of these cases? Why are they grouped together when the subordinate cases were examined one by one?

A final point emerges from the study of the section titles. It concerns the classifier of the two nested enumerations in which we are interested. At first sight, one may think that the titles that we examine—excepting the seventh—are purely functional. They indicate only the position in the tree, and may be replaced by a convenient numbering: $1 - 2 - 2.1 - 2.2 - 2.3 - 2.4$. But it seems to me that the term "case" which appears in these titles is not a word without meaning, required by the grammar in relation to the ordinals "first," "second," etc. Nor does it designate sections of the *text*, but refers to a mathematical truth.

Indeed, Poincaré repeatedly refers to "cases" he has studied, and thus he makes them become objects of his work. Regarding points on the equator, he writes not that we are reduced to the previous study, but "to the cases studied in the beginning of the Chapter." Elsewhere, Poincaré says that "[the fourth] case is more special than the previous ones and [that] it will not occur if X and Y are *the most general polynomials of their degree*" (Poincaré 1881, p. 17).[18] He enters into mathematical considerations about this "case" in this way. The various "cases" are not just sections of the text: their definition is part of the mathematical work through which Poincaré studies the singular points.

The importance to be given to "cases" is reinforced by the existence of another classifier which plays a different role: "types" of singular points. In the middle of the section on "*Special cases left aside*," Poincaré makes an intermediate assessment where he introduces the class of "*singular points of the first kind*." I will return to this class later. For now, I am interested in what he says of this class: "We have seen that there are four types of such points: nodes, saddle points, foci, and centers." But this enumeration does not correspond at all with the sections of the study. In the first subordinate case, we get a node, in the second, a saddle point, in the third, a focus, and in the fourth, a focus or a center. The first of the special cases left aside, meanwhile, gives a node. Thus, certain "types" of singular points occur in two different "cases," and the fourth "case" leads to two different "types." It is therefore necessary to distinguish "cases" and "types of singular points." What is it that separates them? The "types" are the *result* of the analysis: we will see that in each section, Poincaré first studies the behavior of the solutions. He gives the name of the type of singular point which he is dealing with only at the end of the study, just before giving an illustrative example. The "cases," conversely, *organize* the whole study, and correspondingly, they appear in the titles of the sections. Poincaré shows that in order to study the *characteristics* in the neighborood of a point, one can differentiate several "cases" from the problem data. In each "case," he offers an appropriate treatment.

---

[18] Here, as in what follows, I retain the italics of the quoted authors unless stated otherwise.

It is primarily at the level of the separation of the "cases" that Poincaré's thoughts on the degree of specificity of the studied phenomena are expressed. He talks about "special cases" and among them cases that are "more special". About a "case" he says that it will not appear "if X and Y are *the most general polynomials of their degree*" (Poincaré 1881, p. 17). So, what we must analyze to answer the questions posed above and to understand the nature of the distinctive features of certain cases and the way that Poincaré expresses, establishes, and makes use of it, is the process of distinguishing these cases and the way that they are studied.

10.3.2   Singular Points and the Tools Used for Classifying Them

The introduction to the chapter we are studying opens with a bibliographical note where Poincaré places his works as a continuation of those of Cauchy (1842a, c, d 1843) on the one hand, and of Briot and Bouquet (1856b) on the other. It also refers to his earlier work (Poincaré 1878, 1879). We examined the research of Cauchy, Briot and Bouquet, and Poincaré respectively in (Robadey 2006, pp. 53–70). This study allows us to understand how the work of hierarchisation of the cases, which we will now study, connects with the research of Poincaré's predecessors, as well as with his early work on differential equations.

After these preliminaries, Poincaré assumes that one intends to study the *characteristics* near a point with coordinates (α, β), and he sets up notations which will then be used to differentiate the different cases. He introduces a_0, a_1, a_2, b_0, b_1 and b_2, which are the coefficients of the terms of degree 0 and 1 in x and y of the polynomials X and Y expanded around the point (α, β):

$$X = a_0 + a_1(x - \alpha) + a_2(y - \beta) + ...,$$

$$Y = b_0 + b_1(x - \alpha) + b_2(y - \beta) + ...$$

With the notation as specified, the *first case* is determined by the condition that a_0 and b_0 are not both zero. It is then the theorem proved by Cauchy (1842a) which permits us to conclude that one, and only one, *characteristic* passes through the considered point. The *second case case* is also characterized by a condition on the coefficients a_i and b_i: "$a_0 = b_0 = 0$. But a_1, a_2, b_1 and b_2 are not all zero at the same time." This brings to light two complementary forms of the conditions. Some are given by equalities (most often, the cancellation of one or more parameters), others by inequalities (non-nullity). We will come back to this difference.

Having characterized the *second case*, Poincaré comments, saying that the considered point is then a "ordinary singular point," and then immediately states a theorem proved in his thesis. I will refer to it in the sequel as "Poincaré's Theorem".[19] This is how he formulates it here:

---

[19] Poincaré indicates in his thesis that the statement was suggested by Darboux.

If the equation

$$(a_1 - \lambda)(b_2 - \lambda) - b_1 a_2 = 0 \tag{10.1}$$

has two different roots, λ_1 and λ_2;
If the ratio of these roots is positive or imaginary, the general integral of the equation

$$\frac{dx}{X} = \frac{dy}{Y}$$

is of the form

$$z_1^{\lambda_1} z_2^{\lambda_2} = \text{const.},$$

where z_1 and z_2 are ordered series of ascending powers of $x - \alpha$, $y - \beta$, and vanish for

$$x = \alpha, \qquad y = \beta.$$

(Poincaré 1881, p. 14)

In this statement, Poincaré forgets a hypothesis that appeared in his thesis—which is necessary for the theorem to be valid. We still need to ensure that the roots λ_1 and λ_2 are not linked by a relation $p\lambda_1 + q\lambda_2 = 0$ with integer coefficients, now called *resonance*. I will first examine what Poincaré does while ignoring this additional assumption. This will enable us to see how Poincaré proceeded in his analysis of singular points. The conclusions we arrive at will allow us to make a very satisfactory account of the omission of the last assumption.

We see first that the conditions on the a_i and b_i through which Poincaré characterizes the first and second cases are immediately used to formulate theorems. Indeed, it is only when a_0 and b_0 are nonzero that we can apply the theorem of Cauchy. The conditions on the a_i and b_i characterizing the *second case* are also bound by Poincaré to the statement of his theorem. Immediately after having translated them, in saying that "the point (α, β) is then an ordinary singular point," he refers to his theorem as "a theorem relative to *these* singular points" (emphasis mine).

Moreover, this theorem leads directly to the enumeration of subordinate cases. These latter are distinguished by the real or complex values of the λ_i and by their signs, which appear in the statement of the theorem for determining its application conditions: [20]

- First subordinate case: λ_1 and λ_2 real and with a positive ratio.
  Poincaré's theorem applies.
- Second subordinate case: λ_1 and λ_2 real and with a negative ratio.
  Poincaré's theorem does not apply, but the work of Briot and Bouquet allows a conclusion.

---

[20] See Table 10.1 summarizing all cases distinguished by Poincaré and the main points of their study, in Appendix 2.

- Third subordinate case: λ_1 and λ_2 complex and with an imaginary ratio. Poincaré's theorem applies.
- Fourth subordinate case λ_1 and λ_2 complex and with a ratio equal to -1. Poincaré's theorem does not apply.

Thus, the organization of the list of cases formed by Poincaré is based on the hypotheses of the theorems that will be used. This agrees with the use of the classifier "case." The question is to establish that a problem presents only a set of cases and to describe them by applying various theorems according to the data, rather than listing the results. In other words, Poincaré does not seek the conditions under which one has a certain type of singular point, but he examines what can be said about the singular point based on significant parameters; those which appear in the hypotheses of available theorems. The text we study is thus akin to a procedure having several alternatives, rather than to a list of conclusions.

We can go further by noticing that the various theorems which are mentioned are not of the same level of importance. Cauchy's theorem is recalled in order to handle the *first case*, that of nonsingular points. This *first case* is only mentioned at the beginning, but Poincaré does not dwell on it and proceeds immediately to the question of singular points. He refers to the work of Briot and Bouquet where necessary, but it is very much Poincaré's theorem that is at the heart of the classification. It is stated at the opening of the *second case*, and after some geometrical considerations that we will return to in a moment, Poincaré concludes: "Now we can divide the second case into four major subordinate cases." It is thus explicitly Poincaré's theorem, and the geometric considerations that accompany it, which is used for the distribution of the singular points into the four subordinate cases.

Even in the two subordinate cases where Poincaré's theorem does not apply—the second and the fourth—, our author does not content with showing what can be concluded about the shape of the characteristics. He begins by noting that his theorem does not apply. Moreover, the expression of the integral in the form $z_1^{\lambda_1} z_2^{-\lambda_2}$ = const. given by this theorem remains the preferred form in which to indicate the form of the solutions, even in those cases where the theorem does not apply. Even when it does not make it possible to finish the study, the theorem remains the benchmark for the form in which one obtains solutions.

Moreover, in an occurence, Poincaré refers to his theorem using the expression "the general theorem" (p. 16) rather than "the theorem that we recalled in starting" (p. 14), "the theorem we have spoken of" (p. 15), etc. This shows the significance Poincaré gives to this theorem and to the preferred form in which it expresses the integral of the equation.

If the subordinate cases are distinguished by the signs of the roots λ_1 and λ_2, as we have seen, this is not the first thing that Poincaré mentions in their description. The subordinate cases are first characterized by geometric properties—even though Poincaré stresses that they are directly related to the values of λ_i. In modern parlance, the λ_i are the eigenvalues of the linearized equation in a neighborhood of the singular point (α, β) under consideration, but it is the corresponding condition on the eigendirections of the equation that Poincaré gives first.

Here again, Poincaré uses the concepts of projective geometry to reveal what we call eigendirections, and which he called "double lines of the homographic pencil." To a direction defined by its angular coefficient (slope) μ, Poincaré associates the limit direction of the tangent to the characteristics when one approaches the singular point in the direction μ given by the formula $\frac{b_1 + b_2\mu}{a_1 + a_2\mu}$. In other words, he associates with a direction its image under the linear mapping defining the linearized differential equation. The above formula being that of a homography, Poincaré concludes:[21]

> Thus the limit of the line joining the points (x, y) and (α, β) and the limit of the tangent to the characteristic at the point (x, y) form a homographic pencil. (Poincaré 1881, p. 14)

Having given the formula for the homography, Poincaré deduces the equation of the double lines: $\mu(a_1 + a_2\mu) = b_1 + b_2\mu$, where μ is the slope. This equation admits two roots μ_1 and μ_2 about which Poincaré adds: "we easily calculate λ_1 and λ_2 as rational functions of $a_1, a_2, b_1, b_2, \mu_1$ and μ_2." This shows the link between the roots λ_1 and λ_2 involved in the assumptions of the theorem on the one hand, and the double lines which have geometric meaning on the other. Poincaré proposes to deduce the λ_i from the μ_i, and indeed it is effectively in that order that he gives characterizations of each subordinate case: the geometrical condition on the double lines of the homographic pencil first, the condition on λ_i afterwards. Thus, for the first subordinate case—the others are presented in the same way:

> *First subordinate case.*—The double lines of the homographic pencil are real, and any two conjugate lines of the pencil are both in the acute angle formed by the two double lines, or both in the obtuse angle.
> In this case, λ_1 and λ_2 are real and their ration is positive.
> (Poincaré 1881, p. 14)

The geometric characterization that is thus put forward by Poincaré is not a simple illustration, a tacked-on commentary. It is part of the means used to understand the form of solutions of the differential equation. Poincaré uses it directly in his study of the fourth subordinate case, when he remarks: "it is impossible for a branch of characteristic to pass through the point (α, β), since its tangent should be precisely one of the double lines of the involution, which are imaginary." The two characterizations, by the double lines of the homographic pencil and by the values of λ_i,

---

[21] Chasles defines homographic pencils in his *Treatment of higher geometry*: "When two pencils whose lines correspond one to one are such that any four lines of the first have their cross ratio equal to that of the four corresponding lines of the second, we will say that the two pencils are *homographic*." (Chasles 1880, p. 64) It is a special case when the two homographic pencils have the same center. Chasles shows that then there are two special lines "each of which, considered as belonging to the primary pencil, is itself its counterpart in the second." He calls them "*double spokes*" (Chasles 1880, p. 115). The problem considered by Poincaré belongs to this special case: the two pencils are composed of the same lines, which leads Poincaré to speak, in the singular, of "a homographic pencil" and of "double lines of the homographic pencil." The "homographic pencil" of Poincaré thus associates to each line through the point a second line passing through the same point, so that any four lines passing through have the same cross ratio as the four lines associated to them. In modern terms, it is simply a linear mapping. Poincaré then speaks of "two conjugate lines" to describe a straight line and its image under the linear mapping.

are thus jointly exploited by Poincaré, without one appearing as a subordinate to the other. The conditions on the λ_i derive their importance from the centrality of Poincaré's theorem for the study of the subordinate cases. The prominence of the geometric conditions comes from the specificity of Poincaré's *mémoires*, "Sur les courbes", studied by C. Gilain in his thesis. Poincaré's approach contrasts with that of his predecessors in his choice to study the "curves" defined by differential equations, rather than the "functions."

Besides, when Poincaré defines the homographic pencil and its double lines, he presents these geometric objects by linking them to both the geometric shape of the solutions and to the analytical conditions for the application of his theorem. Indeed, on the one hand, the homographic pencil is defined from the tangents to solution curves near the singular point. On the other hand, Poincaré indicates the relationship between the solutions of the equation giving the double lines and the roots, λ_i. Thus it is undoubtedly because it provides an intermediary between the geometry of solutions and the conditions of applicability of the theorem, that the geometric characterization of individual subordinate cases comes first in the description of the subordinate cases.

10.3.3   Several Modes of Hierarchization

From the standpoint of the logical branching structure of the enumeration, the *first case* and *second case* appear at the same level. Likewise for the four *subordinate cases*. But the content of the sections shows that these different cases, and even more so the *special cases left aside*, are not treated in a similar fashion.

10.3.3.1   Singular Points

Poincaré does not emphasize the difference between the first and second cases. In the text of 1881, the only gradation is in the name he gives to the points corresponding to the second case: "ordinary singular points." The term "singular" is clear, however: the second case corresponds to a singular situation compared to the first. The contrast is even more explicit in the summary published the previous year in the *Comptes Rendus*:

> Thus we see: 1° that, by all points of the sphere, except for some singular points, passes one and only one characteristic; 2° that by some singular points pass two characteristics, 3° that by other singular points pass an infinity of characteristics; 4° finally, that a third type of singular point is such that the nearby characteristics turn like spirals around these points without passing through them. I call these three types of singular points the *saddles*, the *nodes*, and the *foci* of the equation. (Poincaré 1880, p. 1)

The three types of singular points listed here, we shall show, correspond with the first three subordinate cases. The numbering puts them on the same level as the points occurring in the framework of the first case, which were mentioned first. But they are distinguished from them in several ways. Their name, "singular" points, reinforces what is already noted by the presentation of the first type of behavior "at all points [...] except [...]." This first behavior is not merely one among others, but that of "all" points, before the exceptions already mentioned. Thus we see overlap, in both the *Note* and the *Mémoire*, between an enumeration that places the first and second cases side by side, and a gradation which contrasts the "singular" points with the others.

A similar phenomenon was already present in Briot and Bouquet (Robadey 2006, pp. 53–70). But what is new in Poincaré is that we find this phenomenon recurring at the next level, among the singular points themselves. The existence of a gradation among different types of singular points is already suggested by the term "ordinary singular points." If some singular points can be called ordinary, then not all are so: the points where $a_1 = a_2 = b_2 = b_1 = 0$ do not fall into the category of ordinary singular points. Furthermore, we will see that Poincaré notices different degrees of generality among the ordinary singular points themselves.

10.3.3.2   More Special Cases, Which do Not Arise if X and Y are the Most General Polynomials of Their Degree

There appears to be a demarcation, from the point of view of generality, beginning with the fourth subordinate case. The first three subordinate cases are treated in parallel fashion according to the following pattern:

- Statement of a condition on the double lines of the homographic pencil.
- Translation into a condition on the values of λ_1 and λ_2.
- Result regarding the applicability of the theorem of Poincaré. In the first and third subordinate cases, it applies and provides the form of the general integral of the equation. In the second, it does not apply and it is the work of Briot and Bouquet which permits the determination of the shape of the solutions.
- Conclusion: behavior of the *characteristics* near the point considered.
- Name given to the singular point obtained.
- Example.

The treatment of the fourth subordinate case begins in a similar way to that of previous cases: Poincaré first gives a geometric characterization in terms of the double lines of the homographic pencil, which are imaginary and in involution. The corresponding analytical condition is that λ_1 and λ_2 are conjugates, with ratio -1. From this condition, Poincaré concludes—as he does in the second subordinate case—that his theorem does not apply.

Then the parallel becomes blurred. Rather than trying to completely resolve this case, Poincaré begins by distinguishing it from the preceding cases from the point of view of the degree of generality:

> This case is more special than the previous ones and it will not arise if *X* and *Y* are *the most general polynomials of their degree*. Let us confine ourselves therefore to a few remarks. (Poincaré 1881, p. 17)

Three points are noteworthy in this quotation. First Poincaré notes that this case is "more special". Somewhere else I propose an elucidation of the elements of analysis that led to this remark, and what he means by "the most general polynomials of their degree" (Robadey 2006, pp. 70–82). We will come back to this. The second point is the consequence that he derives from this observation. Poincaré concludes that he needs not study the fourth case in detail. He simply states the results he has: first that, as the double lines of the involution are imaginary, no branch of a characteristic can pass through the singular point being considered; then, that the characteristics in a neighborhood of the point can only be spirals or cycles, so that the singular point is either a *focus* or a *cycle*. These two results are not obtained by a specific rationale aimed at studying the fourth subordinate case. The first result is shared with the third subordinate case and results from the condition that the double lines of the homographic pencil are imaginary. The second is an explicit anticipation of the results which will be demonstrated later in the *mémoire*. Thus, Poincaré simply recalls at this point what he knows of this fourth subordinate case, but he does not seek to clarify it. The third point that I think is important in this passage is the logical connection established by Poincaré between the first two points. The choice not to look specifically at the fourth subordinate case is expressly justified by the fact that this case is "more special" and "will not arise if X and Y are *the most general polynomials of their degree.*" Such an observation does not come as a remark, a piece of complementary information on the relationship among the different types of singular points. Poincaré makes it an element of the hierarchization, which legitimizes a particular approach in the study of singular points: all of the cases are not of equal importance and one must know where to focus one's efforts of elucidation, rather than studying all cases with equal care, even those that are very special.

We have thus observed three related phenomena by which the fourth subordinate case differs from the first three. First, Poincaré comments on the degree of generality of this case. This observation leads him to give a different treatment, not as fully pursued. And this is translated, from the point of view of the text, by a change in the construction of this section. These three elements are found, even accentuated, in the next section, on "special cases."

The breakdown in the parallelism with the preceding sections is evident from the title of the last section, which is not designated as the fifth "subordinate case." The cases that are considered are immediately classified as "special." They are then characterized only by the conditions on the λ_i corresponding to them, which fall into two classes: $\lambda_1 = \lambda_2$ on the one hand, and $\lambda_1 = 0$ on the other. The geometric characterization has vanished. Finally, in the first lines of the section, Poincaré stresses, as he does for the fourth subordinate case, that such cases "will not arise if X and Y are *the most general polynomials of their degree.*" (Poincaré 1881, p. 18)

Thus we see appearing a first method of distinguishing various types of singular points: the first three subordinates cases include singular points that arise commonly, while the fourth subordinate case and the special cases treated in the last section do not arise if the polynomials are "the most general of their degree." It is this contrast that is adopted in the note announcing the result of the *mémoire* "Sur les courbes", published the previous year in the *Comptes rendus de l'Académie des Sciences* (Poincaré 1880). As shown in the quotation on p. 16, Poincaré only mentions three types of singular points then:

- those through which pass two characteristics, *{i.e.}* the *saddles* obtained in the second subordinate case;
- those through which pass an infinite number of characteristics, the *nodes* of the first subordinate case;
- those around which the characteristics revolve like spirals, the *foci*.

One may note the absence, in this list, of the *centers*, around which the characteristics form concentric closed curves, and which appear in the fourth subordinate case. This circumstance shows that, in this enumeration, Poincaré considers only the singular points falling into the first three subordinate cases. The centers, and more generally all the singular points falling under the fourth subordinate case or special cases left aside, are mentioned only in the last three lines of the *Note*:

> The results which are reported in this summary relate to the most general case, but I had to consider, in the *mémoire*, various exceptional cases, though without being able to think of all of those that arise. (Poincaré 1880, p. 2)

The initial enumeration does not mention these last cases. A second consequence thus emerges, in Poincaré's approach, of the "exceptional" character recognized in these types of singular points. We have seen above that Poincaré considered it legitimate not to analyze in detail the fourth, "more special," subordinate case. We see here that the exceptional nature of these points even justifies ignoring them, at least initially.

10.3.3.3   Special Cases Left Aside

We have just highlighted a separation between the first three subordinate cases on the one hand, and the fourth subordinate case and special cases left aside on the other. Poincaré then explicitly introduces a dichotomy between the "*singular points of the first kind*" and the "*singular points of the second kind.*" The boundary so drawn goes through the section on *special cases left aside*. "*Singular points of the first kind*" groups together the four subordinate cases and the case where $\lambda_1 = \lambda_2$ that opens the last section. We shall soon return to this division. We have therefore two methods of grouping cases: the separation between the two sets is established either at the level of the fourth subordinate case, or inside the final group of cases:

First subordinate case			
Second subordinate case			singular points of first kind
Third subordinate case			
Fourth subordinate case		do not arise if X and Y are the most general of their degree	
Special cases left aside	$\lambda_1 = \lambda_2$		
	$\lambda_1 = 0$		singular points of second kind

This analysis leaves unexplained a textual phenomenon that is nevertheless clear: the division into sections and the titles given to these sections seem to introduce a third method of grouping, which distinguishes the four subordinate cases on the one hand, and the *special cases left aside*, $\lambda_1 = \lambda_2$ and $\lambda_1 = 0$, on the other. This fact encourages examination of the elements by which the fourth subordinate case approaches the first three and differs from those studied in the last section. Indeed, the parallelism that we have previously observed in studying the first three subordinate cases does not disappear at once; it gradually fades when one gets to the fourth subordinate case and then to the special cases left aside. The elements which are lost or modified at each stage are valuable clues for understanding how the different cases being studied are ordered and their relative importance determined.

In fact, from the first lines of the section on the *special cases left aside*, there is a noticeable difference from the four subordinate cases. The characterization of the cases studied in this section is given immediately in the form of conditions on λ_i, without reference to the homographic pencil and its double lines. Moreover, Poincaré no longer mentions his theorem. For the case $\lambda_1 = \lambda_2$, he refers directly to a result of Briot and Bouquet. In contrast, in the second and the fourth subordinate cases, he begins by noting that his theorem does not apply before turning to other means of investigation: the work of Briot and Bouquet for the second subordinate case, and some remarks based on geometry and his subsequent results for the fourth. Finally, he restates the condition $\lambda_1 = \lambda_2$ with respect to only the a_i and b_i.

The last two points, the lack of reference to Poincaré's theorem and the reformulation of the conditions with respect to the a_i and b_i, are obviously linked. We have seen that Poincaré defines the λ_i in the statement of his theorem, at the beginning of the *second case*. The correlation between the lesser importance given to the parameters λ_1 and λ_2 for the characterization of the case, and the lack of mention of the theorem is therefore quite natural. Both points suggest that this theorem is no longer perceived by Poincaré, when he comes to *special cases left aside*, as a central organizing element. However, it is difficult to give a solid interpretation according to which the second and fourth subordinate cases are more deeply bound to the theorem than the first of the *special cases left aside*. Namely, Poincaré's theorem does not apply to any of these cases. Solely from the view of the *mémoire* "Sur les courbes", one can argue that the statement of the theorem contains two conditions, the first may be seen as a prerequisite, while the second is more specifically the validity condition for the result:

If the equation

$$(a_1 - \lambda)(b_2 - \lambda) - b_1 a_2 = 0$$

has two different roots, λ_1 and λ_2;
If the ratio of these roots is positive or imaginary, [...]
(Poincaré 1881, p. 13)

The layout of the text highlights the first condition, while the condition on the sign of the ratio of the roots opens the paragraph that states the result in the form of the general integral. From this perspective, the four subordinate cases satisfy the prerequisite of two distinct roots λ_1 and λ_2. We can then ask the question regarding satisfaction of the second hypothesis, about the sign of the ratio of the roots, and so distinguish the cases where the theorem applies—the first and third—from those where it does not apply—the second and fourth. In contrast, the *special cases left aside* do not meet the prerequisite: in the first case, $\lambda_1 = \lambda_2$, and in the second, $\lambda_1 = 0$, neither of which allows one to consider the ratio of the two roots.

The weakness of this interpretation lies in the fact that the two assumptions, which we put on different levels in view of the presentation of the text of 1881, appear quite parallel in the *Thesis* of Poincaré,[22] from which the theorem is derived:

> *Hypothesis I*—The equation (16) has no multiple roots.
> *Hypothesis II*—If one represents the real and imaginary parts of the λ by the coordinates of n points in a plane, these n points are on the same side of some line passing through the origin.
> (Poincaré 1879, p. CVI)

It is not necessary to go into the details of the formulation of 1879, where Poincaré treats the case of complex partial differential equations in n variables. Suffice it to say that "the equation" (16) corresponds in this context to the equation that defines λ_1 and λ_2 in the *mémoire* "Sur les courbes", and that the condition on the position of the points representing the λ_i in the plane reflects the fact that the ratio of λ_1 and λ_2 will be positive or imaginary.

Thus the theorem, as Poincaré borrows it from his thesis, does not permit a satisfactory accounting for the separation of the four subordinate cases and the special cases left aside. On the other hand, it seems to me that the characterization via the homographic pencil associated by Poincaré, in 1881, with the statement of his theorem, reveals an element of discrimination between these two sets of cases.

When $\lambda_1 = \lambda_2$, in fact, the "homographic pencil," that is, the linear map obtained by linearizing the differential equation, no longer has *two* double lines or eigendirections. Depending on circumstances, either it has only one, or it leaves all the lines invariant. When $\lambda_1 = 0$ on the other hand, all lines have the same image under the linear application, which means we no longer have a homographic pencil. In effect, if the images of four given lines are all identical, their cross ratio is not defined, and it is thus not equal, *a fortiori*, to that of the four given lines. The equation of the double lines as written by Poincaré has a unique solution: the only line which is the image of all lines is naturally its own image, which makes the case $\lambda_1 = 0$ resemble that where $\lambda_1 = \lambda_2$. In all *special cases left aside*, we can note a degeneracy of the "homographic pencil" which was used by Poincaré to characterize the four subordinate cases. Correspondingly, these geometric considerations disappear in this last section. This observation provides a possible interpretation of the contrast between the four *subordinate cases* and the *special cases left aside*, as well as of the grouping of the cases $\lambda_1 = \lambda_2$ and $\lambda_1 = 0$ in the last section.

This hypothesis is reinforced by the fact that the geometrical characterization comes at the beginning of each subordinate case. Furthermore, Poincaré presents

---

[22] As I have already indicated, Poincaré mentions in his *Thesis* a third hypothesis which he seems to forget in 1881.

it just before introducing these four cases: "Now we can divide the second case into four major subordinate cases." The description of the homographic pencil does not appear merely as an illustration of the analytical conditions by which Poincaré distinguishes different cases. It plays a vital role in identifying the degeneracy[23] of *special cases left aside*. It is thus used to hierarchize the cases that Poincaré is led to distinguish by his analysis.

10.3.3.4   Singular Points of the First and Second Kind

After studying the case $\lambda_1 = \lambda_2$, and before examining the case where $\lambda_1 = 0$, Poincaré pauses to summarize the results obtained thus far and to introduce a new method of grouping singular points:

> The five previous cases include all the singular points α, β, such that the two curves
>
> $$X = 0, \quad Y = 0$$
>
> intersect at a single point and not at several coincident points. These singular points will be called *singular points of first kind*, and we have seen that there were four types of such points: nodes, saddles, foci, and centers. (Poincaré 1881, p. 18)

The five cases mentioned here are the four subordinate cases treated in the previous sections, along with the case $\lambda_1 = \lambda_2$. The singular point is a node in the first subordinate case, a saddle in the second subordinate case, a focus in the third, a focus or a center in the fourth, and a node when $\lambda_1 = \lambda_2$. As we saw earlier, this list summarizes the results obtained. However, it does not follow the organization of the study into "cases."

This list further draws attention to an important difference between the singular points of the first kind, and those of the second kind. Poincaré elucidates the geometric behavior of the *characteristics* in the neighborhood of singular points of the first kind in all cases—though less precisely in the latter two cases—and he associates names with these different behaviors. For singular points of the second kind, he gives no information about the shape of the *characteristics*.

He satisfies himself by asserting that "they can always be regarded as the limit of a system of several singular points of the first kind that are coincident." (Poincaré 1881, p. 19) Poincaré justifies this assertion by using a new form of geometric interpretation: the geometry of the curves $X = 0$ and $Y = 0$. The considerations about the homographic pencil at a singular point were directly related to the behavior of the *characteristics* in the neighborhood of this point. Poincaré noted, for example, that there can be a solution through the singular point only if the homographic pencil has a real double line, which is the tangent to the solution curve. The consideration of the algebraic curves $X = 0$ and $Y = 0$ departs from the geometric properties of the

---

[23] This term is not much used by Poincaré, but it is not absent from his vocabulary. It is found, for example, in his work on the three-body problem (Poincaré 1890, p. 360, 363).

characteristics. But this new geometric interpretation will guide the distinction of several cases among the singular points of the second kind.

After finishing the study of singular points of the first kind with the case $\lambda_1 = \lambda_2$, Poincaré deals with the case $\lambda_1 = 0$ through three sub-cases:

- $\lambda_1 = 0$,
- $a_1 = a_2 = 0$,
- $b_1 = b_2 = 0$.

These sub-cases are defined by equalities among the coefficients. But Poincaré uses the geometric-algebraic language that he has introduced to characterize them. This helps explain why he considers precisely these equalities. These three sub-cases correspond, in fact, to the three different ways in which the curves X = 0 and Y = 0 can intersect "at several points coincident with the point (α, β)." Poincaré makes this explicit immediately, reversing the order in which he lists the three sub-cases—we will come back to this:

> In fact, to say that $a_1 = a_2 = 0$ means that $X = 0$ presents a multiple point in (α, β).
> Suppose that $b_1 = b_2 = 0$ means that $Y = 0$ presents a multiple point in (α, β).
> Suppose that $\lambda_1 = 0$ means that $\dfrac{a_1}{a_2} = \dfrac{b_1}{b_2}$, *{i.e.}* that the curves $X = 0$, $Y = 0$ are tangent
> at the point (α, β). (Poincaré 1881, pp. 18–19)

The formulation of these sub-cases presents an interesting peculiarity, which seems characteristic of the way that Poincaré gradually uncovers, throughout this chapter, cases more and more special.

These three sub-cases are first subsumed under the characterization $\lambda_1 = 0$. In fact, if $a_1 = a_2 = 0$, we have $\lambda_1 = 0$. But at the moment where Poincaré decomposes the case $\lambda_1 = 0$ into three sub-cases, he uses again the equality $\lambda_1 = 0$ to characterize the first sub-case. He thus seems to place the case $\lambda_1 = 0$ and the two sub-cases at the same level of the enumeration. Here, $\lambda_1 = 0$ must therefore, according to all evidence, be read as representing the cases where $\lambda_1 = 0$ *with the exception* of those satisfying $a_1 = a_2 = 0$ or $b_1 = b_2 = 0$. Later, Poincaré changes the order in which he refers to them when giving their geometric interpretation, which allows him to list them in a more natural logical order. But the order used first seems to emphasize that $\lambda_1 = 0$ refers in a privileged way to those cases where $\lambda_1 = 0$ *and* where neither of the following is true: (1) $a_1 = a_2 = 0$; (2) $b_1 = b_2 = 0$. These last two sub-cases thus appear as special cases within the case $\lambda_1 = 0$. I will come back to this point when taking stock of the study of this chapter of Poincaré's *mémoire*. I could thus compare it with similar patterns in the way Poincaré introduces more special cases.

10.3.3.5   Completing the Enumeration

One last question will help us better understand the approach followed by Poincaré in this work on singular points: is the enumeration that we just studied exhaustive?

I suggested, in beginning, a way to restore the branching structure to the sections of this chapter of Poincaré's *mémoire*. It consists in seeing the four subordinate cases, as well as the special cases left aside, as sub-parts of the *second case*. Two reasons justify this proposal.

First Poincaré introduces the subordinate cases, stating that "we can divide the second case into four major subordinate cases." This statement clearly shows that the subordinate cases are sub-parts of the *second case*, but also that this *second case* is not exhausted by the four subordinate cases. They are said to constitute the "major" sub-parts, which suggests that the *second case* contains other, complementary, cases.

Moreover, the *special cases left aside* appear to properly belong to the *second case*. This is fairly clear, at least when they are expressed by the conditions $\lambda_1 = \lambda_2$ and $\lambda_1 = 0$. Indeed, the λ_i are defined by an equation defined by the coefficients a_1, a_2, b_1 and b_2. It is necessary that these four coefficients not all be zero so that they define an equation and allow us to determine the λ_i. Now the *second case* is precisely characterized by the fact that the a_i and b_i are not all zero.

If we accept this branching structure, it follows that Poincaré wholly abandons the case where these four coefficients are zero. He is interested only in "regular singular points."

We can suggest a different interpretation. We have seen that Poincaré does not indicate the branching structure of the sections of his chapter except in their titles. Presumably he has taken advantage of the absence of typographical markings to give the final section a special status relative to the branching structure of the previous sections. This section would thus combine under this hypothesis both the sub-cases of the *second case* that were not yet treated, and the cases excluded by the definition of the *second case*, the non "ordinary" singular points.

These last could be included among the singular points of the second kind. The fact that the latter conditions, $a_1 = a_2 = 0$ and $b_1 = b_2 = 0$, are formulated with respect to the a_i and b_i rather than λ_i makes this hypothesis seem quite plausible. The non ordinary singular points are in fact the points of multiple intersection of the curves $X = 0$ and $Y = 0$, and they can be considered as the limit of several singular points of the first kind that would be coincident.

This interpretation would also account for Poincaré's last remark, that "such points cannot exist if X and Y are the most general polynomials of their degree" (Poincaré 1881, p. 19). Indeed, if this remark applies only to singular points of the second kind which are also ordinary singular points, it is redundant with the similar remark at the beginning of the section, which concerns the special cases $\lambda_1 = \lambda_2$ and $\lambda_1 = 0$. But if one considers that at the end of the section, Poincaré no longer restricts himself to cases where a_1, a_2, b_1 and b_2 are not all zero, this last remark contains new information in that the singular points of the second kind, *in which we include* the non ordinary singular points, do not arise when X and Y are the most general of their degree.

10.3.4 A Keen Interest in That Which is Essential

We have seen that Poincaré introduces, in his study of singular points of differential equations, various types of gradation in the different cases which he studied. All this leads him to label some cases as "singular," "more special," "exceptional," "of the second kind" in comparison with others.

This gradation is correlated with an important aspect of Poincaré's approach. He seeks what is essential in order to concentrate his efforts there, rather than trying to present the most comprehensive study possible. Several indications help to highlight the priority he gives to the study of the most general cases.

This procedure of Poincaré first manifests itself in the way that he presents the cases that he will handle. On many occasions, he announces initially only the principal cases, even if he subsequently studies and indicates supplementary cases.

Thus, when he introduces the study of the subordinate cases, he mentions only "four major subordinate cases" into which the second case is divided. The other cases are then called, significantly, "special cases left aside."

We saw a similar phenomenon in the *Note* published in the CRAS. Poincaré first mentions only three types of singular points: the *saddles*, the *foci* and the *nodes*. These three types correspond to the first three subordinate cases. All other types of singular points, being more special, are ignored until the last paragraph of the *Note*, where Poincaré reports the existence of "exceptional cases."

The imprecision that we have emphasized regarding the scope of the last section is similar. We cannot know if he saw his list of singular points as exhaustive or not, but at least it is clear that he does not explicitly emphasize its completeness. The existence of residual cases which he would not have treated does not seem to worry him, once he has established that they are more special cases.

Correspondingly, the order chosen by Poincaré for examining the different cases consists of systematically disregarding the most special cases until the end. Thus, the major cases still appear first, and the order of treatment of the cases conforms simultaneously with all the gradations that Poincaré introduces among the groups of cases.

Thus all those cases that, as Poincaré emphasizes, do not manifest themselves in the most general polynomials of their degree are collected together after the fourth subordinate case. Among them, the singular points of the second kind are treated last, and within those, the cases where $a_1 = a_2 = 0$ and $b_1 = b_2 = 0$ are listed last.

This is reflected in a particular textual phenomenon that we find repeatedly. The special sub-cases, such as $a_2 = a_1 = 0$ in relation to $\lambda_1 = 0$, or such as $\lambda_1 = \lambda_2$ relative to the case where the roots are real and of the same sign, are systematically indexed and studied *after* the cases to which they belong at the upper level of the tree. However, the case $\lambda_1 = \lambda_2$ requires special treatment, different from the case where the roots are real and of the same sign—of which it is a sub-case. The case of equality is excluded by the assumptions of Poincaré's theorem, but this is not recalled when studying the first subordinate case, where the two roots are real and of the same sign. Similarly, when Poincaré lists the three cases $\lambda_1 = 0$, $a_1 = a_2 = 0$, and

$b_1 = b_2 = 0$, we have seen that we must understand that the two special sub-cases $a_1 = a_2 = 0$ and $b_1 = b_2 = 0$ are excluded from the case $\lambda_1 = 0$, although nothing indicates this when one reads $\lambda_1 = 0$: the reader must take into consideration the entire hierarchy to interpret correctly the characterization $\lambda_1 = 0$.

This phenomenon manifests itself in two ways in the *Note* (Poincaré 1880). One encounters it at the level of a sentence when Poincaré describes the different behaviors of the *characteristics* in the neighborhood of a point. He introduces the first case, when the point is not singular, by these words: "through all the points, except for some singular points, passes one characteristic and one only." The ordinary points of which he speaks here are first designated as "all the points"— and not "most points" or a similar phrase—and the "singular" points which escape this rule are reported and studied only later. The same method of exposition is found on the scale of the whole note: Poincaré writes it as if the three types of singular points that he first surveyed were the only ones. He only mentions at the end that there may be other, exceptional cases.

Poincaré therefore adopts in this work a specific approach which consists of studying the possible cases by beginning with the most general before considering more and more special cases. These are initially assumed to be excluded, but this sometimes remains implicit.

We have identified two elements in the *form* of the discourse, through which Poincaré shows an interest in the most general case. The first is an explicit emphasis on the principal cases. The second is a standard method of organizing the text. It consists of making what is essential appear first. The same interest is also observed at the level of the *contents* of the study.

When he gradually rolls out the originally isolated special cases, Poincaré gives, in fact, a more and more concise treatment. The economy of means employed to treat the most special cases is explicitly justified by their lesser importance. We have seen that Poincaré satisfies himself with "some remarks" about the fourth subordinate case. The case $\lambda_1 = \lambda_2$ is also studied very quickly, without distinguishing two cases as Briot and Bouquet did, and as we do today.[24] Finally, for the singular points of the second kind, Poincaré merely says that they can be obtained as the limit of several singular points of the first kind which become coincident.

10.3.5 A Significant Omission

This analysis can account for an important omission of Poincaré. I will show that this circumstance reinforces the conclusion that our author focuses on the essential.

I have pointed out in stating Poincaré's theorem that he "forgets" a hypothesis when he states, in his *mémoire* "Sur les courbes", the theorem proved in his thesis. One must indeed ensure that one of the roots λ_1 and λ_2 is not a multiple of the other

---

[24] In modern terms, these two cases are those where the matrix of the linearized equation is diagonalizable and those where it is not.

to conclude as Poincaré does. This requirement is reflected in the thesis of 1878, but Poincaré fails to mention it in 1881.

We can certainly rule out the hypothesis of unscrupulous omission. Poincaré had no reason to hide these special cases, which one calls "resonant" today. Indeed, in exactly the same way as in the case of two equal roots, they do not arise if the polynomials are the most general of their degree. Poincaré could have treated them the same way.

But because these exceptions to the theorem of Poincaré are more special, it seems probable that Poincaré simply forgot the corresponding hypothesis when quoting the statement of his theorem. Doubtless he did not reread his thesis. He states in the *mémoire* the result as he remembered it. The really crucial assumption is the one which is reflected, in the case of two variables, by the fact that λ_1 and λ_2 must be of the same sign or imaginary. This assumption is truly restrictive: it excludes all singular points for which the ratio of the two roots is negative. The assumption of non-resonance, in turn, excludes only the most special cases. Moreover, when Poincaré restates the results of his thesis in the *Analyse de ses travaux scientifiques* (Poincaré 1921), he mentions only the hypothesis which corresponds to the sign of the ratio λ_1 / λ_2. He omits the assumption of non-resonance as well as one requiring the roots to be distinct. This is consistent with the conclusion to which we have been led by the analysis of the work of 1881: Poincaré gets to the essential.

So why does he forget the assumption of non-resonance, but not that which states that the roots must be distinct? I think we can give a satisfactory response by noting the importance held by linear equations in Poincaré's discussion. Poincaré's theorem provides, under suitable assumptions—among which non-resonance should be included—, that we can linearize the equation by an analytic change of variables. The classification of singular points following the statement of the theorem aligns with the classification of the linear equations. In fact, all of the examples given by Poincaré are examples of linear equations. Thus, Poincaré seems more interested in the qualitative description, over the domain of the reals, of the solutions of linear equations in different cases than in the linearization process itself. Now, the resonances that occur when roots are multiples of each other have no effect on the behavior of solutions of the linearized equation. They only cause obstructions to linearization. Moreover, in the real domain, the resonant equations—nonlinear—have a behavior similar to that of their linearization. Thus, only the case where the two roots are equal is qualitatively remarkable, and appears as such in the study of the linearized equation. This may explain how Poincaré forgets the assumption of non-resonance, but not that requiring that the roots be distinct. It is his study of linear equations, to which his theorem reduces the general problem, that recalls to him his assumption. The assumption of non-resonance, however, was probably immediately recognized by Poincaré as a minor hypothesis since it excludes only those cases he would call "more special." Consequently, it was less likely to be present in his memory, so that he may have forgotten it (Arnold 1980).

10.3.6   *Grading Criteria*

The analysis we have made of the various gradations by which Poincaré prioritizes the cases he studies also allows us to speculate on the criteria he uses for assessing the significance of different cases. It seems that Poincaré articulates two types of information for defining and prioritizing cases.

We have already noticed the role played by the *conditions of equality* and inequality among the coefficients defining the differential equation—or between the λ_i that are deduced from it. Equalities define more special cases, while inequalities do not influence the degree of generality of the cases they define.

Thus the second case, which includes the regular singular points, differs from the first by the conditions on a_0 and b_0. We are in the first case when a_0 and b_0 are not both zero, whereas we are in the second case when $a_0 = b_0 = 0$. The *ordinary* singular points are characterized by the additional condition—of inequality—that a_1, a_2, b_1 and b_2 are not all zero. Poincaré is not very explicit, but it is quite clear that the singular point is *no longer ordinary* if these four coefficients are zero.

The first three subordinate cases are then defined only by inequalities.[25] The fourth subordinate case, which is called "more special," is characterized by an equality joined with an inequality: λ_1 and λ_2 are supposed imaginary, and their ratio is *equal* to -1.

The three cases among which Poincaré divides the singular points of the second kind are also characterized by equalities. More precisely, the first, $\lambda_1 = 0$, is defined by *one* equality, and the next two by *two* equalities. Now we have seen how the last two actually form a sub-case, mentioned only when studying the singular points of the second kind.

We observe therefore a correlation between the number of equalities defining a case and its identification as more special.

But these conditions are not the only factors used to characterize the different cases. They are almost always associated with some *geometrical conditions*. Initially, these are geometrical conditions on the properties of the homographic pencil associated with the singular point. Later, when there is no longer such a homographic pencil, Poincaré interprets the conditions on the coefficients by relating them to the geometric properties of the intersection of the curves $X = 0$ and $Y = 0$.

These geometrical conditions are not mere illustrations of analytical conditions on the coefficients. Together with the hypotheses of the theorems available in the first part of the study, and alone in the second part, they show the equalities and inequalities relevant for distinguishing the cases. This is particularly clear for the singular points of the second kind, where only the interpretation in terms of the intersection of $X = 0$ and $Y = 0$ justifies singling out the cases $a_1 = a_2 = 0$ and $b_1 = b_2 = 0$, rather than some other equality among the coefficients. In the first part of the study, it is clear that the differentiation between subordinate cases relates

---

[25] This distinction between cases where the are real and those where they are imaginary is related, algebraically, to an inequality: it depends on the sign of the determinant of the Eq. 10.1 of the second degree defining these roots.

to the hypotheses of Poincaré's theorem. But our author highlights the geometric interpretation of the conditions thus defined. The fact of giving such prominence to geometric interpretation for distinguishing cases may explain in part the omission of the hypothesis of non-resonance: the latter corresponds to nothing geometrically.

Furthermore, we have seen that geometry seems to also play a role in the ranking of the cases. Such is the interpretation that seems most likely to explain the distinction between the subordinate cases and special cases left aside. The boundary between the singular points of the first kind and those of the second kind is explicitly defined geometrically: the five cases consolidated under the name "singular points of the first kind" are united by the property that "the curves $X = 0$ and $Y = 0$ intersect at a single point and not at several points that would be coincident." The singular points of the second kind are further characterized by the property that $\lambda_1 = 0$. But this property, as a condition of equality, does not justify seeing them as secondary to those of the first kind. Among these, there are indeed points belonging to the fourth subordinate case, and therefore the case $\lambda_1 = \lambda_2$, both defined by an equality added to $a_0 = b_0 = 0$. The relationship between the singular points of the first kind and those of the second kind is that the latter can be regarded as "the limit of several singular points of the first kind that would be coincident." Poincaré relates this property to the geometric characterization of the points of the second kind by using the same expression: "coincident points." So it is that the geometric characterization plays a key role in distinguishing—and articulating—singular points of first and second kind.

The importance of geometrical considerations does not diminish the role of the characterization by conditions on the coefficients. Indeed, the latter provide unity and coherence for the definitions of all of the cases, over and above the use of the two different geometrical interpretations. They permit, therefore, the construction of a tree that includes the various cases identified by Poincaré. The importance of these conditions on the coefficients may be particularly clear with regards to the case $\lambda_1 = \lambda_2$: this characterization brings together two different degeneracies of the homographic pencil (only one double line/all the lines invariant). Poincaré does not attempt to distinguish them, but merely notes that there were always an infinite number of *characteristics* passing through the singular point in question, which is therefore a *node*.

Conclusion

Thus Poincaré enlists various approaches for defining cases and prioritizing them in a very complex fashion. This hierarchy is the expression of a significant work on the degree of particularity of the considered cases. Poincaré does not content himself with placing a general case and some special cases in opposition, but establishes some means of differentiation among the special cases. This thought is not *thematized*. The criteria used to highlight the particularity of certain cases seem to be various. They are not made explicit except by their implementation: Poincaré does not define a mathematical notion apt for measuring the particularity of a case, let alone to uniformly measure the different cases that he distinguishes. Despite this, the ranking and organization achieved by Poincaré are of a finesse altogether striking.

In later chapters of his *mémoire*, we see that Poincaré uses this hierarchy to define a "general case," in which the singular points belong only to the first three subordinate cases. The gradation of the special cases developed early in the *mémoire* plays an important role for the remainder of the study (Robadey 2006, pp. 84–91).

In the *mémoire* "Sur les courbes", Poincaré introduces a special form of conditions for establishing a result. Between universal conditions—what is true "in all cases"—and restrictive conditions—what is true "in such and such cases"—Poincaré inserts a new type of condition that we could be called generic: what is true in "the general case". It is not, however, a matter of an imprecise *generic reasoning* such as thoses pointed out by Hawkins in his research on the history of matrix theory (Hawkins 1975a, b, 1977a, b) (see also Robadey 2006, pp. 77–82). In 1881, Poincaré defines accurately, even before developing his analysis, the "general case's" limits. However these limits must be distinguished from restrictive conditions, for they "they do not undermine the generality".

The possibility of defining a "general case" appears to be correlated with Poincaré's specific way of working: the definition of hierarchized lists of cases. Through this privileged approach to problems, Poincaré develops a reflection on the degree of generality of the different cases he is led to consider. From the generality point of view, this gradation between cases forms an important prerequisite to the "general case" definition.

The transition from a hierarchized list of cases to a "general case" proves to be particularly fruitful in the context of establishing a global result based on a list of possible local behaviors. This is the 1881 study's framework, and again that of Poincaré's study of periodic solutions (Robadey 2006, pp. 91–97).

When Poincaré inaugurates in his *mémoire* "Sur les courbes" a new, global approach to the study of differential equations, he relies on a technique he introduced in his first article (Poincaré 1879): the organisation of the cases into a hierarchy (Robadey 2006). Thus, my analysis allows us to shed light on the process through which Poincaré distinguished himself from his predecessors and formulated new questions and new methods.

Appendix 1

One will find here all quotations in their original language in order of appearance.

10.2 Text and context, from Briot and Bouquet (1856b, p. 133):

> Les cas où l'on peut intégrer une équation différentielle sont extrêmement rares et doivent être regardés comme des exceptions. Mais on peut considérer une équation différentielle comme définissant une fonction, et se proposer d'étudier les propriétés de cette fonction sur l'équation différentielle elle-même.

10.2 Text and context, from Poincaré (1881, p. 3):

> Malheureusement, il est évident que, dans la grande généralité des cas qui se présentent, on ne peut intégrer [les équations différentielles] à l'aide des fonctions déjà connues, par exemple à l'aide des fonctions définies par les quadratures. Si l'on voulait donc se restreindre aux cas que l'on peut étudier avec des intégrales définies ou indéfinies, le champ de nos recherches serait singulièrement diminué, et l'immense majorité des questions qui se présentent dans les applications demeureraient insolubles.
>
> Il est donc nécessaire d'étudier les fonctions définies par des équations différentielles en elles-mêmes et sans chercher à les ramener à des fonctions plus simples.

10.2 Text and context, from Riemann (1854, pp. 246–247). cited by R. Chorlay:

> Les travaux que nous avons signalés sur cette question avaient pour but de démontrer la série de Fourier pour les fonctions que l'on rencontre en Physique mathématique [...]. Dans notre problème, la seule condition que nous imposerons aux fonctions, c'est de pouvoir être représentées par une série trigonométrique; nous chercherons donc les conditions nécessaires et suffisantes pour un tel mode de développement des fonctions. Tandis que les travaux antérieurs établissaient des propositions de ce genre: »si une fonction jouit de telle et telle propriété, elle peut être développée en série de Fourier«, nous nous proposerons la question inverse: »si une fonction est développable en une série de Fourier, que résulte-t-il de l sur la marche de cette fonction, sur la variation de sa valeur, quand l'argument varie de manière continue ?«

10.3.2 Singular points and the tools used for classifying them, from Poincaré (1881, p. 14):

Si l'équation

$$(a_1 - \lambda)(b_2 - \lambda) - b_1 a_2 = 0 \tag{10.2}$$

a deux racines différentes, λ_1 et λ_2;
Si le rapport de ces racines est positif ou imaginaire, l'intégrale générale de l'équation

$$\frac{dx}{X} = \frac{dy}{Y}$$

est de la forme

$$z_1^{\lambda_1} z_2^{\lambda_2} = \text{const.},$$

où z_1 et z_2 sont des séries ordonnées suivant les puissances croissantes de $x - \alpha$, $y - \beta$ et s'annulant pour

$$x = \alpha, \qquad y = \beta.$$

10.3.2 Singular points and the tools used for classifying them, footnote from Chasles (1880, p. 64):

> Quand deux faisceaux dont les droites se correspondent une à une sont tels que quatre droites quelconques du premier aient leur rapport anharmonique égal celui des quatre droites correspondantes du second, nous dirons que les deux faisceaux sont *homographiques*.

10.3.2 Singular points and the tools used for classifying them, from Poincaré (1881, p. 14):

Donc la limite de la droite qui joint les points (x, y) et (α, β) et la limite de la tangente à la caractéristique au point (x, y) forment un faisceau homographique.

10.3.2 Singular points and the tools used for classifying them, from Poincaré (1881, p. 14):

Premier cas subordonné.—Les droites doubles du faisceau homographique sont réelles, et deux droites conjuguées quelconques du faisceau sont ou toutes deux dans l'angle aigu formé par les deux droites doubles, ou toutes deux dans l'angle obtus.
Dans ce cas, λ_1 et λ_2 sont réels et leur rapport est positif.

10.3.3.1 Singular points, from Poincaré (1880, p. 1):

On voit ainsi: $1°$ que, par tous les points de la sphère, sauf par certains points singuliers, passe une caractéristique et une seule; $2°$ que, par certains points singuliers, passent deux caractéristiques; $3°$ que, par d'autres points singuliers, passent une infinité de caractéristiques; $4°$ enfin, qu'une troisième sorte de points singuliers est telle, que les caractéristiques voisines tournent comme des spirales autour de ces points sans qu'aucune d'elles aille y passer. J'appelle ces trois sortes de points singuliers les *cols*, les *nœuds* et les *foyers* de l'équation donnée.

10.3.3.2 More special cases, which do not arise if X and Y are the most general polynomials of their degree, from Poincaré (1881, p. 17)

Ce cas est plus particulier que les précédents et il ne se présentera pas si X et Y sont *les polynomes les plus généraux de leur degré*. Bornons-nous donc à quelques remarques.

10.3.3.2 More special cases, which do not arise if X and Y are the most general polynomials of their degree, from Poincaré (1880, p. 2)

Les résultats qui sont rapportés dans ce résumé se rapportent au cas le plus général; mais j'ai dû examiner, dans le Mémoire, différents cas exceptionnels, sans pouvoir pourtant envisager tous ceux qui se présentent.

10.3.3.4 Singular points of the first and second kind, from Poincaré (1881, p. 18)

Les cinq cas précédents comprennent tous les points singuliers α, β, tels que les deux courbes

$$X = 0, \qquad Y = 0$$

s'y coupent en un seul point et non en plusieurs points confondus. Ces points singuliers s'appelleront *points singuliers de première espèce*, et l'on a vu qu'il y avait quatre sortes de pareils points: les noeuds, les cols, les foyers et les centres.

10.3.3.4 Singular points of the first and second kind, from Poincaré (1881, pp. 18–19)

En effet, dire que $a_1 = a_2 = 0$, c'est dire que $X = 0$ offre un point multiple en (α, β).
Dire que $b_1 = b_2 = 0$, c'est dire que $Y = 0$ offre un point multiple en (α, β).
Dire que $\lambda_1 = 0$, c'est dire que $\dfrac{a_1}{a_2} = \dfrac{b_1}{b_2}$, c'est-à-dire que les courbes $X = 0$, $Y = 0$ sont tangentes au point (α, β).

Appendix 2

Study of the equation $\frac{dx}{X} = \frac{dy}{Y}$ to the point p oint (α, β)

$$X = a_0 + a_1(x - \alpha) + a_2(y - \beta) + \ldots$$
$$Y = b_0 + b_1(x - \alpha) + b_2(y - \beta) + \ldots$$

1ˢᵗ case	$a_0 \neq 0$ *ou* $b_0 \neq 0$.	Cauchy's thm	Only one *characteristic* passes by (α, β).
2ⁿᵈ case	$a_0 = b_0 = 0$, a_1, a_2, b_1, b_2 *not* all zero.	[Statement of Poincaré's theorem]	

[Singular points of the first kind:]

1ˢᵗ case	*sub.*	$\lambda_1, \lambda_2 \in \mathbb{R}$, $\frac{\lambda_1}{\lambda_2} > 0$ [but $\lambda_1 \neq \lambda_2$].	Poincaré's thm.	All the *characteristics* pass by (α, β). $\boxed{\text{node}}$
2ⁿᵈ case	*sub.*	$\lambda_1, \lambda_2 \in \mathbb{R}$, $\frac{\lambda_1}{\lambda_2} < 0$.	Briot and Bouquet's result	Two *characteristics* pass by (α, β). $\boxed{\text{saddle}}$
3ʳᵈ case	*sub.*	$\lambda_1, \lambda_2 \in \mathbb{C}$, $\frac{\lambda_1}{\lambda_2} \in \mathbb{C}$.	Poincaré's thm.	*Characteristics* are spirals around (α, β). $\boxed{\text{foci}}$
4ᵗʰ case	*sub.*	$\lambda_1, \lambda_2 \in \mathbb{C}$, $\frac{\lambda_1}{\lambda_2} = -1$.	"some remarks"	*Characteristics* are spirals $\boxed{\text{foci}}$ or closed concentrics curves $\boxed{\text{center}}$ around (α, β).

Particular cases left aside

$\lambda_1 = \lambda_2$.	Briot and Bouquet's result	All *characteristics* pass by (α, β). $\boxed{\text{node}}$

[Singular points of the second kind:]

$\lambda_1 = 0$. $\lambda_1 = 0$ *or* $a_1 = a_2 = 0$ *or* $b_1 = b_2 = 0$	The singular point can be considered as the limit of a system of several singular points coinciding with one another.

Table 10.1 Organisation of Chapter II of Poincaré (1881)

References

Arnold, Vladimir. 1980. *Chapitres supplémentaires de la théorie des équations différentielles.* Moscow: Mir Publishers.

Barrow-Green, June. 1997. *Poincaré and the three body problem.* Providence, RI:AMS-LMS.

Briot and Bouquet. 1856a. Étude des fonctions d'une variable imaginaire. *Journal de l'école Polytechnique* XXXVI:85–131. (Gallica.bnf.fr).

Briot and Bouquet. 1856b. Recherches sur les propriétés des fonctions définies par des équations différentielles. *Journal de l'école Polytechnique* XXXVI:133–198. (*Mémoire* submitted to the Institut de France, on August 21, 1854. Available online at Gallica.bnf.fr, following the *mémoire* Briot and Bouquet (1856a)).

Cauchy, Augustin-Louis. 1842a. Mémoire sur un théorème fondamental, dans le calcul intégral. *Comptes rendus des séances de l'Académie des Sciences* XIV:1020–1023. (Session of June 27, 1842. Œuvres Cauchy (1882–1974), s. 1, t. 6, 461–467 (excerpt No. 167)).

Cauchy, Augustin-Louis. 1842b. Mémoire sur l'emploi d'un nouveau calcul, appelé calcul des limites, dans l'intégration d'un système d'équations différentielles. *Comptes rendus des séances de l'Académie des Sciences* XV:14. (Session of July 4, 1842. Œuvres Cauchy (1882–1974), s. 1, t. 7, 5–17 (excerpt No. 169)).

Cauchy, Augustin-Louis. 1842c. Mémoire sur l'emploi du calcul des limites dans l'intégration des équations aux dérivées partielles. *Comptes rendus des séances de l'Académie des Sciences* XV:44–58. (Session of July 11, 1842. Œuvres Cauchy (1882–1974), s. 1, t. 7, 17–33 (excerpt No. 170)).

Cauchy, Augustin-Louis. 1842d. Mémoire sur l'application du calcul des limites à l'intégration d'un système d'équations aux dérivées partielles. *Comptes rendus des séances de l'Académie des Sciences* XV: 85–101. (Session of July 18, 1842. Œuvres Cauchy (1882–1974), s. 1, t. 7, 33–49 (excerpt No. 171)).

Cauchy, Augustin-Louis. 1843. Remarques sur les intégrales des équations aux dérivées partielles, et sur l'emploi de ces intégrales dans les questions de Physique mathématique. *Comptes rendus des séances de l'Académie des Sciences* XVI:572. (Session of March 13, 1843. Œuvres Cauchy (1882–1974), s. 1, t. 7, 308–325 (excerpt No. 206)).

Cauchy, Augustin-Louis. 1882–1974. *OE uvres Complètes d'Augustin Cauchy. Publié sous la direction scientifique de l'Académie des Sciences.* Paris: Gauthier-Villars. Première série: *mémoires* extraits des recueils de l'Académie des sciences de l'Institut de France, 12 volumes et un volume de table générale. Deuxième série: mémoires divers et ouvrages, 15 volumes. (Gallica.bnf.fr).

Chasles, Michel. 1880. *Traité de géométrie supérieure.* 2nd ed. Paris: Gauthier–Villars. http://name.umdl.umich.edu/abn2684. Accessed 19 April 2015.

Dirichlet, Peter Gustav Lejeune. 1829. Sur la convergence des séries trigonométriques qui servent à représenter une fonction arbitraire entre des limites données. *Journal für die Reine und Angewandte Mathematik* 4:157–169.

Gilain, Christian. 1977. *La théorie géométrique des équations différentielles de Poincaré et l'histoire de l'analyse.* PhD thesis, Université Paris I.

Hawkins, Thomas. 1975a. Cauchy and the spectral theory of matrices. *Historia Mathematica* 2:1–29.

Hawkins, Thomas. 1975b. The theory of matrices in the 19th century. In *Proceedings of the International Congress of Mathematicians, Vancouver, 1974.* vol 2, 561–570. Vancouver: Canadian Math. Congress.

Hawkins, Thomas. 1977a. Another look at Cayley and the theory of matrices. *Archives internationales d'histoire des sciences* 27:82–112.

Hawkins, Thomas. 1977b. Weierstrass and the theory of matrices. *Archive for History of Exact Sciences* 17:119–163.

Luc, C., M. Mohajid, M.-P. Péry-Woodley, and J. Virbel. 2000. Les énumérations: structures visuelles, syntaxiques et rhétoriques. In *Document électronique dynamique: Actes du troisième* colloque internationnal sur le document électronique, *CIDE'2000, Lyon, 4–6 juillet 2000*, 21–40. Lyon: Europia production.

Luc, Christophe. 2001. Une typologie des énumérations basée sur les structures rhétoriques et architecturales du texte. In *Huitième conférence annuelle sur le traitement automatique des langues naturelles, TALN–RECITAL 2001, Tours, 2–5 juillet* http://www.atala.org/doc/actes_taln/AC_0045.pdf. Accessed 31 July 2014.

Poincaré, Henri. 1878. Note sur les propriétés des fonctions définies par les équations différentielles. *Journal de l'école Polytechnique* 45:13–26. (Œuvres Poincaré (1916–1954), t. 1, xxxvi–xlviii).

Poincaré, Henri. 1879. *Sur les propriétés des fonctions définies par les équations aux différences partielles*. PhD thesis, Faculté des sciences de Paris, August 1, 1879. Œuvres Poincaré (1916–1954), t. 1, il–cxxxii.

Poincaré, Henri. 1880. Sur les courbes définies par une équation différentielle. *Comptes rendus de l'Académie des Sciences* 90:673–675. (Œuvres Poincaré (1916–1954), t. 1, 1–2).

Poincaré, Henri. 1881. Sur les courbes définies par une équation différentielle *(first part)*. *Journal de Liouville (third series)* 7:375–422. (Œuvres Poincaré (1916–1954), t. 1, 3 44).

Poincaré, Henri. 1882. Sur les courbes définies par une équation différentielle *(second part)*. *Journal de Liouville (third series)* 8:251–296. (Œuvres Poincaré (1916–1954), t. 1, 44–84).

Poincaré, Henri. 1885. Sur les courbes définies par les équations différentielles *(troisième partie)*. *Journal de Liouville (fourth series)* 1,167–244. (Œuvres Poincaré (1916–1954), t. 1, 90–158).

Poincaré, Henri. 1886. Sur les courbes définies par les équations différentielles *(quatrième partie)*. *Journal de Liouville (fourth série)* 2:151–217. (Œuvres Poincaré (1916–1954), t. 1, 167–222).

Poincaré, Henri. 1889. Sur le problème des trois corps et les équations de la Dynamique. *Mémoire couronné du prix de S.M. le roi Oscar II de Suède*. This *mémoire*, already completed by several long notes comparatively to the *mémoire* that indeed received the king Oscar's Prize, was printed by *Acta mathematica*, but withdrawn before publication, in order to be replaced by Poincaré (1890), where the text is very largely reworked (see Barrow-Green (1997) for further details about the two versions).

Poincaré, Henri. 1890. Sur le problème des trois corps et les équations de la Dynamique. *Mémoire couronné du prix de S. M. le roi Oscar II de Suède. Acta Mathematica* 13:1–270. (Œuvres Poincaré (1916–1954), t. 7, 262–479. In fact, this *mémoire* is a largely reworked version of Poincaré (1889)).

Poincaré, Henri. *Œuvres*. 1916–1954. Paris: Gauthier-Villars.

Poincaré, Henri. 1921. Analyse de ses travaux scientifiques. *Acta Mathematica* 38:1–135. (This *note*, though published only in 1921, was written by Poincaré in 1901).

Riemann, Bernhard. 1854. Ueber die Darstellbarkeit einer Function durch eine trigonometrische Reihe. *Habilitationschrift* submitted to the faculty of philosophy of the University of Göttingen. Göttingen: *Abhandlungen der Koeniglichen Gesellschaft der Wissenschaften zu Goettingen, 13*. *Opera*: Riemann (1898), 225–272 and Riemann (1876), 227–265.

Riemann, Bernhard. 1876. *Bernhard Riemann's Gesammelte mathematische Werke und wissenschaftlicher Nachlass, H. Weber and R. Dedekind, editors*. Teubner. Quoted from the second edition of 1892, with the supplement by M. Noether and W. Wirtinger (1902), reprinted in 1990 by Springer Verlag and Teubner jointly, under the Title *Gesammelte mathematische Werke, Wissenschaftlicher Nachlass und Nachträge. Collected papers*.

Riemann, Bernhard. 1898. *Œuvres mathématiques*. Paris: Gauthier-Villars. Translated by L. Laugel from the German second edition, with a preface by M. Hermite and a speech by M. Felix Klein.

Robadey, Anne. 2004. Exploration d'un mode d'écriture de la généralité: l'article de Poincaré sur les lignes géodésiques des surfaces convexes (1905). *Revue d'histoire des mathématiques* 10:257–318.

Robadey, Anne. 2006. *Différentes modalités de travail dans les recherches de Poincaré sur les systèmes dynamiques*. PhD thesis, Université Paris 7—Denis Diderot, Department of History and Philosophy of Science, Paris. http://tel.archives-ouvertes.fr/tel-00011380/. Accessed 31 July 2014.

Index

A

Ab, See Āryabhaṭīya

Absolutive, 198, 202

Abstract, 151, 268, 282, 341, 349, 355

Acceptance, 73, 75, 143, 161, 257

Act of communication, 2, 6, 51, 52, 204

Acting, 2, 13, 116, 311

Action, 6, 11, 14, 23, 51, 63, 64, 67–69, 73, 76, 79, 80–83, 123, 127, 153, 194, 198–203, 207, 208, 210

Actor, 6, 7, 9, 39, 165, 365

Adjustment, 143, 163–165

Advice, 49, 62, 64, 65, 71, 72, 94, 99–102, 111, 112, 123

Agreement, 162

Aid, 31, 63, 64, 70, 71, 77, 252, 259, 308, 323

Ailment, 12

Akkadian, 283–284, 306

Algorithm
   execution of an algorithm, 184, 189, 192, 199, 202, 205, 206, 208, 211, 317, 325, 330, 333, 335
   giving an algorithm, 24
   number of steps of an algorithm, 201
   order of the steps, 23, 24, 181, 202
   statement of an algorithm, 25, 184, 188, 190, 197–199, 201, 203, 206–208, 210–211, 288–297
   text on algorithm, 317–381, *See also* Text

Amateur, 18

Analytic, *See* Function

Anaphora, 243

Animal, 152, 145, 147–150, 155, 156, 158, 160

Anscombe, G.E.M., 45, 79, 81, 84, 246, 263

Argument, 36, 51, 125, 127, 203, 222, 233, 248, 274, 302–304, 310, 332, 350, 353, 356, 357, 390

Argumentative structure, *See* Structure

Aristotle, 156, 227, 234, 235, 252–257

Arrangement, 39, 144, 206, 234, 235, 259, 262

Article (of journal), 53, 63, 88, 89, 94, 151, 189, 190, 197, 388

Aryabhaṭa, 22, 24, 194, 197, 199, 201, 203, 204, 206, 208, 211

Āryabhaṭīya (Ab), 22, 183–185, 198, 200, 202, 209

Āryabhaṭīyabhāṣya (BAB), 185, 203, 209, 213–215

Assertion, 2–4, 28, 51, 55, 57, 58, 62, 64, 73, 78, 79, 196, 199, 224, 247, 297, 322–323, 331, 346, 354, 367–368, 379, 406

Assertive, *See* Speech act

Attitude, 62, 82, 149, 199

Austin, J.L., 2–4, 7–8, 20–21, 23–24, 28, 37, 45, 50–52, 62, 68, 79, 81, 83, 84, 116, 119, 143–144, 153, 157, 184, 188, 204, 210, 218, 232–233, 237, 242

Author, 9, 10, 12, 13, 15, 18, 21, 22, 35, 89, 91, 92, 100, 101, 108–111, 117, 119, 123, 125, 127, 151, 153–156, 161–164, 185, 191, 211, 227, 234, 235, 251, 254

Authority, 15, 18, 21, 71, 72, 83, 118, 119, 123, 127, 149, 151, 163–165, 224

Availability, 159

B

BAB, See Āryabhaṭīyabhāṣya

Bhāskara, 22–25, 195, 197, 200, 203, 205–211

Bhāskarācārya, *See* Bhāskara

Bhaṭaprakāśikā (SYAB), 22–25, 185, 200, 209

Bible, 144

Binomen (nomenclature), 147, 152, 153, 163

© Springer International Publishing Switzerland 2015

K. Chemla, J. Virbel (eds.), *Texts, Textual Acts and the History of Science,*
Archimedes 42, DOI 10.1007/978-3-319-16444-1

Printed in the United States
By Bookmasters

Printed in the United States
By Bookmasters